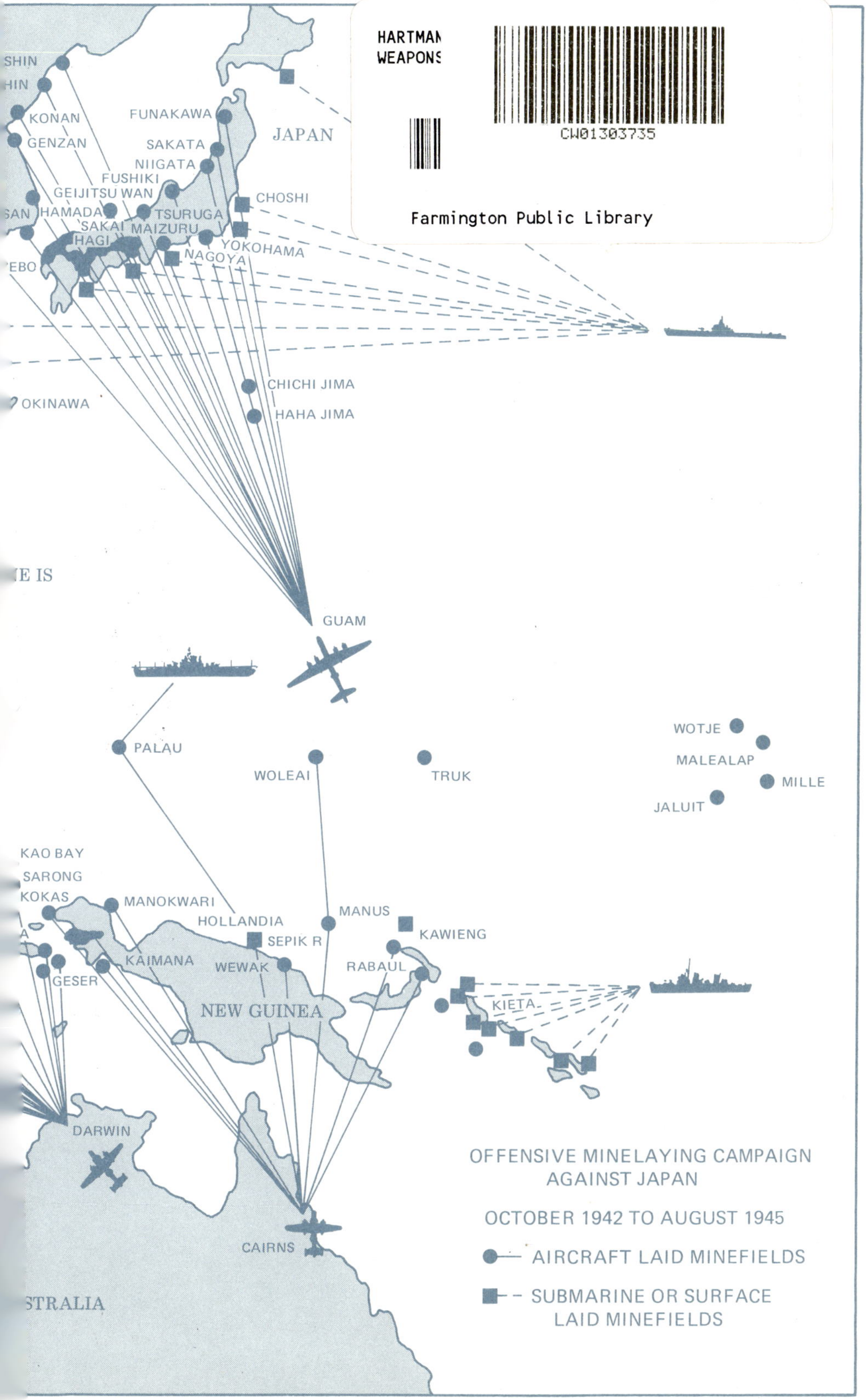
JAPAN
KONAN
GENZAN
FUNAKAWA
SAKATA
NIIGATA
FUSHIKI
GEIJITSU WAN
HAMADA
TSURUGA
SAKAI
MAIZURU
HAGI
CHOSHI
YOKOHAMA
NAGOYA
CHICHI JIMA
HAHA JIMA
OKINAWA
GUAM
PALAU
WOLEAI
TRUK
WOTJE
MALEALAP
MILLE
JALUIT
KAO BAY
SARONG
KOKAS
MANOKWARI
HOLLANDIA
MANUS
KAWIENG
SEPIK R
KAIMANA
WEWAK
RABAUL
GESER
KIETA
NEW GUINEA
DARWIN
CAIRNS
OFFENSIVE MINELAYING CAMPAIGN
AGAINST JAPAN
OCTOBER 1942 TO AUGUST 1945
AIRCRAFT LAID MINEFIELDS
SUBMARINE OR SURFACE
LAID MINEFIELDS

Weapons
That Wait

America's greatest liquid assets are the oceans on either side of the continent.

—A quote from *A Cartoon History of United States Foreign Policy 1776–1976*

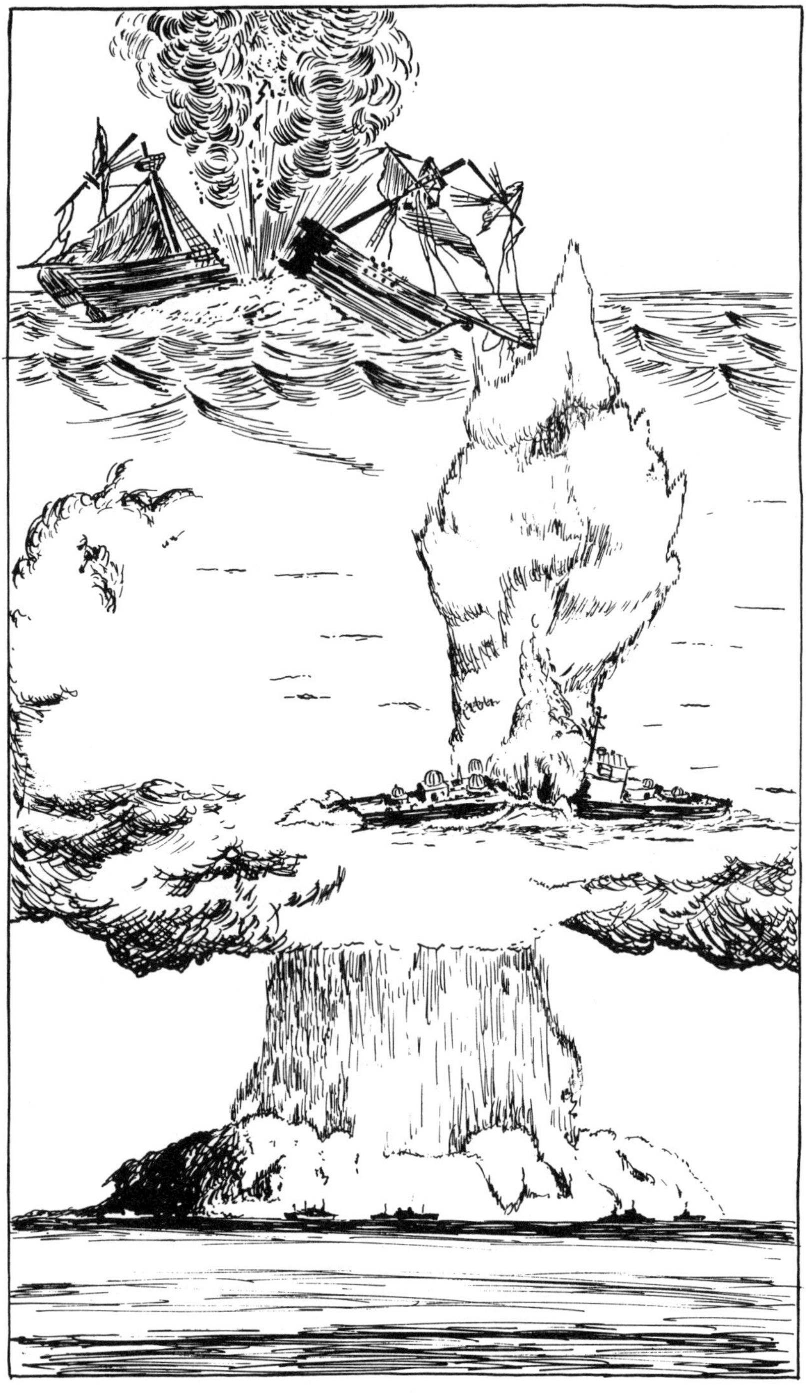

Weapons That Wait

Mine Warfare
in the U.S. Navy

Updated Edition

*Gregory K. Hartmann
with Scott C. Truver*

Naval Institute Press
Annapolis, Maryland

Copyright © 1979, 1991
by the United States Naval Institute
Annapolis, Maryland

All rights reserved. No part of this book
may be reproduced without written permission from
the publisher.

Library of Congress Cataloging-in-Publication Data

Hartmann, Gregory Kemenyi.
    Weapons that wait : mine warfare in the U.S. Navy / Gregory K. Hartmann with
Scott C. Truver. — Updated ed.
      p. cm.
    Includes bibliographical references and index.
    ISBN: 1-55750-350-8
    1. Mines, Submarine—United States. 2. United States. Navy—Equipment.
I. Truver, Scott C. II. Title.
V856.5.U6H37   1991
359.9'382—dc20   90-42356

Printed in the United States of America on acid-free paper ∞

9 8 7 6 5 4 3 2

First printing

Frontispiece: Underwater Explosions against Ships.
(*Top*) Fulton's sketch of the brig *Dorothea,* a 200-ton ship sunk in 1805 by two 170-pound gunpowder charges exploded underwater. (The two charges were bridled together.)
(*Middle*) Ex-USS *Moore* (DE 240). Sunk by Mk 48 torpedo in underbottom explosion of approximately 1,200 pounds of TNT equivalent. (Experiment on 12 June 1975.)
(*Bottom*) Baker shot at Bikini Atoll in 1946, 20,000 tons of TNT equivalent. Many ships sunk and damaged.

Contents

List of Illustrations	vii
List of Tables	viii
Foreword to the Updated Edition	ix
Preface to the Updated Edition	xi
Preface to the First Edition	xiii
Acknowledgments to the First Edition	xv
1. Introduction	3

    Mines and attitudes, 3
    What kind of war? 9
    Threat evaluation—Soviet Navy, others? 12
    Importance of mine warfare, 14

2. History	17

    Beginnings, 17
    From Civil War to World War I, 32
    World War I, 42
    Mines come of age, 55
    Important lessons and questions, 84

3. Mine Targets and Types	86

    Influence fields, 86
    Vulnerability, 98
    The kinds of mines, 103

4. Technology and Mine Design	106

    Definitions and discussion, 106
    Examples of the influence of technology on mine warfare, 112
    Fields of science related to mine warfare, 116
    Mine warfare as a system, 116

5. Countermeasures ... 121
   The problem, 121
   The simulation of ships' fields, 130
   Mine hunting, 133
   Vessels for mine countermeasures, 141
   The deep-water countermeasure, 146
   Sweeping and clearing in Haiphong and Suez, 148
   Recent countermeasure efforts in Western Europe, 155

6. Measures of Effectiveness ... 162
   Operations research, 162
   Simulation, 167
   The effects of damage, delay, and dismay, 169

7. Readiness and Delivery ... 174
   Ready for what? 174
   The obsolescent stockpile and training, 176
   Vehicles and ordnance, 179
   Full circle? 182
   Does the unexpected always happen? 187

8. Strategy and Tactics (or Vice Versa) ... 190
   Purpose, 190
   Matching mines to targets; the calculation of threat, 192
   Observation and replenishment, 195
   Law of the sea and other matters, 196

9. A Catalogue of Specialties ... 200
   Harbor defense, 200
   Riverine and land mining, 201
   Mobile mine, 203
   Deep-water mines, 204
   Under-ice mines, 207
   Miscellaneous, 209

10. R&D Policy ... 211
    The best policy is honest, 211
    Requirements and invention, 215
    The sophisticated buyer, 219
    Evaluation—How much and who says? 222
    Producibility and competition, 226

11. Limited War and Terror ... 229
    What's past is prologue? 230
    A new twist: Terror and "low-intensity conflict," 243

12. Into the Twenty-First Century ... 263
    Image and reality: Mine warfare in a superpower navy, 264
    Final observations, 287

Appendix A. Mine Chronology ... 291

Appendix B. U.S. Navy Mine Warfare Platforms and Systems ... 296

Appendix C. The Pressure Drop under a Moving Ship ... 329

Bibliography ... 331

Index ... 337

Illustrations

Endpapers. The Offensive Minelaying Campaign against Japan.
Frontispiece. Underwater Explosions against Ships.   ii
1. David Bushnell or Joseph Plowman?   18
2. Mine Mark 6 Planting Sequence.   39
3. Submarine Mine-laying Techniques.   44
4. Mine Mark 6.   49
5. Circuits of the K-Device.   50
6. The North Sea Barrage.   54
7. Functional Diagram of the Induction-Type Magnetic Mine.   64
8. The Oyster Mechanism.   68
9. Japanese Traffic in March 1945.   79
10. Japanese Traffic in August 1945.   79
11. Starvation Campaign, April–August 1945.   80
12. Typical Ship's Magnetic Signatures (a) and Induced Current (b).   90
13. Typical Ship's Pressure Signature.   93
14. Spectra of Submarine Noise at Two Speeds.   95
15. Sound Pressure Contours.   95
16. Damage Zones of a Typical Vessel.   101
17. Probability of Actuation for Influence Mines.   102
18. Mine Warfare as a System.   119
19. British and American Minefields 1914–18.   246
20. British Minefields in Europe 1939–45.   247
21. A "Backwater" Mission?   276

Tables

1. German Submarine Mine Laying, Western Atlantic, World War II. 70
2. U.S. Mine Types. 104
3. Mine Warfare Needs Related to Navy Basic Research. 110
4. Linkage of System Components to Technologies and Needs. 117
5. Economic Cost per Enemy Ton Casualty: WWII, Pacific. 237
6. Mines and Damage in Two World Wars. 242
7. World War II in Two Theatres. 243
8. Mines Used in the Twentieth Century. 244
9. Noteworthy Mining Efforts. 245
10. Red Sea/Gulf of Suez "Mines of August" Incidents (1984). 251

Foreword to the Updated Edition

Mines are becoming more and more a weapon of choice. They are a powerful political as well as military option. At the same time, they are relatively humane, in that once a minefield has been announced, as required by the rules of the Geneva Convention, ships enter the field at their own risk; accordingly, mines do not have to sink ships to be strategically and tactically effective. Mines also lend themselves well to modern technology such as microchip electronics, wherein they can be the smartest of weapons. Mines, together with their counterpoint, mine countermeasures, make up a body of naval warfare called mine warfare.

While there are perhaps eight or so major categories of naval warfare, few have the "leverage" and "force multiplier" effect of mine warfare. Mine warfare can be a major factor in circumstances ranging from low-intensity conflict with a terrorist nation to all-out strategic survival. On the other hand, few forms of naval warfare have been so little understood and so underutilized by military professionals and statesmen alike.

Accordingly, it is most appropriate and worthwhile that mine warfare be studied, understood, appreciated, and utilized in all its defensive and offensive subtleties—much as the knight, once studied, can be so valuable in chess. This is particularly true for the United States, which, as an island nation and leader of the free world, must continue to be the premier maritime sea power throughout the twenty-first century.

Dr. Hartmann's first edition of *Weapons That Wait* was a textbook on mine warfare on which I relied heavily as Commander of the Mine Warfare Command from 1979 to 1984. I encouraged all that I could to read it, both within the military and without. *Weapons That Wait*, in my view, was and remains the best of the mere handful of unclassified works on this subject. It is broad, incisive, thorough, and illuminating.

In his updated edition of *Weapons That Wait* Dr. Hartmann, ably assisted by Dr. Truver, has retained all the solid underpinning of the original work, while adding the recently unfolding aspects that are so pertinent for our study and grasp. The Falkland Islands War, the "Mines of August," and the Persian Gulf conflict are addressed with special insight as Dr. Hartmann and Dr. Truver describe how mine warfare is achieving center-stage status, in both utilization and impact, in modern conflict.

The essential point in modern conflict is not how well a handful of mine specialists understand mine warfare, offensive and defensive. The key is how well the general line practitioners of all designators—from the Chairman of the Joint Chiefs of Staff to the action officers, of all services, on staffs and in command centers—understand and use their knowledge of mine warfare. It is worthy of note that Admiral William Crowe's unique understanding and appreciation of mine warfare during his tenure as Chairman contributed significantly to the successful outcome of the 1988–89 Persian Gulf "Earnest Will" operation.

In this light, I encourage both military and nonmilitary students of warfare to read and reread this timely and thought-provoking update of Dr. Hartmann's classic *Weapons That Wait*. I may not agree with all of its points, but I do agree fully with its emphasis on this important capability, or lack thereof. Mine warfare will wait no longer for the maritime nations of the world to understand it and assimilate it into their day-to-day thinking.

Charles F. Horne III, Rear Admiral, USN (Ret.)

Preface to the Updated Edition

It was suggested that an updated edition of this book be issued, not only because the first edition has been sold out for some time but, more important, because interest in the subject of mine warfare has been increasing in the some ten years since the book's first publication.

We cannot report on everything that has been going on in the area of mine warfare. We do not know everything, and even if we did we would be unable to reveal it all; new operational or technical items are likely to be highly classified. Consequently, this updated account is necessarily fragmentary and incomplete. Unclassified accounts, however, have become more numerous and more readily available than in earlier decades. For example, the *Proceedings* of the U.S. Naval Institute have carried many articles in recent years about countermeasures and mine developments in NATO and other countries, as well as some U.S. efforts. Other sources are the *International Defense Review, Jane's Defence Weekly, Jane's Weapon Systems,* and Norman Polmar's excellent *Ships and Aircraft of the U.S. Fleet* and *Guide to the Soviet Navy,* published by the Naval Institute Press. I am much indebted to Sylvia and Richard Humphrey of the Naval Surface Warfare Center at White Oak, Maryland, for their work in collecting many of the relevant articles and making them available to me.

This updated edition attempts to describe the events of the past decade in terms of mine operations, developments, and countermeasures, and to merge the relevant current information with chapters 11 and 12 of the first edition without unnecessary duplication. For this difficult editorial feat I am much indebted not only to Paul Wilderson (Naval Institute Press), who motivated it, but also to Dr. Scott C. Truver (Director of National Security Studies at Information Spectrum, Inc.), whose access to the U.S. Naval Institute Military Database has been most helpful and whose scholarly approach to the problem is well appreciated. The extensive new appendix B describing current U.S. mine warfare platforms and systems has been compiled through the efforts of Dr. Truver.

Gregory K. Hartmann

Preface to the First Edition

Mines are of vital importance in the waging of war. Either in our hands or as used by an enemy they can make the difference between winning and losing. They are, moreover, relatively cheap to have and relatively safe to use. Mines have been used in growing numbers in successive wars since their first effective use in the American Civil War. Mines blockaded the German submarines between the North Sea and the English Channel in 1918. In 1946 the Japanese stated that the war would have been over sooner if the mine operations against them had started sooner. The atom bomb then would not have been used, and certainly things would be different today—but in what way it is difficult to say. Perhaps there would have been an all-out nuclear demonstration in the interim, if the world had not seen what two small bombs could do to two cities. In May 1972 when the harbor at Haiphong was finally mined, all shipping in or out came to a halt, and a peace agreement was finally negotiated within a few months. The evidence is that North Vietnam could not have gone on without replenishing its supplies of war material.

Many of these facts are unknown to the general public. They are also unknown to many responsible officials in the government and even in the Navy. I believe it to be of the utmost importance that current interest and experience in the subject of mine warfare be made available to a broad readership and that immediate plans be made to ensure the preservation and use of this knowledge. The consequences of neglect are far too devastating to allow indifference or parochial or selfish interests to consign the mine warfare readiness of the United States again to the limbo it has periodically endured.

Some may say that mine warfare is so valuable that we must bury it in secrecy, a concept that leads to ignorance and neglect. Others may feel that the less said about it the better—if it is good, let's keep it to ourselves; if it is not, then let it fade quietly away. Secrecy is clearly not the way to condition either the public or the Navy to the advantages of mine warfare. Every informed citizen is rightly concerned about the expense of nuclear weapons and the horrendous risk in their proliferation. It is long past the time when security lies in this direction. It is high time that a lesser expenditure should give greater security at lower risk. We need to measure our defenses in terms of effectiveness in real and likely situations, not in terms of overspending to achieve parities which will never be put to the test. Such expenditures create more risk, are inflationary, and are useless in the continuing real world. The real world exists somewhere between holocaust and absolute peace. The resources of the real world include diplomacy and capabilities for graduated force (limited war) as well as hope in the brotherhood of man. Since World War II, the United States has spent two thousand billion dollars for defense, and about one billion of this for mine warfare. I believe that the priorities and expenditures should be realigned. We could get a lot more defense for a lot less money, and this is what everybody is trying to do or at least so they say.

A glance through the bibliography at the end of this book will show that the open literature on mine warfare is small, but growing. Some comments about other sources of information are given at the beginning of the bibliography. The numbers that appear in brackets in the text are references to the like-numbered items in the bibliography. When a direct quotation appears in the text, the page numbers are given as well.

It would seem that every five years on the average a book on mine warfare appears—not an excessive rate. This is not a book for specialists. It is my hope that when the whole matter is now put forward again, the story and its significance will reach an ever-wider audience. This new statement should let us understand what our future readiness will depend upon. It should let us see what is to be done now in order to be in a desirable state later. As Edmund Burke said, "The public interest requires doing today those things that men of intelligence and goodwill would wish, five or ten years hence, had been done."

Acknowledgments to the First Edition

The preparation of this book would not have been possible without the help and assistance of many people. The support provided by the Naval Surface Weapons Center under J. E. Colvard, technical director, and the White Oak Laboratory under E. H. Beach, deputy technical director, is gratefully acknowledged. I would like also to thank the many individuals who have helped me in one way or another. For technical advice: present or former members of the White Oak staff C. J. Aronson, D. Brumbaugh, W. C. Chamberlin, L. C. Fisher, R. E. Hightower, S. Humphrey, J. Martin, C. Rowzee, E. Schuchart, J. P. Smaldone, W. C. Wineland. For logistic support: editorial services, George Hamlin; typing, Nancy Lanham, Marj Fox Grey, Julia Carson, Kandy Coppinger, Christine Fusco; library, Helen Conoley; illustrations, Betty Jew. For help of various kinds I am also indebted to many others, among them Lee Hunt of the Naval Studies Board of the National Academy of Sciences; Philip K. Lundeberg, Curator of Naval History, Smithsonian Institution; G. G. Gould of the Naval Coastal System Laboratory; C. M. Schoman of the David W. Taylor Naval Ship R&D Center; and Jan Snouck-Hurgronje and Carol Swartz of the Naval Institute.

But my greatest indebtedness throughout the years is to the many outstanding and dedicated men both at the Naval Ordnance Laboratory and outside in the greater mine community, with whom it has been my privilege and honor to work. In spite of these advantages, or perhaps because of them, it is important to say that any opinions expressed in this book are solely those of the author.

Weapons
That Wait

1 Introduction

Mines and attitudes

Naval mine warfare is difficult to understand. It is both obscure and controversial. Yet it is an important adjunct of the capacity of the United States to wage war. Since World War I, the United States has made increasing use of naval mines, and has also emphasized its own work to provide countermeasures against the possible enemy use of mines. These two specialities—the development of mines and the development of countermeasures, in all their aspects of concept and use—constitute the subject of mine warfare.

This book is about naval mine warfare and does not consider the use of mines on land. Nevertheless, the general principles of mine warfare apply equally well whether the mines are deployed under ground or under water. In fact, the main idea of mine warfare is to let the enemy run across the weapon rather than to let the weapon seek the enemy. This, of course, can be done equally as well on land as in water, presuming in each case that it is not easy to detect or locate the weapon. Thus, mine warfare removes from consideration many of the conventional aspects of warfare—i.e., face-to-face combat, and the pursuit or capture of the enemy. Instead, mine warfare gives the enemy the option of doing nothing—i.e., not advancing, not occupying territory, not moving his men or materials by land or sea—or of risking severe losses by attempting these things. It is indeed a more pleasant thing to establish a blockade by laying a minefield that will maintain the watch, than to patrol the same area without the minefield and be exposed to enemy action the whole time. But now it must be said that there is always a chance that a vehicle can safely pass through a minefield. If the chances are good, we say that the threat of the minefield is low. If the chances of getting through safely are poor, then the minefield threat is high. The threat of sinking or of damage can simply be equated to the chance of sinking or of damage associated with passage through the field.

One of the difficulties in understanding mine warfare is the fact that the assessment of its value is a statistical exercise. For example, if the mines could be exactly located and exactly avoided, every vehicle could avoid every mine provided there weren't too many, and the threat would be zero. Even in such a case, however, the net speed of transit through the minefield would be decreased, which would have the effect of robbing the enemy of some of his potential mobility. But in practice it is not possible to eliminate error. Therefore, the threat of a mine field is the result of a complex calculation in which location errors, navigation errors, ship influence fields, mine responses, and elements of surprise, urgency, and countermeasure activity all play a role. In such calculations it doesn't matter whether a ship makes a safe transit. Any ship that takes the risk and escapes disaster does not guarantee the safe transit of others. The practical matter is that if the threat is perhaps one in a hundred, a commander would take the risk if he felt he had no other choice. Consider, for example, Admiral Farragut at the entrance to Mobile Bay. If the field has a 10 percent threat, however, the chances are that nobody will try it. Some say that a 25 percent threat is the critical level. It is also clear that once it is known that mines have been laid, it is necessary to try to determine the threat. In the absence of information on how many mines there are or what kind or how effective they are, there is always one way to determine the threat and that is to undertake passage of the field. If you get away with it nine times and fail on the tenth, you have an estimate of the threat at 10 percent. In such an undertaking, however, it would be better to send out a specially designed vessel for the purpose rather than to risk a valuable ship or one with a valuable cargo.

Such special vessels are called minesweepers, and really what one would like them to do is to present to the mine such a sequence of physical appearances—in terms of magnetic, acoustic, or pressure fields, or combinations of these, increasing and decreasing with time and with distance—that the mine is deceived and will proceed to detonate. It would then also be desirable if the sweep were in a place where the mine explosion would not damage it, or if the sweep were so cunningly constructed that it was immune to such damage. If all these things could be accomplished, then it would be a simple matter not only to determine the minefield threat, but to continue using the sweepers until that threat had been reduced to the acceptable level. Even if these goals could be achieved, it must be remembered that during the time when the threat assessment and sweeping activity is going on to nullify the minefield, the field is performing a main function in preventing the passage of the ships or cargoes which are waiting for safe transit. If these supplies or war vessels are urgently needed in another theater of war, their absence could be crucial. The minefield then should be credited with contributing to a successful action elsewhere even though it actually may have sunk or damaged no ship at all.

Thus we see that the evaluation of minefield effectiveness in war depends almost on a state of mind, on an accounting after the fact, on evidence which appears after the war, such as the number of ships that were sunk which were not observed by our side. When actions and the results of those actions are long separated in time, the connection between them has to become a matter of faith and belief rather than a matter of direct observation. Herein also lies a difficulty for mine warfare. Protagonists claim it as the ultimate in warfare. Others regard it as a dangerous and at the same time unglamorous game which is made possible by diverting funds from more worthwhile pursuits—like building faster airplanes or deeper-diving submarines. Clearly such extreme views on either side are incorrect. An agreement on some middle ground is desirable and perhaps can be achieved by a consideration of the facts, although when one brings in facts about money it is hard to be reasonable and even harder to be objective.

The Mine Advisory Committee of the National Academy of Sciences in 1970 published a report entitled "The Present and Future Role of the Mine in Naval Warfare" [33]. The series of studies leading to this report was called Project Nimrod. In Genesis Chapter 10, Verses 8 and 9, Nimrod is said to be "a mighty hunter before the Lord." Perhaps the title reflects the thought that mines should play a mighty role in warfare. Their role, however, whether mighty or not, is essentially the passive one of waiting rather than the active one of hunting. Even this distinction is becoming blurred, as we shall see, with the introduction of mines which can indulge in a limited pursuit of the target once they have been triggered by its approach.

After reviewing the history of mine warfare, the committee concludes that mines are extremely cost-effective weapons of attrition. They can, in effect, reduce the extent of navigable areas. The report goes on to say that many responsible persons in the mine-warfare community believe that mine warfare does not receive the attention or emphasis it should. It further states that many persons in decision-making positions, even within the Navy, do not understand the characteristics or principles of mine warfare. Why should this be so? A number of factors undoubtedly contribute to the situation. For example, military people are "vehicle oriented." Mines can be delivered by many vehicles, but are not ordinarily associated with any particular one. Nevertheless, the discovery in World War II of the overwhelming practicality of delivering mines by air was coupled with the fact that the Army Air Corps and not the Navy had the planes which were large enough to deliver significant payloads at great distances. This situation led to a significant delay in the final agreements to carry out the mining campaign against Japan.

Second, as already mentioned, the results of mine laying are hidden or delayed—in short there is no "bang" appeal. More important, perhaps, is the fact that credit for an action is hard to assess and assign. In the delivery or sweeping of mines, there are, clearly, dangerous situations to be encountered which call for courage, good training, and devotion to duty, and which should be duly recognized and rewarded.

Third, there are some who feel that mine warfare is perhaps the weapon of inferior forces (and hence that only inferior forces will use it); that it is covert or ungentlemanly, and, as has been said of it, not quite cricket. These attitudes are somewhat hard to understand in this century when advances in technology have so broadened the possibilities for mutual destruction that in retrospect mine warfare looks like a very gentlemanly way to apply restricted persuasive measures in a limited and controlled manner.

Fourth, there are a few miscellaneous points—such as the claim that an interest in mine warfare does not contribute to one's personal advancement in our Navy, although it is believed that in other navies —such as the Soviet Navy—mine warfare is a popular speciality. There, for example, ship efficiency prizes are given in mining as well as in the conventional areas of gunnery, torpedo firing, and ASW. This indicates the relative importance given to mine warfare in the Soviet Navy and it is quite appropriate, because the great Russian tradition has been in defensive mining.

It is important early in our study to stress the difference between defensive and offensive mine warfare. The first ideas for control mining were clearly defensive; the fields were designed to keep out enemy ships and to provide protection for ours. The installation of such fields was made in waters under our own control. After the fields were established, our own forces in theory were able to operate without constraint except to stay clear of contact mines which could not be controlled. Defensive mines were to keep invaders out. If the enemy did not attempt a transit, there would be no losses. The defensive field that sinks no ships because none appear has accomplished its purpose perfectly. It is a mistake to fault a defensive minefield for failing to sink ships. It is therefore very obfuscating to lump together all mines planted in a war, both defensive and offensive, and then to list their effectiveness by counting up the number of mines per ship casualty.

Offensive mining, on the other hand, is designed to attack enemy shipping which is going about its own business (not necessarily on war missions), in waters largely under enemy control. In order to place mines in such waters in the days before submarines and airplanes were developed, it was necessary to have control of the surface. But if you have control of the surface, it was argued, there is a positive disadvantage in placing mines in your way. Therefore, mines were unnecessary as offensive weapons and were thought of only as defensive—i.e., laid in our own waters to keep out enemy ships. Offensive naval blockade was to be done by surface warships. This doctrine expressed by Admiral Mahan in 1890 [29], was similar in its conclusion to the British Admiralty view expressed nearly a century earlier by Lord St. Vincent when he said of Pitt's interest in Fulton's ideas about mines, "Pitt was the greatest fool that ever existed to encourage a mode of warfare which those that commanded the seas did not want, and which if successful would deprive them of it." [10, P. 2]

The argument that offensive mines could only be laid if we had control of the surface and that in this case they were not only unnecessary but a positive hindrance, was completely undercut by the appearance of mine-laying submarines and mine-laying aircraft. The new technological facts of the twentieth century—namely, that mines can be used as offensive weapons when laid by submarine or aircraft (which did not exist when Mahan wrote his books)—serve to confirm his views about the value of naval blockade. Mahan was not against offensive mine warfare *per se*, although this is what his doctrine was held to be. He was actually opposed to encumbering tactical surface forces with mines which would be unnecessary to their operation. The new delivery methods now offer a means of achieving an objective with naval mines that was not available in Mahan's time.

Some have said that the glamor of new technology in the fields of propulsion, missiles, and surveillance has decreased interest or emphasis in mine development. This is, however, not necessarily so. The technical resources of solid-state electronics, information processing, new materials, and weapon propulsion and control are of vital application to the many problems of a new mine warfare capability. This capability is currently under development in both the mine and the countermeasure areas.

It has been said that the lack of readiness in the present mine stockpile is a major cause of disinterest in the use of mines. Some have maintained that it takes too long to assemble and check out present mine material and that it should be redesigned to be a "ready round." Such goals where realistic are desirable. Clearly, however, it will be impossible to do away with all maintenance or trouble. Things will deteriorate in storage, reliability will have to be checked, safety standards will have to be kept. The relative emphasis on redesign, or on new design, will depend on many factors—the extent of the existing stockpile, the cost of maintenance, the availability of R&D dollars, and the assignment of personnel to these many functions. The main point here, perhaps, is that these conditions are not the cause of a lack of emphasis on mine readiness; rather they are the result of such a lack. First must come the understanding and the appreciation of the result which is wanted; then will come the direction to achieve this desired result.

Finally, mines must be recognized as a supplement to other forces, not as a substitute or replacement for them. In other words, the use of mines with other forces is a matter of balance and degree and cannot be substantiated as an argument to do away with other forces. There can always be circumstances when a blockade or a waiting policy is not sufficient and a positive action, such as an attack or an invasion, is necessary. For these cases mines may play a supporting or a preparatory role, but not a principal role. Nevertheless, mines are expected to be used more, as their cost effectiveness is realized and as economic priorities limit the growth of military budgets, not only in this country but in poorer countries of the world. Furthermore, the new technology is opening up added possibilities for mine use. One objection to the use of mines which has been voiced is that once a channel or an area is mined, it is extremely difficult to change your mind about it. Once the harbor at Hanoi was filled with mines it was largely a matter of waiting until the sterilizers, mostly set for a six-month period, had the opportunity to function. The reliability of the sterilizer, a device to render the mine inert after a chosen interval, can be set at any desired level by adequate redundancy in design. It is also possible that a "command control device" can be provided which, on appropriate coded signals, can turn a mine off before its sterilizer is programmed to function—or even turn it on again. Thus, with such embellishments the minefield could enter more into the tactical domain of warfare and perhaps better meet some of the criticisms of tacticians. War cannot be made into a riskless activity, but the risk in using our own material can be set at any acceptable level by a proper program of technical development and environmental testing. Such a program would indeed make it possible to pass safely through our own minefield after it had been operating against the passage of others.

In summary, ships are necessary for the movement of goods and materials from place to place. Ships generally must leave port and return to port. Mines can prevent this by merely covering a channel or an area, not finding their targets at the time they are laid. The targets later unwittingly find the mines. The mine once laid is in constant readiness awaiting the opportunity to attack. The mine remains on station waiting, withholding its fire—at no additional cost—until its maximum opportunity arrives. It thus achieves maximum effectiveness at minimum cost and at minimum risk to its own forces. Mines do not go to sleep unless they are programmed to.

What kind of war?

At the beginning of this account it may be worthwhile to attempt a glimpse into the future. In this way perhaps the relevance or direction of mine warfare as a part of that future can be guessed. It is obviously very risky to talk about a future that may be determined by things that are at present unknown. However, there are some things which are known and which are not likely to change; for example, five-sixths of the earth's surface is covered by oceans, and the general configuration of the land is not likely to change much—even allowing for continental drift or the melting of the polar icecaps. It appears inescapable that goods and materials will have to be moved from place to place and that this will involve transportation by sea. The cargoes in question are bulky and heavy—for example, oil, ore, wheat, and machinery—and it is assumed that they cannot be transported in adequate quantities by air.

A similar assumption made before World War II about the transportation of people would not have held up. Ocean liners have all but disappeared now except perhaps as vacation resorts. The transport of troops to and from Vietnam, for example, was almost exclusively by air. It remains to be seen whether passenger ships will come back because of the growing fuel shortage. The necessity of transporting bulk supplies by sea, however, if even to supply fuel so the airplanes can fly home, remains an assumption which it is difficult to see changing with time. The continuing function, then, of the Navy must be to ensure the safety of such transits for us or to prevent such transits for our enemies in war. This function is described by Admiral Zumwalt and others as *sea control*. Another function of the Navy is described by the term *projection of force*. This covers such diverse situations as showing the flag, the evacuation of U.S. citizens, amphibious operations, riverine warfare, air support, conventional bombing, or engagements with naval forces at sea. These functions, perhaps in varying degrees, are likely to remain in the future as long as there is a national navy. And, of course, the two main functions mentioned here are intertwined in that it might be very difficult to carry out the sea-control function without exercising the projection-of-force function.

It is very important to realize that there has been a most significant change in undersea warfare since the end of World War II and that this change has not been tested or used in warfare at all. This began with the advent in 1956 of the *Nautilus*, the first nuclear-powered submarine. Today, twenty-four years later, virtually all U.S. submarines are nuclear powered, and the Soviets have a matching number of strategic and tactical nuclear submarines. The sea-control problem of the future lies with the tactical nuclear submarines. In World Wars I and II the submarines of the time were formidable opponents even though they were not truly submarines in the modern sense of the word. They had to come up for air from time to time, to look around, and to receive or send messages. Their speed when submerged was limited in both magnitude and endurance. It was therefore possible to thwart submarines by bunching together a lot of merchant or cargo vessels in convoys and by providing naval escort vessels which could carry out counterattacks. Since the search and detection range of a submarine was relatively short (say ten miles), it was almost as easy to miss a whole convoy as it was to miss a single ship. If a convoy was spotted, however, it was difficult to attack many ships before the whole thing had passed over. The submarine, if submerged, could not keep up. If she came to the surface, she was fair game for the escorts. Further, her weapons were relatively short-range torpedoes so that the escorts only needed to search relatively close to the targets. Today an attack submarine has none of these disadvantages. She has, if anything, a speed advantage over the cargo ship; she can stay submerged indefinitely; she has long standoff weapons (i.e., the Soviets do); she has better communications and intelligence information telling her where her targets are. She also has better sonar, both active and passive, for detection and for fire control. Thus, it is by no means clear that the convoy system is an automatic answer to the modern (true) submarine as it was to the old diesel-electric boats with or without snorkels. It would seem that it is absolutely essential to develop a highly effective quick-reaction system to detect, localize, and destroy a nuclear attack submarine once it has identified itself by launching its attack. Such a system would tend to make the exchange ratio of ships sunk per submarine sunk too low for the enemy continuance of submarine warfare. Even so, the initial advantage of surprise and initiative would remain with the submarine.

Some thoughtful people in the ASW community—such as the late William McLean, inventor of Sidewinder, Technical Director of Naval Ordance Test Station, China Lake, and later of Naval Undersea Center, San Diego—have advocated the abandonment altogether of surface ships. If submarines have such an advantage because they cannot be found, why not confer this advantage on cargo ships? Why not have a fleet of submersible tankers or other cargo vessels and thus make it so difficult for submarines to find targets that their effectiveness would become negligible? This is a complex question. One of the standard replies is that submarine cargo vessels would be more expensive than ships to carry the same cargoes and yet would have to be used in peacetime if they were to be used during a war. Whether they would be more expensive or not, it is certainly true that a transition away from surface shipping would take a long time, because presumably present surface ships would not be abandoned until they were worn out.

The idea of solving the submarine menace by eliminating readily detectable surface ship targets is not only technically debatable as to how safe from detection a submarine tanker would be, but it is economically uncertain and of course emotionally repugnant to many of those who joyfully go down to the sea in *ships*. The idea, however, should not be written off. It should be given careful and long-range consideration. But in the meanwhile, what is to be done about the sea-control mission, and particularly the part that lets us use the sea against enemy opposition? Presuming that we must continue to use surface vessels, we must continue to develop an antisubmarine system in order to reduce the ship-to-sub exchange ratio. Whatever that ratio is, the number of our ships lost will depend on the number of enemy submarines that get into action and are thus attacked. If this number can be reduced, then our shipping losses will be reduced in the same ratio. It is thus very cost-effective indeed to reduce the number of submarines on station by blockading their passage to the areas of operation of our own shipping. Such interdiction as we shall see can be established by the proper planning, development, and use of sea mines. The development of such mining capability in deeper water is fortunately coincident with the increased necessity to have something to contain the submarine menace which in wartime would now be so much more dangerous than it ever has been in the past.

Another viewpoint, which also owes its origin to the discovery of nuclear fission, is sometimes expressed—namely, that a future world war, if it occurs at all, will be over quickly. Therefore, we will not have to worry about blockades, attrition, or antisubmarine warfare. Such an unattractive possibility cannot be completely excluded, of course, although a sudden trip to total extinction or to the Stone Age must be numbered among the unlikely events. Much more likely, it would seem, is the sort of self-serving restraint in this era of deterrence and assured second-strike capability (incidentally provided by Polaris-type submarine systems) that would lead to controlled hostilities and to the very situation in which submarine warfare would be the highest manifestation—a kind of guerrilla war of the sea with massive civilian populations suffering shortages, not bombs.

Another possibility which may not be as remote as all-out nuclear war is an agreement of the major nations to abandon nuclear weapons altogether with ironclad guarantees, inspections, and so forth. Such a utopian state of affairs would still leave the way open for attempts at sea control by one side or another, unless all war were ruled out and a world state were decreed to settle all disputes. It is of course possible that such peaceful solutions will become popular if it is clearly demonstrated that no one can gain from an opposite course of action. It is thus conceivable that the atom bomb has made land warfare too dangerous for anyone to succeed at it, and that the response of appropriate measures including naval mine warfare will do the same for the war at sea. But in order to beat our swords into plowshares it will be necessary first to have the swords.

Threat evaluation—Soviet Navy, others?

Because the Soviets now have the world's largest submarine force, it is clear that the Soviet Navy poses a major threat to our freedom of the seas. It must also be remembered that the Soviets have the world's largest mine force, which is largely devoted to the defenses of their own ports and coastal waters. The combination of mines and submarines used as minelayers for offensive purposes is clearly a possibility that must not be ignored. In fact, such a possibility is much more likely in the Soviet Navy than it is in ours and, among other things, will depend on how successful our tactical antisubmarine measures are. The relative safety of covert mine laying may then make this an appealing way to carry out offensive warfare. Russia was the first nation to develop a submarine exclusively for mine laying. Although this took place before World War I and was not pursued, nevertheless the idea, further developed by the Germans and the British, could have an important rebirth as applied to the new era in submarine design. If such a thing were to occur and our ports and approaches were being mined, there would be at least two possible responses. First, a countermeasure or sweeping effort would have to be made to try to find and remove the offensive mines. Second, it should be possible to prevent the mine laying itself or to move it farther out to sea by planting our own antisubmarine deep-water mines. These mines would pose no threat to surface traffic and would be hazardous only to submarines which either were afraid to surface or were ignorant of the minefield plan. In other words, one way to fight a threat of covert submarine-laid mining is by the use of a covert antisubmarine mine.

It is indeed difficult to imagine that hostilities of this sort will ever again be considered preferable to a peaceful and rational settlement of differences among great powers. Nevertheless, it is the job of governments to be prepared for whatever eventualities can be visualized. A war at sea and all of its variations does appear to be more rational for this country than massive land wars and in any case will have to precede any such later actions. We are also assuming that because it seems more rational it is also more likely!

There is another interesting corollary to the use of deep-sea antisubmarine mines. It should be possible further to protect our own nuclear deterrent submarine forces (Polaris, etc.) by keeping them at times within a mine barrier which would prevent intrusion by hostile submarines. Such mining in peacetime would of course be unacceptable, because it could result in the sinking of foreign submarines on or in the high seas. The law of the sea, however, is subject to change and evolution. The elaborate international conferences on the use of the sea bed for minerals, while producing very little progress, are a case in point. If Byron* were alive today he might have written:

Roll on, thou deep and dark blue ocean—roll,
Ten thousand fleets sweep over thee in vain;
Man marks the earth with ruin; his control
Stops with the edge of the continental shelf!
Or the 300-mile limit, whichever is greater.

---

* Roll on, thou deep and dark blue ocean—roll!
Ten thousand fleets sweep over thee in vain;
Man marks the earth with ruin—his control
Stops with the shore;—upon the watery plain
The wrecks are all thy deed, nor doth remain
A shadow of man's ravage, save his own,
When, for a moment, like a drop of rain,
He sinks into thy depths with bubbling groan,
Without a grave, unknell'd uncoffin'd and unknown.
                Childe Harold Canto IV Stanza 179.

One reason for the difficulties encountered in Law of the Sea conferences to date is the involvement of so many parties and in particular of the so-called third-world nations. And this brings to mind that mine warfare has traditionally been thought of as a logical tool of the underdog. Not only is it true that it is used as a weapon of last resort or as a weapon in lieu of others, but even more as a weapon which costs less and is affordable by a poor nation. It is therefore not difficult to understand why a poor nation might embark on such a course. Further, the guerrilla or terrorist option is something that has occurred in the past in other contexts and should certainly not be ruled out as totally unlikely for the future. Certainly the technology is available, and the situation may indeed arise where blackmail or banditry is a "viable option" for some objectives. Consider, for example, the difficulties that could be caused by a mining venture in the Persian Gulf to prevent the sale or delivery of oil from the Middle East. This and other scenarios give emphasis to the necessity for having a rapidly deployable ability to sweep mines even in remote parts of the globe.

Importance of mine warfare

Why should mine warfare be studied when we are hoping for generations of peace? Mine warfare is important because it is likely to be used in the future even if strategic deterrence successfully prevents nuclear war. Mines have been used in every major U.S. war since the American Revolution—i.e., both before and after the arrival of the nuclear age. It is therefore of interest to review the development and use of mines on a worldwide basis and to see how the growth of technology has influenced that development. This review will be taken up in the next chapter, where we will see, in hindsight, how technology—and a very simple technology at that—has influenced tactics.

The growth of nuclear weapons in size and number between the superpowers has led to the concept of deterrence. This is the view that it would be mutually suicidal for anyone to resort to the strategic use of nuclear weapons. It is everyone's hope and expectation that the existence of world-destroying weapons will prevent their use, at least by rational people (defining these as people who do not wish to commit suicide). It is disturbing to recall that at least one world leader did commit suicide, and this was just before the dawn of the nuclear age. A further hope which is open to everyone is that if the nuclear powers can come to believe that they will never use their nuclear weapons, they might try to phase them out as unnecessary expenses. However that may be, the existence of nuclear weapons has not prevented the usual mix of large and small conflicts since 1945—wars, civil or international, declared or undeclared, which L.F. Richardson has studied in his interesting book *Statistics of Deadly Quarrels* [41]. In at least two of these wars, in which the United States was involved (Korea and Vietnam), the use of naval mines played a significant role—in the former in the hands of the enemy, in the latter in ours. The existence of nuclear weapons has limited the extent to which a non-nuclear or limited war may be pushed. If one of the contestants is driven to the point where he believes he has nothing more to lose, he may give up his nuclear restraint and use nuclear weapons if he has them. The other contestant would not want to push things so far that he might become the victim of his own actions. Consequently, the physics of Einstein may be the most important force for peace after all, or at least for restraint in war. Therefore, the limited war and the war with naval mines remain as very likely possibilities.

The study of mine warfare is a relatively unknown subject outside the mine community. Most of the U.S. mine development and technology for the last half century has been centered in the Naval Ordance Laboratory (now the White Oak Laboratory of the Naval Surface Weapons Center), which has consistently devoted approximately 20 percent of its effort to projects oriented toward mine warfare. There is, therefore, a tradition and body of knowledge here which has led to a few publications in the open literature listed in the bibliography. These and others are explicitly referred to from time to time and still others are listed as general references.

It seems clear that mine warfare is important because, as we have discussed, it is likely to be with us in the future. The second claim to importance lies in the noteworthy accomplishments of the past. Duncan [12] says that in World War I a total of approximately 240,000 mines were planted by all parties. The Allies lost to mines 586 merchant ships and 87 warships, not counting 152 small patrol boats and minesweepers. The Central Powers' losses to mines included 129 warships not counting an unknown list of submarines and an unknown number of merchant ships. For a listing of mines used and ships lost in the two World Wars, see Table 6. The total ship damage in the first war from mines was greater than that by gunfire and torpedoes! It is interesting that Great Britain and Germany both had great war fleets but did not actually have battle contact sufficient to prove which was superior. The U.S. Navy, also "with a great fleet of warships, did not engage the enemy with them, but probably performed its greatest war service by building 100,000 and planting 56,000 mines (Mk 6) on which no money had been spent before the declaration of war" in

1917. [12, P. 163] In World War II the U.S. and the United Kingdom laid down about 300,000 mines, of which approximately 200,000 were defensive, and sank virtually no ships. The remaining 100,000 mines in Europe and in the Pacific sank or severely damaged a total of 2,665 Axis and Japanese ships. In the last four and one-half months of the mining campaign against Japan, 670 ships including 65 warships were sunk or severely damaged, and this accounted for just about all the ships the Japanese had. They were thus completely cut off from necessary supplies of oil, raw materials, and food from Asia. Later, Admiral Tamura (who was in charge of Japanese mine countermeasures) said that had this campaign started three months earlier, or in January 1945 instead of April, Japan would probably have capitulated before August. It would have been unnecessary to inaugurate the nuclear age at Hiroshima!

There is much interesting information on this mining campaign, including Japanese interviews and opinions, in the U.S. Strategic Bombing Survey, [49]. In the last four and one-half months of the war the Japanese loss of shipping to mines was about equal to the U.S. loss of ships to German submarines in the Atlantic during the entire war! Most of the U.S. ship loss occurred in 1942 and the first half of 1943 when about 400 ships were torpedoed. The highest U.S. monthly ship loss was forty in May 1942 [34]. (The U.S. ships then averaged about 5,500 tons compared with the Japanese ship average of about 2,000 tons at war's end.)

From such statistics we conclude that mine warfare is important. Many may be surprised at these numbers. But, as noted in Reference 16, "the greater surprise should be that we are surprised; that we have not heard about this before. The public relations organization of the mine community is rather ineffective, or more accurately, rather nonexistent. As we have learned, 'there is nothing good or bad, but thinking makes it so' (Hamlet, Act 2). But one must at least have heard of it before thinking can begin." [P. 7]

2 History

Beginnings

The first example of the use of an unattended underwater explosive device against ships occurred in the Delaware River at Philadelphia in 1777. The idea was quite straightforward. The British warships that were the targets for this action were anchored in the river. If kegs of gunpowder could be set afloat upstream without anyone noticing it, they would drift down and might by chance run into one or more of the ships. It would then only be necessary to have a device which would ignite the gunpowder when the keg bumped into the ship. In 1777, the method for setting off gunpowder was well known and quite simple. A spark generated by the action of a flintlock, such as in an ordinary musket, was allowed to fall into the powder. Chemistry took care of the rest. It was only necessary to cock the hammer so that an impact would dislodge it.

David Bushnell, while a student at Yale, had been experimenting with the explosion of gunpowder underwater and was convinced that the explosion of a charge of gunpowder against the side or bottom of a ship of that date, even a man-of-war, would damage it seriously or cause it to sink. Bushnell was authorized by General Washington to proceed with his scheme against the British ships. This newly invented sea mine, which incidentally Bushnell referred to as a torpedo, consisted of a charge of powder loaded into a watertight keg which was suspended a few feet below a float on the surface. The flintlock trigger arrangement was assembled inside the keg with the powder and "adjusted so that a light shock would release the hammer and fire the powder" [12, P. 3]. It would appear, however, from other sources that the flintlock was operated by a lever on the outside of the keg which was connected possibly by a wire or rod to the lockplate on the inside and which had to be pushed rather than shocked in order to operate.

Philip K. Lundeberg's book on *Samuel Colt's Submarine Battery* [28, P. 2] contains a picture entitled "One of the kegs celebrated in the time of the Revolution" and also shows a view of the underside of the keg's cover where a portion of the lockplate of a Brown Bess flintlock musket (manufactured in England by Galton) is mounted. The keg looks more like a milk churn than a beer keg and has a volume of about one cubic foot. The keg mine illustrated on page 2 of Lundeberg's book was deposited in the Peale Museum of Philadelphia in 1793 by a Major George Fleming and was said to have been fabricated at Bordentown, New Jersey, in the cooper shop of Colonel Joseph Borden and designed by an inventive pin maker, Joseph Plowman. (See figure 1 and photograph courtesy Smithsonian Institution.) These mines were said to represent a community effort. The detonator element "consisting of a spring-lock arrangement fabricated by gunsmith Robert Jackaway, was apparently triggered upon disturbance of a wooden firing arm by a passing vessel." The Fleming keg

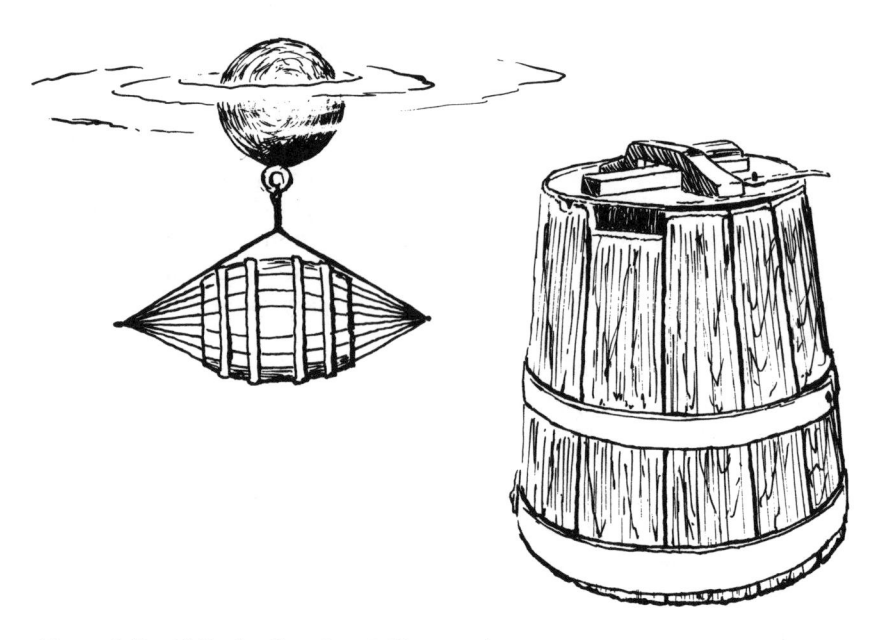

Figure 1. David Bushnell or Joseph Plowman?

mine has been in the Division of Naval History of the National Museum of History and Technology since 1972. At present it is the oldest known example of a sea mine. From this, it is not clear whether this was really Bushnell's original mine or a somewhat later and improved version. Since there were several mines (Lundeberg says a score) released by Bushnell in one night in December 1777, by his own account, there must have been a place for their manufacture. Probably this was Bordentown, which was conveniently near by.

Whether or not Bushnell actually used the beer keg or the butter-churn mines at Philadelphia, the tactical operation was unique. That there was no further use of this invention in the Revolutionary War may be due to the lack of success in the operation conducted, or to the lack of opportunity or of material for a second attempt. The initial failure of the operation does illustrate in a modest way the interaction of tactics and technology as well as the role of chance in determining the outcome of military events. In the first place, by modern standards or really by any sensible standards, the gadgets which Bushnell launched in the river that night were hideously unsafe. Days later one of the kegs, which had been spotted by two boys, exploded when they were trying to retrieve it. This unfortunate accident not only killed the boys but alerted the British to their danger and caused them to destroy by gunfire the rest of the kegs when they drifted by. Second, it was of course necessary to set the mines in secret, and this was done in the dark of night. Bushnell and his helpers, however, were not familiar with the river above the anchorage and as a result set the mines afloat farther upstream than security really demanded. As a result the mines drifted back and forth with the tide and became further delayed by ice in the river. If they were launched as a Christmas present, it was literally Twelfth Night before they arrived. By this time the British frigates had been relocated nearer shore to avoid the ice. If it had not been for the unfortunate accident, it is possible that the whole flotilla of mines would have floated out to sea and there would have been no "Battle of the Kegs" for Lord Howe to write home about. The principal value of the whole affair, then, may be from the historical point of view—that this was after all the first mine wartime operation in history and that its date was "late" December 1777 or, if one prefers, 5 January 1778.

This affair was unique in several respects. The idea of floating the mines until they encountered the target was atypical; in almost all later history, mines were stationary or anchored and the target had to come to them. But far more unusual was the method for igniting the gunpowder charge. This was unique to these particular mines and although tried later by Fulton in some of his experiments, was not

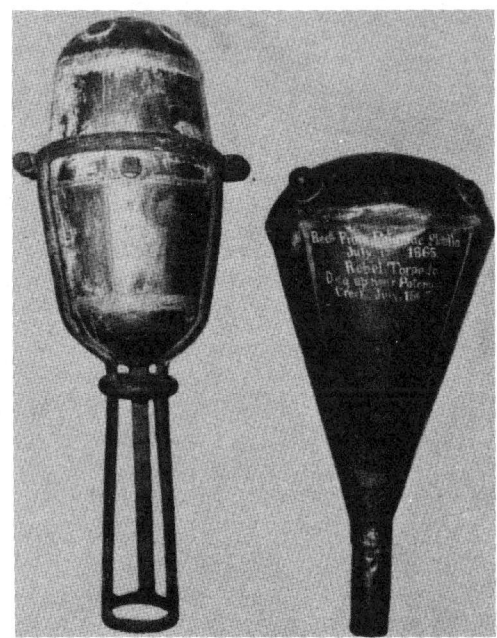

Revolutionary powder keg (top left); Civil War mine, Singer type (top right). Spar torpedo (bottom left); Brooks type mine (bottom right). Note that this mine is labeled "Rebel Torpedo."

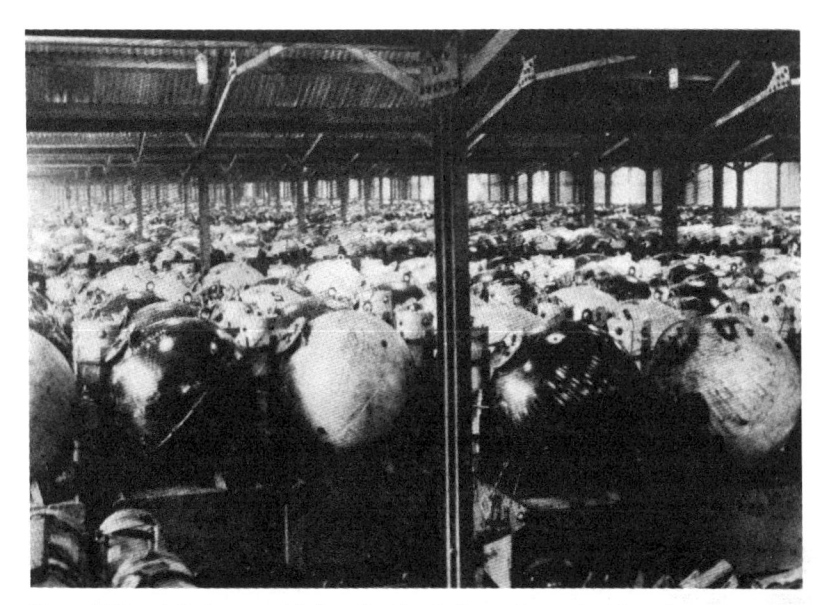

Moored Mine, Mk 6, assembled in the Ready Issue Store awaiting the return of the Mine Squadron prior to being embarked for the next laying operation at Inverness, Scotland, 1918.

Mine stockpile on Tinian, June 1945. Mines were sometimes loaded aboard aircraft within 48 hours after arrival during the height of Operation Starvation.

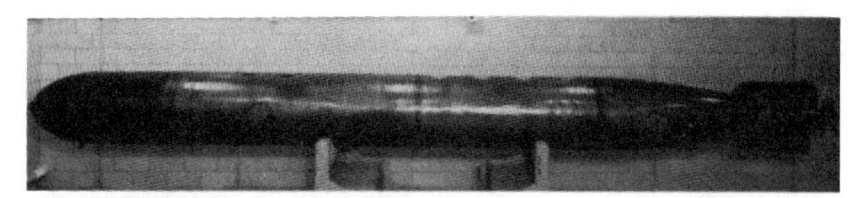

Mk 27 Mobile Mine.

Mk 25 Mine with components exposed.

Post-WWII mine entering pressure test chamber at White Oak (1952).

"Working the stockpile." Mk 55 Mine on fork lift.

Test-fitting mines on B-52 aircraft.

23

very successful. The reliable underwater initiation of gunpowder was going to have to wait for the science of electricity.

Although gunpowder, or what was later called gunpowder, was described in the thirteenth century by Roger Bacon (1214–1294)— and he was aware of its explosive property—the use of this mixture to propel objects from "firearms" was first made by Berthold Schwarz, a Franciscan monk of Freiburg, well after Bacon's death. According to Read's *Explosives* [40], there is no evidence that Roger Bacon was acquainted with the propellant power of gunpowder or with its use in guns. Read shows an illustration of a primitive but large gun in a manuscript of 1325 (From 'De officiis regum' by Walter de Millemete, in the Library of Christ Church, Oxford). The picture shows a flask- or bottle-shaped horizontal vessel (not unlike Dahlgren's gun of 1850) with a large arrow emerging from its elongated neck and an ordnanceman holding a long cord of burning fuse leading to a hole in the top of the barrel. This appears to be the prototype of cannons, bombards, or mortars for the next several centuries.* The scheme of igniting gunpowder in small arms by means of a spark coming from a flint and steel is attributed to Schwarz. This became the prototype for handguns of all sorts leading to the flintlock musket which came into use in the mid seventeenth century and a hundred years later was the principal hand weapon of the American Revolutionary War. It was this spark-making mechanism that Bushnell adapted for his impact-initiated mines.

There was, however, another method for igniting gunpowder, which was invented and demonstrated by Benjamin Franklin and described by him in a letter to one of his London correspondents, Peter Collinson of the Royal Academy, in 1751. In his own words:

> I have not heard that any of your European Electricians have hitherto been able to fire Gunpowder by the Electric Flame. We do it here in this Manner.
>
> A small Cartridge is filled with Dry powder, hard rammed, so as to bruise some of the Grains. Two pointed Wires are then thrust In,

---

* According to Read there was a powderworks at Augsburg, Germany, in 1340. Both gunpowder and cannon were being manufactured in England in 1344 and probably earlier. In English military operations, it is said that guns and gunpowder were first used at the Battle of Crécy in 1346, although it is also claimed that some sort of cannon was used in 1327 by Edward III against the Scots. Although the use of guns, mortars, and cannon proliferated rapidly after the initial invention, it is perhaps amazing that no significant improvement was made in either the gunpowder or the method of initiating it for about five hundred years! Gun cotton and also nitroglycerine was invented in 1846 exactly 500 years after the Battle of Crécy.

one at Each end, the points approaching Each other in the Middle of the Cartridge, till within the distance of half an Inch: Then the Cartridge being placed in the Circle [circuit], when the Four Jarrs are discharged, the Electric Flame leaping from the point of one Wire to the point of the other, within the Cartridge, among the powder, fires It, and the Explosion of the powder is at the same Instant with the Crack of the Discharge. [Quoted by Lundeberg, Reference 28].

The "Jarrs" referred to are the Leyden jars first described by Cunaeus of Leyden in 1745. One can only marvel at the speed with which information and application moved around the Western world at a time when a one-way trip across the Atlantic could take over two months. This exploit of Franklin's, although of little practical value in a military application because of the necessarily cumbersome problem of handling and charging Leyden jars, nevertheless may have influenced him some twenty-four years later to stop off and visit the workshop of David Bushnell near Saybrook, Connecticut, in 1775. Bushnell was working on a scheme by which a gunpowder mine could be conveyed in a hand-propelled one-man submarine to be attached underwater to the hull of an enemy vessel. Bushnell's submarine, called the *Turtle*, was actually used on 7 September 1776 in an abortive attempt to fasten the first limpet mine in naval warfare to the hull of Vice Admiral Richard Lord Howe's flagship *Eagle* in the Hudson River at New York. It is not entirely unlikely that General Washington's permission to try out the mine schemes in New York and Philadelphia rested to some degree on the scientific respectability lent Bushnell by Franklin's visit.

The schemes for igniting gunpowder by flame or sparks originated in the Middle Ages. The use of the electric discharge between two sharp points imbedded in the powder was new, depending as it did on the high electric potential which could be accumulated in a Leyden jar. It was also transitory because of the discovery of a new source of electricity, initially by Galvani in 1791 and expanded by Volta around 1800 into a practical source of continuous electrical current at relatively low voltage. It is interesting to see how this could be exploited to fire gunpowder. The voltage available was too low to make sparks and so something else had to be done. The answer, of course, was to replace the spark gap by a fine wire which would heat up when the voltage was applied and which, when hot enough, would ignite the powder surrounding it. This idea, actually using a thin platinum wire which would get red hot without melting, was first used, however, as late as 1839 when British Colonel Charles Pasley removed a wreck lying off Portsmouth Harbor, England, by an underwater demolition.

This operation, by the way, was an all-British one, with Sir Charles Wheatstone and Michael Faraday providing scientific advice and a Daniell's cell being used instead of Volta's. This successful event—using insulated electric cables, batteries, hot-wire initiators, and gunpowder in watertight containers—provided a demonstration to the world that this was a reliable, safe, reproducible system that anybody could build and operate if he took adequate care. It was no longer a field for genius, good luck, or black magic. It is perhaps surprising, however, that it took forty years to get from the battery to the hot-wire igniter. Let us go back a moment to see what else had been happening.

Even before Volta's battery, Tiberius Cavallo, a friend of Volta, had tried to initiate gunpowder by a spark, even as Franklin had, but at a considerable distance. Perhaps the principal aim of electrical experimenters in these first days was to transmit some sort of message over a distance. One of the greatest problems was to get conductors properly insulated so that the electrical effects could be transmitted without leaking off. It was obvious that if one wished to transmit a message under water, as for example from one side of a river to the other, the wires could not be suspended from insulating threads as was the case on land; they would instead have to be wrapped in an insulating material. Before insulated wire was developed, however, Cavallo reported in his "Complete Treatise on Electricity," London 1795, that he had "some years ago" fired gunpowder at a distance. Quoting from Lundeberg's quote, "At first I made a circuit with a very long brass wire, the two ends of which returned to the same place, whilst the middle of the wire stood at a great distance. In this middle an interruption was made, in which a cartridge of gunpowder mixed with steel filings was placed. Then by applying a charged Leyden phial to the two extremities of the wire, (viz. by touching one wire with the knob of the phial, whilst the other was connected with the outside coating) the cartridge was fired. In this manner I could fire gunpowder from the distance of three hundred feet and upwards." [28, P. 3] The iron filings were in fact a step toward the idea of a hot filament because they reduced the spark-gap resistance and supposedly made it possible to have longer lead wires.

Between Volta's battery and Col. Pasley's system, several developments occurred in Germany and in Russia. In Germany, Samuel Thomas Sömmering experimented with telegraph apparatus, and encouraged by visits from diplomats and other government officials as well as the renowned Baron Alexander von Humboldt, he continued his experiments with wire insulated with India rubber and varnish. He succeeded in 1812 in telegraphing through some 10,000 feet of cable. Among his diplomatic visitors was one Baron Pavel L'vovich Schilling von Cannstadt, a member of the Russian diplomatic mission in the Bavarian capital (Munich). With the approach of hostilities between Russia and France in 1812, Schilling intensified his efforts toward the possible use of these insulated wires not only to send messages but to detonate gunpowder charges planted on the opposite bank of a defended river by a controlling observer. He insulated his cables with tarred hemp and copper tubing (for protection) and also devised yet another scheme for initiating the explosion. Perhaps following a lead from Humphry Davy, who had demonstrated an electric arc using pointed carbon electrodes as early as 1800 (also using a battery source), Schilling devised a carbon-arc fuse consisting of two pieces of pointed carbon in contact and embedded in the powder.

In the fall of 1812 Schilling carried out a demonstration of his system at St. Petersburg where he detonated powder charges that were controlled from the opposite bank of the Neva River. As Lundeberg says, "Although Russian military engineers do not appear to have adopted his galvanic system at that critical juncture in the Napoleonic invasion, Baron Schilling retained a lively interest in its development." In fact, in 1814, when the various allies were in Paris, Baron Schilling, with the Russian forces there, repeated on several occasions the demonstration across the Seine he had previously made across the Neva, to the delight and amazement of the local inhabitants. In fact, the Baron had a lot of the showman in him, in this manner anticipating the antics of both Fulton and Colt and perhaps of anyone else who is trying to interest, inform, or persuade a patron in something new. For example, in demonstrating his carbon-arc igniters to Alexander I at the tsar's summer camp near St. Petersburg, "Once Baron Schilling had the honor to present a wire to the Emperor in his tent. He begged his Majesty to touch it with another wire, whilst looking through the door of the tent in the direction of a very distant mine. A cloud of smoke rose from this exploding mine at the moment the Emperor, with his hands, made the contact. This caused great surprise, and provoked expressions of satisfaction and applause." [28, P. 5]

With these preparations, it is perhaps not pure chance that after Schilling's death in 1837 his efforts in galvanic mine development were reviewed at the Imperial Academy of Sciences. The chief reviewer was a young Prussian émigré, Moritz von Jacobi, who two years later was appointed by Nicholas I as scientific leader of a joint services Committee on Underwater Experiments. This was a working group destined to carry out a program on galvanic mine developments for some fifteen years before the Crimean War (1854–56). No doubt because of this preparation, the Russians did actually deploy some mines during that war at Kronstadt as a precaution to guard St. Petersburg, although the hostilities actually took place a thousand miles away on the Crimean peninsula. Although he was a distinguished member of the Imperial Academy, Jacobi is best known for the invention and development of yet another device for firing gunpowder, which does not depend on either long wires or hot wires and yet which was used to initiate mines by impact. The device is a chemical method of generating heat and consists of a glass tube partly filled with sulfuric acid and imbedded in a mixture of sugar and potassium chlorate powder. For example, if the glass tube is protected in a lead sheath from ordinary impacts, it can nevertheless be broken if a ship smashes into it, bending the lead. The acid mixes then with the potassium chlorate and sugar and generates a hot mixture which is sufficient to ignite gunpowder.

While progress in mine warfare occurred slowly in Russia from the Napoleonic to the Crimean War, a certain progress was also developing in Germany principally at the hands of Werner von Siemens. Siemens joined the Prussian army in his youth, and in 1846, at the age of twenty-seven, was appointed to a commission to establish an underground telegraph system in Prussia. Lundeberg [28, P. 13] says, "During the Schleswig-Holstein War in 1848 von Siemens collaborated with his brother-in-law, Professor Karl Himly of the University of Kiel, in designing and laying a field of galvanically controlled mines in the approaches to Kiel that effectively discouraged Danish naval bombardment of that seaport. Following these early achievements, von Siemens emerged rapidly as an international entrepreneur of telegraphic systems, completing numerous major lines in both Prussia and Russia and ultimately attaining stature as a giant of early German scientific industry." Patterson in his Nimrod Report History [33] says these mines were wine cask mines, and so it can be presumed that this was the first controlled moored minefield in history, and also the first installed in time of war.

Any remarks on the history of mine warfare should note that Robert Fulton contributed many ideas and much work to the subject without gaining the measure of acceptance that he sought. As early as 1797 he proposed the use of various floating charges to the British to use against the French, and thereafter upon rejection went to the French to try to sell the idea of submarine-laid mines to Napoleon as a means of attacking Britain. During this time he built the *Nautilus* in which he remained submerged at a depth of twenty-five feet for two hours. In 1803 the French decided against further experiments, and Fulton returned to England where for a time he was paid a salary and given funds and facilities for experiments. It was at this time that he accomplished the successful destruction of the 200-ton brig *Dorothea*. After the Battle of Trafalgar (in which his schemes played no part) in 1805, he returned to the United States and proceeded to Washington, D.C., where he represented his submarine and his underwater bomb (the first moored mine) to Mr. Madison, the Secretary of State, and Mr. Smith, the Secretary of the Navy. These gentlemen were much interested and granted him funds for continuing the tests.

In 1807 he succeeded in blowing up a brig in New York harbor but only after several failures. He was probably using two mines bridled together as in the *Dorothea* experiment, but the mines were improperly balanced and turned over, spilling out the priming powder so that the gun-lock spark had no effect. These failures made the government skeptical of the whole scheme. Nevertheless, in 1810 he again attempted to destroy a sloop, but the captain of the sloop was permitted to deploy a protective net around the ship so that the mines could not get close enough. In these experiments attempts were also made to harpoon the target ship from a small boat, as a method of getting the mine charge, which was tied to the harpoon, close to the ship. After the harpoon was lodged in the ship, the tidal current was relied on to drift the mine alongside. Counterfire from the ship made this a hazardous method, and the nets defeated the mines anyway. Mr. Fulton admitted that he could not destroy a ship so protected, but pointed out that the provision of such equipment for all enemy ships would be a burden and a hindrance to mobility. An impartial government committee of experts would not recommend adoption of any of Mr. Fulton's proposals. However, they overlooked the device that really had possibilities for development, namely, the *moored mine*. Against powder casks anchored in a harbor and floating at or near the surface of the water, equipped with firing locks which would explode the powder on contact with a moving ship, the naval officers could devise no defense. Such weapons could apparently close the harbors of the United States against the vessels of an enemy. The problem of getting the explosive charge to the ship was solved by simply having the ship run into the explosive charge. It will be noticed that up to this point the mine had been conceived as a device for attacking an anchored ship. If the ship were moving, however, the mine could be anchored and would thus assume its true role of lying in wait until its victim approached.

During the War of 1812, Fulton made proposals to the U.S. government to plant moored sea mines in harbors and to use the harpoon system with mines for attack. Nothing much came of this. Fulton died in 1815, but as late as 1814 he was at the Navy Yard in Washington getting casks and other gear for experiments which were never done. Fulton's experiments were largely designed to show over and over again that an explosion would sink a ship, rather than concentrating on more reliable ways to get an explosion to occur. They were designed to convince the skeptics rather than to advance the art.

The Fulton story is remarkable in that for a period of eighteen years with access to the highest levels in three countries, he was unable to get any real acceptance of his ideas beyond token support for experimentation. He did quite well at this and his promotion techniques were superb. However, the technical input to the process all this time was minimal. He was stuck with the flintlock initiating mechanism, which was highly uncertain and unsafe. One suspects that even the best salesmanship will not succeed with a faulty product. But we have seen that there were other factors at work. If the product had been perfect, it would have been shunned by a sensible navy—at least in the year 1805. The successful destruction of *Dorothea* just six days before the Battle of Trafalgar was ill-timed. Although the result—the ship was broken in two—was, as Duncan says, "highly gratifying to Fulton, it greatly alarmed the British naval authorities." [12, P. 19] After Nelson destroyed the French and Spanish Fleets, however, "England had no need of submarines, or of mines, or of Fulton."

Fourteen years after Fulton's death, Samuel Colt, later of revolver fame, began to experiment with underwater explosions fired by electrical means. He used an electric current (from a galvanic battery) flowing through a fine wire to heat the powder and cause the explosion. In this he was contemporaneous with Colonel Pasley. Whether information passed between Colt and Pasley or whether each conceived of this scheme independently is not determined. As early as 1831 Robert Hare, professor of chemistry at the University of Pennsylvania, had proposed a method for igniting blasting powder in rock crevasses, which was safer and more reliable than using the Leyden jar spark. He suggested igniting "explosive gaseous mixtures" using a battery as the source of energy. Whether he used a hot wire or sharpened carbons is not clear. He extended his suggestion, however, to the idea of using remotely controlled underwater mines in conjunction with coastal defense and as a part of the planned installations going with specific forts. This idea was taken up and pressed very hard by Samuel Colt. (Colt says he thought of it in 1829.) Colt conceived of the moored minefield as a defensive controlled-weapon system and was very concerned about not only firing the moored mines, but about firing each at the right time when a target vessel would be over a particular mine or within damage range. In other words, he was concerned about the fire-control aspects of the system and had a scheme whereby an image of the harbor and of any ships in it could be superimposed by means of a concave reflecting mirror onto a chart showing the locations of the separately firable mines. He was so secretive about his arrangements that even in his patent applications he was not at pains to reveal what he had developed.

This among other things, in spite of a successful demonstration with a moving ship in the Potomac River, led to the government decision in 1844 to terminate support and interest in his proposals. According to Lundeberg the best account of Colt's firing system is contained in a newspaper account rather than in his patent, as follows:

> ... the Battery consists of a light sheet-iron box filled with gunpowder, and having two copper wires wound around with cotton, then varnished with a mixture of gun shellack, alcohol and Venice Turpentine, and extending through tight corks in one side of the box, having a piece of platina wire extending between them in the box amongst the gunpowder, and the two copper wires extending off from this box (which may be anchored in the channel of a river) to a large one of Grant's Electricity collecting (connecting) machines, electrified by a large Galvanic Battery, which may be seven or eight miles distant from the box and where the operation (operator) is, having one of the wires in his hand ready to attach them to the collectors (connectors) the instant the signal is given to explode the box. [28, P. 32]

Drawings attest to the fact that "anchored" means "moored", although unless there are air spaces or other floats not mentioned it is clear that boxes "filled with gunpowder" would immediately sink to the bottom. Colt spared no effort to promote his ideas, but he refused to let anyone know precisely how they were to work. He combined excessive claims as to the effectiveness of his devices with excessive secrecy as to how they in fact did work. The result of this behavior was very soon to turn off all official interest in his proposals, and he had to achieve success in this world by inventing and manufacturing the Colt revolver, which received considerable stimulus from the Mexican-American War, 1846–48. Or as Lundeberg says, "That the United States did not, as in the case of Russia, integrate observation mines with its Third System of coastal fortifications during the two decades prior to the Civil War appears to stem in no small part from the fact that Samuel Colt, as principal American proponent of galvanic harbor mines in that era, represented their utility in a manner that antagonized those military officials charged with long-range development of the nation's systems of coastal fortifications, a system then approaching the climax of its technical development." [28, P. 12]

None of the experiments or claims resulted in any support or action on the part of the government, and so no further thought was given to mine warfare in this country until the Civil War. As we have seen, the preparations made in Europe during these years did lead to operational use of mines in Kiel, and in the Crimean War. Patterson mentions somewhat cryptically a use of electrically controlled mines in the Harbor of Venice during the French-Austrian War in 1859. He says, "The position of the mines relative to ships was ascertained by means of a *camera obscura* in which a chart marked with the mine plants could be compared with the reflected images of ships on the surface." [33, P. 17] This indeed seems like an application of Colt's fire-control idea, but if so it is without attribution. The idea disappeared without a trace later when the influence fields of ships were used to indicate their presence in the vicinity of a mine for either a warning signal or a firing signal. This stage of development, however, did not occur until about three quarters of a century later.

Appendix A is a somewhat detailed chronology which lists the technical events in the development of explosives, and of electricity and physics which have made possible some of the applications for mine warfare which are also identified.

From Civil War to World War I

Although developments up to this time were of basic importance, they had had little practical significance. They did, however, provide the background for the first effective use of mines in warfare. In 1861, Matthew Fontaine Maury, the creative chief of the Navy's Depot of Charts and Instruments and founder of naval oceanography, decided at the outbreak of the Civil War to throw his lot in with his native South. Convinced that the only possible defense of a nation with a vast coastline but without a navy lay in the use of mines, he established a Torpedo Bureau and a Torpedo Corps in Richmond and proceeded to design, develop, and deploy a number of mines of various types. In those days the term *torpedo* covered almost all kinds of uncontrolled or manually controlled underwater weapons, i.e., what we now call mines. The power-driven mobile torpedo had not yet been invented. When that event did occur (the first Whitehead torpedo was built in Fiume in 1866), it became customary to designate the older nonpropelled items as *mines* and use the term *torpedo* for newer self-propelled items. In most recent times the distinction has again become blurred by the combination of mine and torpedo characteristics in a single weapon (Captor).

Now as Patterson notes, in a war marked by the first large-scale construction of ironclads, not one of these vessels succeeded in sinking another. Several of them were sunk by mines, however. A table given by Patterson [33, P. 19] shows the record of ships sunk or damaged by mines during the Civil War. The table lists thirty-six ships of which only one, the Confederate ram *Albemarle,* was sunk by a Union mine (in Plymouth, North Carolina). All the rest were ships sunk or damaged by Confederate mines in southern waters, the James River, Charleston, Mobile Bay, Cape Fear River, the Yazoo River (near Vicksburg), and so on. Three of these were Confederate ships —the victims of Confederate mines which had accidentally broken loose from their moorings. The remaining thirty-two ships, all under the Union flag, were armored ships, gunboats, and transports. In their encounters with Confederate mines, the score goes like this: twenty-three destroyed, one wrecked, four seriously damaged, one disabled, one internally damaged, one slightly damaged, and one not damaged. These statistics illustrate a point which has caused some concern about the measurements of the effectiveness of a mining effort. A very crude measure would be simply to count the number of sinkings. However, this would certainly underestimate the consequences of the mining effort. It would ignore the diversion of effort due to repairing damaged vessels. One could measure the effectiveness by the ratio of the cost of the losses to the cost of building and installing the minefields, although this would be more a way of comparing one kind of mine with another. The main effect of a mine operation, however, should lie in the importance to both attackers and defenders of the sinkings produced and the delays thereby caused, the failures to deliver, arrive, receive, or attack, and the further deprivation to the loser of those assets which he formerly possessed. These are matters which are very difficult to measure. It would appear, however, that the Confederate mining efforts were remarkably successful in terms of the damage done for the effort expended.

The so-called "Singer" mine (invented by a brother of the sewing-machine manufacturer) is credited as being the most successful Confederate mine. It was a hand-planted moored mine with a charge of about sixty pounds of black powder. It carried a heavy cap on its upper surface which would be knocked off by a ship running into it. When it came off, it released a spring-driven plunger which struck a fulminating charge—thus exploding the mine. This method of initiating the charge was doubtless adapted from the new method of initiating a firearm, which originated from the discovery of fulminate of mercury in 1800 by Edward Charles Howard (brother of the twelfth Duke of Norfolk), and the patent in 1814 by Thomas Shaw of Philadelphia of a percussion cap using fulminate and potassium chlorate. This finally led to the replacement of the spark-producing flintlock mechanism and made it possible for the desperate and inventive Southerners to adapt the percussion cap quickly to mines in 1861.

The Torpedo Bureau of Richmond also used the identical chemical initiator with its frangible glass tube, sulfuric acid, potassium chlorate, and sugar that Jacobi had invented before the Crimean War. Is it possible that Maury had learned of this device in 1853 in Brussels where he had organized an international conference on oceanographic and naval subjects? The Singer mine initiator, however, was held to be more reliable than the chemical scheme, except that a new difficulty arose—the firing spring which was exposed to the seawater tended to get clogged up by sea growth over a period of time, especially in tropical waters. It could have been this emerging defect which protected Admiral Farragut at Mobile Bay when he chose to ignore mines which had been in place for some time and sail on through the field.

Patterson's table [33, P. 19] also shows that two Union ships were attacked by "Davids," the *New Ironsides* damaged in Charleston in 1863, and the USS *Housatonic* sunk on 17 February 1864 also in Charleston. "David" was a generic term given to an underwater manned vessel used to attach a spar torpedo (or as we now would say, limpet mine) to the hull of a moored ship. The David was so called possibly because it attacked a Goliath, rather than because of its rather remote connection with David Bushnell and his *Turtle*.

Today in New Orleans in the State Museum across the street from the St. Louis Cathedral is a curious iron vessel with a deep keel, about twenty feet long, with a four-foot beam, and designated C.S.S. *Pioneer*, "the first successful submarine." This vessel, designed by Baxter Watson, Jr., James R. McClintock, and Horace Lawson Hunley, was built in New Orleans in 1861 in McClintock's Machine Shop. The vessel could hold three men. She was made of quarter-inch sheet iron and propelled by a hand crank attached to a shaft that held a four-bladed propeller at the stern. She could make up to about $3\frac{1}{2}$ knots. Along her keel was a ballast tank that could be filled with water to submerge. To surface, the crew had to pump the water out by hand. When submerged, air reached the cramped compartment through a long rubber tube extending from a float on the surface to the inside of the cabin. She had diving planes on the bow and rudders fore and aft. She carried a "spar torpedo." In the bow there was machined a hole large enough to run a long lance through. At the end of this lance was a detachable barbed iron point, and attached to the point was a 75-pound charge of gunpowder containing a bottle of acid and a "triggering mechanism." The idea was that when the *Pioneer* would stealthily approach an enemy ship, the lance would be thrust firmly into the ship's wooden hull. Then the submarine would back off leaving the iron point and the charge, the lance remaining in the submarine. The charge was connected to the submarine only by a long line. As the *Pioneer* backed out of range to safety, the line would grow taut and trip the mechanism in the torpedo that broke the bottle of acid and exploded the charge.

So much for the concept. Farragut came into New Orleans before the *Pioneer* could be used. The owners decided to scuttle her in Lake Pontchartrain rather than risk compromising her design by capture, and this was done in 1862. They decided to move to Mobile to build a new larger submarine. There, Hunley and others built another submarine, still hand cranked, using an old boiler for a hull. Once submerged, the crew of nine had only the air in the hull to breathe. On trials she drowned her crew and Hunley, for whom she had been named. After four more ill-fated trials, permission was granted to Lieutenant George Dixon of the Alabama 21st Infantry to undertake further operations. It is not clear how he got this vessel to Charleston. On the night of February 17, 1864, however, he rammed the *Hunley*'s spar torpedo into the side of the steam sloop of war USS *Housatonic*. The sloop sank with no loss of life; the *Hunley* also sank, for the last time, with the loss of her crew of five. This Pyrrhic victory is said to be the first submarine sinking of a ship. It is also the first sinking of a submarine by a ship!

The original submarine, *Pioneer*, which never saw action, was located and recovered in 1878 and presented to the museum in New Orleans where nearly a century later it excites the wonder and admiration of the curious. (Some of this information comes from *Historic Ships Afloat*, [11]).

Before leaving the Civil War, mention must be made of Admiral Farragut's entrance into Mobile Bay in 1864. The Confederates had planted a field of eighty or ninety moored mines, not all identical. As Duncan tells it [12], fortunately for Admiral Farragut they had been planted some time before he attacked Mobile in 1864. Nevertheless, the story is incredible. The *Tecumseh*, an ironclad vessel of 1,034 tons, led the attack. Just as she reached a point where her guns had the range of the defending forts, she struck a mine which exploded and she sank in a very few minutes. The *Brooklyn* was following the *Tecumseh*; her captain saw the *Tecumseh* sink and saw other mines in the water. He altered his course and signaled to Admiral Farragut on his flagship USS *Hartford* that torpedoes were present. Farragut was furious and signaled to the *Brooklyn*, "Damn the torpedoes [mines], Captain Drayton, go ahead." No more mines fired. Later it was discovered that the mines were inert due to long immersion (corrosion) and wave action.

This must certainly be the first and last time that a commander could safely ignore a minefield! But a successful commander, of course, is one whose decisions lead to success, whatever the reasons. Admiral Farragut, incidentally, believed that the use of mines was unethical and that they were "sneak" weapons and "devilish" devices. During the Civil War he wrote, "I have never considered it [the use of mines] worthy of a chivalrous nation." [12, P. 13]

Another account of this same opinion by Admiral Farragut is more complete and changes his meaning: Arnold S. Lott [27, P. 10] quotes him from a statement made on 25 March 1864 to the Secretary of the Navy as follows: "Torpedoes [mines] are not so agreeable when used by both sides; therefore, I have reluctantly brought myself to it. I have always deemed it unworthy of a chivalrous nation, but it does not do to give your enemy such a decided superiority over you."

According to the Phoenix Museum in Mobile, the 1,034-ton monitor ironclad *Tecumseh* was found in 1966 almost buried in sand in thirty-six feet of water 200 yards offshore, i.e., off Fort Morgan, which guarded the entrance to the bay from its eastern side. The story was that the Smithsonian Institution was going to raise the old warship, but recently decided to give it back to the state of Alabama. They are still debating over it, while the price of salvage has gone up into the millions.

Between the end of the Civil War and the beginning of World War I very little happened in the United States to advance the state of the mining art. It would seem that it took an active war to unleash American inventive genius and to generate the energy which is summoned only when there is a crisis. The Bureau of Ordnance was proceeding in its own way with the problem of mine readiness. In 1898, a Bureau of Ordnance Annual Report refers to mine design work at the Naval Torpedo Station, Newport, R.I., (here "Torpedo" means torpedo and not mine!) and states that gun cotton mines were prepared and issued." This was a technological or engineering advance—to replace the old monopoly of gunpowder with the more powerful nitrated cellulose, first prepared in 1838 by Pelouze and improved in safety by further processing advances in 1865 by Sir Frederick Abel. As late as 1909 the Bureau was describing its "Mk 2 Naval Defense Mine" which was designed and manufactured for it by a French firm, Sauter-Harlé of Paris. This mine was essentially the Elia mine, having an automatic anchor and a mechanical firing arm. The Elia mine was originally designed in Italy by Commandante Elia. As far back as 1901, the British had examined this design. Since the original inertial firing device had a tendency to explode prematurely in a seaway, they replaced it with a mechanical arm which could release a cocked spring and so fire a percussion detonator. This, then, became the "British Elia."

About 1915, the Bureau of Ordance began to manufacture its own mines using a design owned by Vickers Company of England. This mine also had an automatic anchor and was loaded with TNT instead of gun cotton. (The first nitrated benzene ring explosive, picric acid [nitrated phenol], was not known as an explosive until 1871. It was introduced as a filling for shells in 1885, and from that time the coal-tar industry rapidly developed other explosives, including TNT, which became plentiful in World War I.) In 1917 when we entered World War I, we discovered that the mines of the British Elia design which we were manufacturing had been found to be quite unsatisfactory in practice by the British. The difficulty lay in the mechanical arm for impact firing of these mines. The British were loath to use the Hertz horn scheme for initiation because the glass they made for the electrolyte was quite sensitive to countermining shock (i.e., the shock from a nearby exploding mine). It was not until late 1916 that the British, by improved glass-manufacturing techniques, had remedied these defects and produced the reliable so-called H-2 mine fitted with Hertz horns. It was the H-2 mines that finally made the Dover minefields successful. In 1917, however, the United States was in much the same position with respect to mine warfare that it had been in at the conclusion of the Civil War; it had no usefully functioning mines.

Although the United States had been doing very little on the mine front between the Civil War and World War I except buying obsolete mines abroad, other nations had been busy improving their mine capabilities. The Hertz horn, which was to prove so valuable for firing contact mines, evolved right after the German-Austrian War of 1866. This 1868 invention consisted of an electrolyte (potassium bichromate solution) in a glass tube sheathed in a soft metal horn, usually lead, which stuck outside the mine. When bent by contact with a ship, the glass would break and the electrolyte would complete the circuit in a battery which would then fire the electric detonator. This became standard for contact mines. (Heinrich Hertz, 1857–1894, of electromagnetic fame could hardly be the author of this, for he was only eleven years old at the time.)

The automatic anchor, developed in the late nineteenth century at HMS Vernon at Portsmouth, made it possible to moor mines quickly at a predetermined depth without measuring the water depth, provided it did not exceed the maximum cable length available. Figure 2 shows how this scheme works. A plummet weight descends faster than the anchor until it is a distance below equal to the depth eventually desired for the mine case. At this point plummet and anchor sink together leaving the case on the surface. When the plummet hits the bottom, the loss of tension locks the cable, and the anchor pulls the mine down below the surface an amount equal to the distance the plummet lay below the anchor. It is not necessary to know the depth of the water!

Another improvement made by HMS Vernon consisted of a new firing device, namely, an inertial switch which would connect to the detonator the voltage from a battery placed in the mine itself. This is the first instance of a mine carrying its own battery.

Although in this interim period the Russians, Germans, French, and British were the principal proponents of mine warfare, even the Spanish apparently had mine capabilities in their depleted arsenals, at least according to Patterson. He says that "without arguing the unanswerable question of whether the battleship *Maine* was mined, it is still known that the Spanish employed mines off Cuba and that they likewise emplaced them at Manila and Subic Bays in the Philippines." [33, P. 22] These mines indeed were either avoided or ineffective. They produced no casualties. But the believable proposition that the Spanish had somehow subversively blown up a U.S. battleship moored peacefully in Havana Harbor had a great deal to do with the U.S. entry into the Spanish-American war in 1898. Admiral Rickover, seventy-eight years later, is able to argue fairly conclusively, with the help of analysis by R. S. Price and I. S. Hansen (explosive and structural damage specialists of the Naval laboratory establishment) that the *Maine* blew herself up by igniting her powder magazines through the spontaneous combustion of adjacent coal storage bunkers. [42] The unanswerable question seems to have been answered. Thus is seen the international importance of mine warfare even when later it is shown that there were no mines involved!

The major use of mines and their first use at sea in naval actions occurred in the Russo-Japanese War of 1904. Mines played a decisive role in this war. It is not surprising that the Russians would rely on defensive mining outside their Pacific ports, as they had been devel-

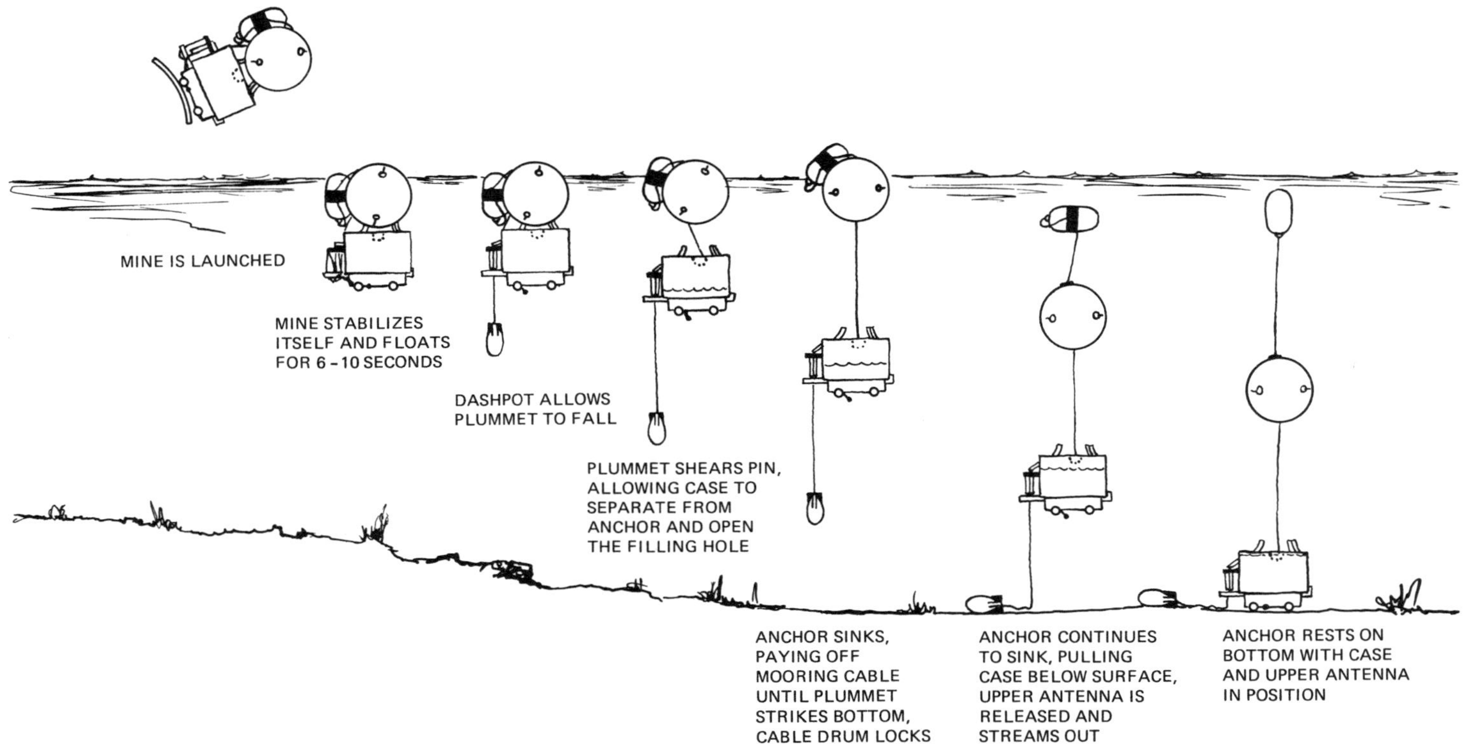

Figure 2. Mine Mark 6 Planting Sequence.

oping mine concepts and uses since the days of Jacobi and before. But it is perhaps surprising that the Japanese emerged at that time as a major naval power, complete with the most modern mines, which they used offensively. They had also learned to cast-load them with picric acid and hence were credited with having a new secret super-explosive. The Japanese were able to lure a Russian force over their minefield with serious consequences for the Russians. The Russians on their own part used mines in a defensive manner to extend their own coastlines to sea, preventing Japanese ships from coming in to give fire support to their troops on land. In this war, counting both sides, three battleships, five cruisers, four destroyers, two torpedo boats, one minelayer, and one gunboat were sunk by mines while other ships were severely damaged. The Russians laid the first mines in that war. There are two stories worth mentioning about it. One is that the Japanese minelayer *Yenisei* of 2,500 tons blundered into her own minefield and was blown up by a mine she had just laid. This event stressed the importance of good navigation or, failing that, of a delayed arming mechanism so that minelayers could get clear. The other story relates to the Russian loss of their 11,000-ton battleship *Petropavlovsk* which hit two mines and went down while crossing a known Japanese minefield. The Russian Admiral Makaroff lost his life in that event. He refused to consider mines dangerous and would not change course! Perhaps he was following Admiral Farragut's example without his luck.

According to Captain Cowie [10], Japan lost two battleships, four cruisers, two destroyers, one torpedo boat, and one minelayer—the latter the *Yenisei* which struck her own mines. The Russians lost one battleship, one cruiser, two destroyers, one torpedo boat, and one gunboat. All these casualties were to mines laid in the open sea and constituted about two thirds of all ship casualties. There were no sinkings due to torpedoes.

The Russo-Japanese War showed that mines were formidable weapons. The Russians sank more Japanese ships by mines than by any other form of attack (perhaps a poor commentary on their gunnery). They also were able to affect operations ashore by the use of mines at sea. This was also the last war in which only military forces were the main aggressors and the main targets. Thereafter, the sinking of commercial shipping by any means available would become simply a facet of war by attrition. However, many mines used in this war had broken loose from their moorings and were to be found in later months floating into shipping lanes with subsequent casualties to noncombatants on the high seas. Such a state of affairs was not to be viewed with apathy by shippers or by insurance companies. Concerns thus aroused were influential in bringing about the Hague Convention of 1907 in which mines were put under a series of fairly ambiguous restrictions which were intended to protect merchant shipping. The meeting was attended by forty-four nations. Cowie [10, Ch. 9] discusses at some length the issues which led to the final Convention. To begin with the British were unable to persuade the Russians and the Germans to agree to give up mining altogether. Russia claimed that mines were essential for her defense since her coasts were so extensive and her fleet so limited. The Germans felt that mines should be permitted in the actual and prospective theaters of naval operations. The British felt that such a ruling would in fact make possible the secret dissemination of submarine mines on the outbreak of war and could lead to the promiscuous destruction of noncombatant shipping in areas like the English Channel, the Straits of Gibraltar, and elsewhere. They were against it. The British also noted that apart from the designation of limited areas for mining, and the absolute prohibition of unanchored floating mines, it should be possible to provide a technical means whereby a moored mine which broke loose would automatically destroy or scuttle itself after a certain period of time.

The conferees finally agreed to a number of provisions which were ambiguous in fact and which were unenforceable even if there had been no loopholes. It was agreed that it was forbidden to lay drifting mines unless "they are so constructed as to become harmless one hour at most after those who have laid them have lost control over them." It was also forbidden to lay "automatic contact mines which do not become harmless as soon as they have broken loose from their moorings." Also, it was agreed that it was forbidden to lay automatic contact mines off the coasts and ports of the enemy with the sole object of intercepting commercial navigation! Another article states that when anchored contact mines are used, every possible precaution must be taken for the security of peaceful navigation. The belligerents will undertake to provide, as far as possible, for these mines to become harmless after a limited time has elapsed and "to notify the danger zones as military exigencies permit by a notice to mariners which must also be communicated to the governments through the diplomatic channel." Finally the convention agrees that if some of the contracting powers do not at present own perfected mines with self-destruct or sterilizing features and hence could not comply with the agreed-on rules, then they will undertake to convert the materiel of their mines as soon as possible so as to bring it into conformity with the foregoing requirements.

In looking at this 1907 International Agreement, one is struck with the notion that the same sort of reasoning is exhibited here as Orwell depicted so masterfully in *Animal Farm*. The original commandments written on the barn were modified slightly as time went on so that they contradicted their original meanings. Thus, bad effects of mines on innocent bystanders or neutral traders were easily dealt with by declaring the mined area to be a blockade. It then became more legal as it became more enforceable. When World War I broke out seven years later, both Germany and Britain largely ignored the Convention, each claiming that the other had broken the rules. The United States and some of the other participating nations had not ratified the Convention in the first place. A main point of interest, as Cowie says, is that the Convention was due for reconsideration in 1914, but the First World War intervened. It thus remains as the sole instrument for the conduct of mine warfare.

The Convention contains no rules forbidding the laying of mines on the high seas, nor against any sort of mine other than "automatic contact mines." The use of influence mines was still in the future. Further, the use of mines lodged on the bottom was not imagined. Such use would of course deemphasize the fear of mines breaking loose and drifting about in the world's shipping lanes. However, the Convention did promote a discussion of this underhand and cowardly tool of warfare (mining) and did attempt to describe the circumstances under which this form of warfare would be considered legitimate. Perhaps it can be said that this Convention on the one hand, and the later proliferation of more devastating tactics—such as mass bombing of cities—on the other, have succeeded in converting mine warfare from an underhand and nonchivalrous activity to a restrained and humanitarian form of persuasion.

World War I

At the outbreak of World War I the Russians, the Germans, and the British were equipped—each after his fashion—to use mines in the conduct of the war. Initially, the Russians laid some 2,200 moored mines at the entrance to the Gulf of Finland to protect St. Petersburg from attack from the sea. These fields were supplemented by additional mines off the Latvian coast and along the ore route from Sweden to Germany. As a result of these actions the Germans lost four cruisers and seven torpedo boats. Rear Admiral Friedrich Ruge [43], in charge of German mine warfare in the Second World War, has called these Russian mine operations the most clearly conceived and ably executed of the entire war. They were not, of course, able to tip the balance of such Russian military disasters as Tannenburg or the effects of the Revolution of 1917.

Although the Germans laid several large defensive minefields west of Heligoland, which were effective in keeping out British submarines, the major German use of mines was offensive against British shipping in British estuaries and around British ports. The German defensive mines were, however, a major reason for the British decision not to go into the Heligoland Bight after the German Fleet. According to Patterson [33, P. 36], when the Battle of Jutland took place the decision not to follow the German Fleet was strongly related to the British fear of mines. He quotes Admiral Jellicoe as having said earlier before the Admiralty that "If, for instance, the enemy battle-fleet were to turn away from an advancing fleet, I should assume that the intention was to lead us over mines and submarines; and should decline to be so drawn."

Beyond such defensive successes, however, the most significant development of the Germans in their North Sea mine operations was the use of submarines as minelayers (see figure 3). These boats were dedicated to this purpose. They were capable of carrying and launching eighteen, and later thirty-four, Type IV mines through six mine tubes which were inclined aft from top to bottom to prevent mines from hanging up when the submarine was moving. A soluble plug was used to delay the release of the mine from the anchor by ten to twenty minutes so that the mine-laying submarine could get clear. Later in the war there was developed a delayed rising or a variably delayed rising mine (from one to four days) so that mines would appear later in fields that the British had already swept, [25, 33]

In the first two years of the war, the British had experienced great difficulty with the reliability of their Elia-(Italian) designed stockpile, principally because of the mechanical firing arm, a British designed "improvement." They replaced these by late 1916 with Hertz Horn mechanisms and were thus finally able to make an effective mine blockade in the English channel.

One of the British developments portended great future potential although its use in World War I was minimal. This was in effect the first magnetic influence ground mine called the M-sinker. The mechanism consisted of two bar magnets with different moments of inertia pivoted one over the other. These aligned themselves in a magnetic north-south direction but when a steel ship came by, its increasing magnetic field caused one magnet to turn more quickly than the other. This relative deflection would cause a contact to be made completing a firing circuit through a detonator. The mechanism was provided with a clockwork delay to allow the hemispherically shaped mine to stabilize itself on the sea bottom before arming took place. The actual use of the M-sinker was very limited and its effectiveness

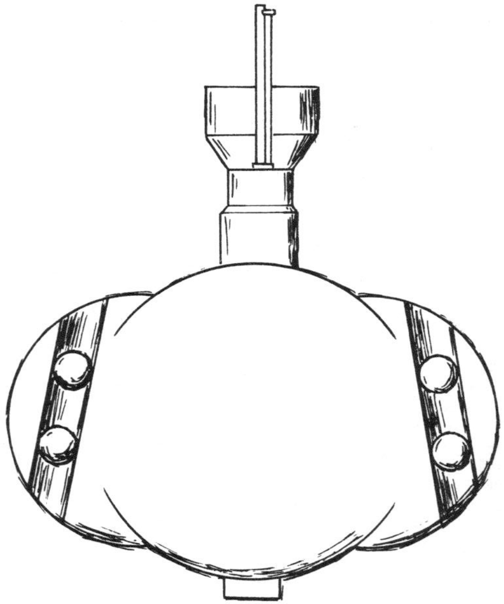

BRITISH E CLASS, 1915

GERMAN UC BOAT, 1915

Figure 3. Submarine Mine-laying Techniques: German UC Boats, 1915; British E Class, 1915; Russian *Krab*, 1915; American *Argonaut*, 1927

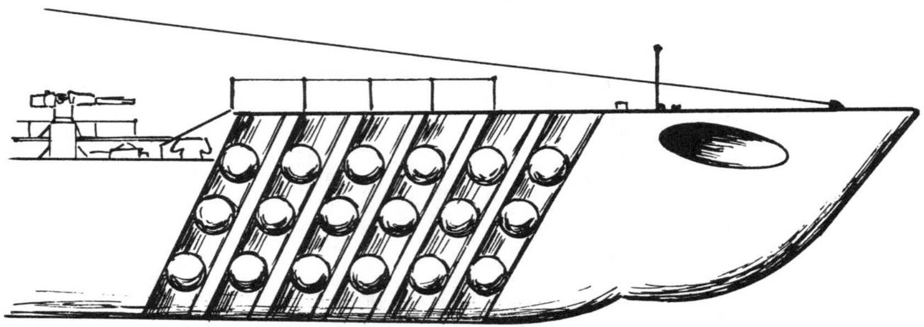

German UC was capable of launching 18 Type IV mines through six inclined mines tubes; design was later improved to give capability of 34 mines. Six British E-class submarines were converted into minelayers having vertical tubes in saddle tanks; 32 mines could be carried in 16 tubes. Below, the Russian "Krab" was the first submarine minelayer to be undertaken. Mines were planted through two stern-facing tubes and conveyed rearward by belt. American "Argonaut" carried ambitious minehandling systems including revolving cages to move mines from storage racks (3,4,5,6) into launching tubes (1,2); traversing truck shuttled between lines 1 and 2 for initial loading.

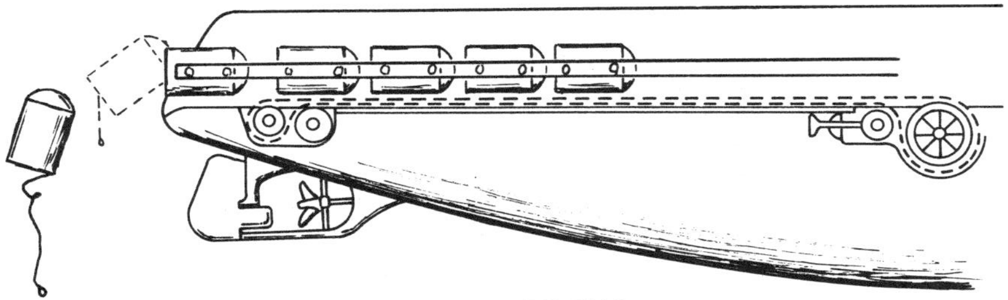

REVOLVING LOADING CAGE (2)

1-2 = USED TO LOAD MINES INTO LAUNCHING TUBES
3-6 = USED TO LOAD MINES INTO STOWAGE AREAS
7 = COMPENSATING TANK

RUSSIAN "KRAB", 1915

LAUNCH TUBE (2)

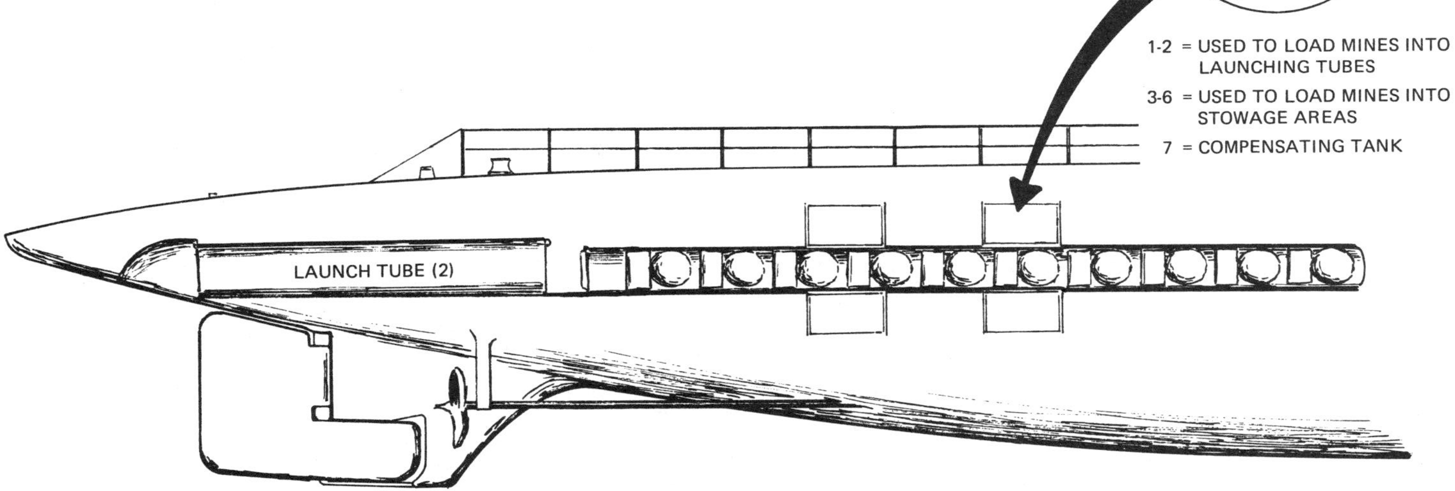

AMERICAN "ARGONAUT", 1927

unassessed. It was, in fact, a very finicky device and in the subsequent period between world wars caused a certain amount of learned calculation to be done in the various mine establishments to figure out how to improve it. The Germans copied the idea and used it in the early days of World War II. The United States tried to improve upon it but gave up and instead copied a single magnet design which was recovered by the British from a German mine drop in November 1939 on the flats at Shoeburyness (in the Thames estuary). It is fair to say the real development of the magnetic ground mine and in fact of other influence ground mines had to wait until World War II, although, as we have said, the earliest laboratory beginnings occurred before 1918. Ground influence mines were used in World War II and, significantly, were laid by aircraft as well as by ships and submarines.

The mining of the Dardanelles by the Turks starting in August 1914 was a most fateful event, and perhaps unexpected. Patterson says, "How and where the Turks got their mines and how they managed to lay them is not known. The mines are thought to have been a mixed bag from many countries, strung with salvaged anchor wire, all of doubtful longevity. But of their effect there is no doubt—Russian wheat did not leave the Ukraine for France, Turkey was not surrounded from the north and west by Allied Forces, nor was a naval wedge driven between the Central powers of Turkey and Bulgaria." [33, P. 31] My hunch is that they had some help and advice from German staffers who were in the country.

As Barbara Tuchman describes in a fascinating account [46], the Germans were most anxious to obtain an active alliance with the Turks in order to sever the connection between Russian ports in the Black Sea and the Allied fleets in the Mediterranean. At the outbreak of war, there were two German warships, the battle cruiser *Goeben* and the light cruiser *Breslau* in the Mediterranean. These ships evaded the British and French forces in the Mediterranean and arrived in a few days at Constantinople after demanding and receiving from the Turks a safe conduct through the Dardanelles. In the Tuchman account, there is no mention of mines in the Dardanelles in August, although Meacham [32] cites Turkish records which show that the first mines were laid on 3 August. The same sources show, however, that 150 of the mines in the Kephoz barrier were laid after 1 January 1915 and that the other 200 were laid between August and January, i.e., after the German warships had passed through. In later October, these same ships, flying Turkish flags but still under the command of German Admiral Souchon, steamed into the Black Sea and shelled Odessa and other ports, sinking a Russian gunboat. This calculated action caused Russia to declare war on Turkey and thus finally brought Turkey actively into the war on the side of the Central powers. Britain and France declared war on Turkey on 5 November and immediately began plans to force the Dardanelles, capture Constantinople, and again open the Black Sea to Russia. This they were unable to do. Tuchman says, "With the Black Sea closed, her [Russia's] exports dropped by 98 percent and her imports by 95 percent. The cutting off of Russia with all its consequences, the vain and sanguinary tragedy of Gallipoli, the diversion of Allied strength in the campaigns of Mesopotamia, Suez, and Palestine, the ultimate breakup of the Ottoman Empire, the subsequent history of the Middle East, followed from the voyage of the *Goeben*." [46, P. 168]

We should note that these events would not have followed if the British and French fleets had been able to force the Dardanelles on 18 March 1915, the date of their maximum effort. They were unable to do so because of a very small number of mines which on that day sank the French battleship *Bouvet* with a loss of 638 men out of a complement of 709; sank two British cruisers, HMS *Irresistible* and HMS *Ocean*; and damaged HMS *Inflexible* so that it had to be beached. These ships were sunk by a line of twenty mines which were secretly laid during the night of 7–8 March several miles south of the known Kephoz minefield. Meacham quotes the conclusion reached by the admirals at the close of day: "The battleships could not force the straits until the minefield had been cleared—the minefield could not be cleared until the concealed guns which defended them could be destroyed, and they could not be destroyed until the Peninsula Gallipoli was in our hands; hence we should have to seize it with the Army." [32]

Thus began one of the most miserable land campaigns in history, resulting in failure. Meacham says, "The enterprise was finally given up—but the mines were still in place at the end of the war and sweepers had to precede the triumphal entry of British warships past the silent guns into Constantinople on 12 November 1918." [32] There is an ironic footnote to this story. Cowie [10] states that on 20 January 1918 the ex-German battle cruiser *Goeben* was damaged and the cruiser *Breslau* was sunk in the Dardanelles minefield while operating under Turkish control!

As another example of effective mining, let us move from the Dardanelles in 1915 to the English Channel in 1917. Many German submarines based in ports in Flanders at that time passed south through the Channel en route to their Atlantic hunting grounds. In the Dover area, they were opposed by various means including minefields, nets, and surface patrols—none of which was very effective. The unrestricted submarine warfare against British shipping had become critical for Great Britain and gave new emphasis to plans for closing the Dover Straits to submarines. Admiral Keyes, a veteran of the defeat at Kephoz, had a great belief in the value of mining. This belief was not shared by the senior naval officer at Dover, a Vice Admiral Bacon. The argument was settled eventually by the assignment of Keyes to relieve Bacon. The Keyes plan consisted of a deep moored minefield of over 5,000 mines extending from Folkestone to Cap Gris Nez plus a large surface patrol using searchlights and flares at night. We are told by Meacham that up to this time, the British mines were both ineffective and unreliable. "However, a new mine (designated H2) using chemical horns copied from captured German weapons became available for the project commencing in November 1917. Planting started in December and was completed before the new year. This field, with a calculated threat of between .35 and .45, forced the Germans after several losses to give up the Dover route and to seek access to the Atlantic through the longer northern route."

These successes led to a wide-scale use of mines, the North Sea Barrage. This was a last-minute effort on the part of the United States and obviously would have been far more effective if it had been done at the beginning, in 1914, instead of at the end, in 1918! As has been said, the United States was manufacturing an obsolete British moored mine during the first years of the war, while the British were busy convincing themselves that their mine was really not a suitable device —especially for so ambitious a project as blocking off the German submarines' North Sea exit to the Atlantic. However, in 1917 when the United States got into the war, a great enthusiasm for combating the terrible submarine menace developed. Inventors all over the country proposed various schemes, and one of these made use of the fact that a steel ship coming in contact with a copper wire in salt water could produce a galvanic current which could be made to fire a mine. The virtue of the copper wire or antenna was that it could be held up by an auxiliary float of its own (submerged), so that the ship or submarine would not have to hit the mine. It could merely run into the wire but would still be within lethal range of the explosion. This extended the cross-sectional area within which passage was unsafe by a factor of at least three, so that the number of mines necessary to seal off the North Sea became much more manageable.

A great deal of enthusiasm developed in the Bureau of Ordnance for this project, which was an enormous undertaking starting from scratch with minimal time—although no one then knew that the war would be over in eighteen months. The conviction that the project could be undertaken successfully stemmed from the state of the technology at the time. The electrolytic firing device was a bit risky, because it had not been tried out on what we today would call an evaluation lot. However, the newly designed mines could be laid automatically just by steaming along and dropping them over the fantail from built-in rails. The automatic anchor took care of the depth setting. The charge consisted of 300 pounds of TNT. Calculations showed that approximately 100,000 mines in the North Sea would substantially inhibit submarine passage. The Admiralty somewhat reluctantly gave its approval for the mine barrage in October 1917, and the Secretary of the Navy authorized the Bureau of Ordnance to proceed with the procurement of 100,000 mines, designated Mk 6, at a cost of $40 million, or $400 per mine. In today's dollars, I suppose this would be of the order of $4,000 each. Figure 4 is a drawing of the famous Mk 6 complete with anchor, plummet, and antenna float.

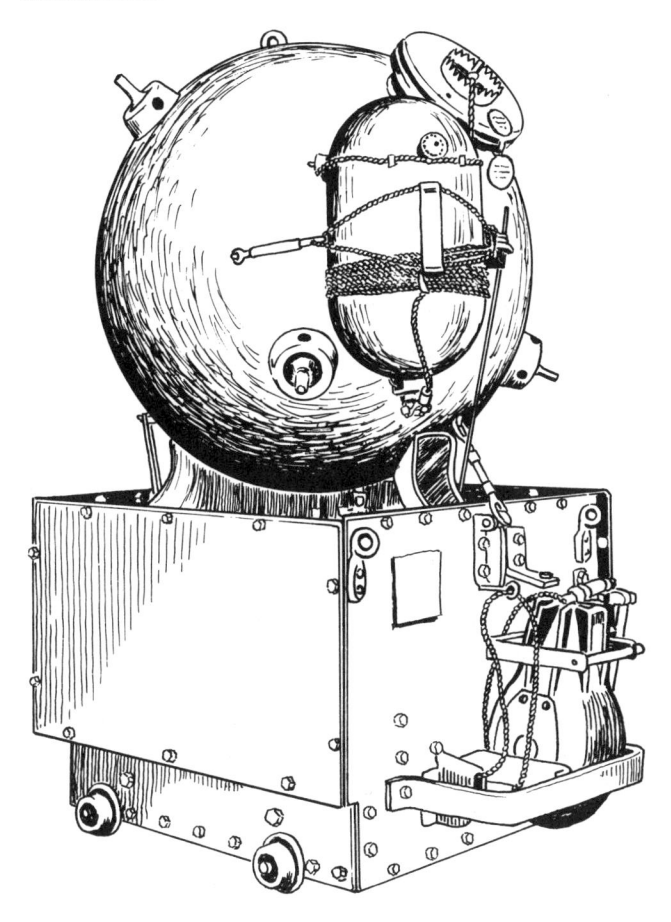

Figure 4. Mine Mark 6

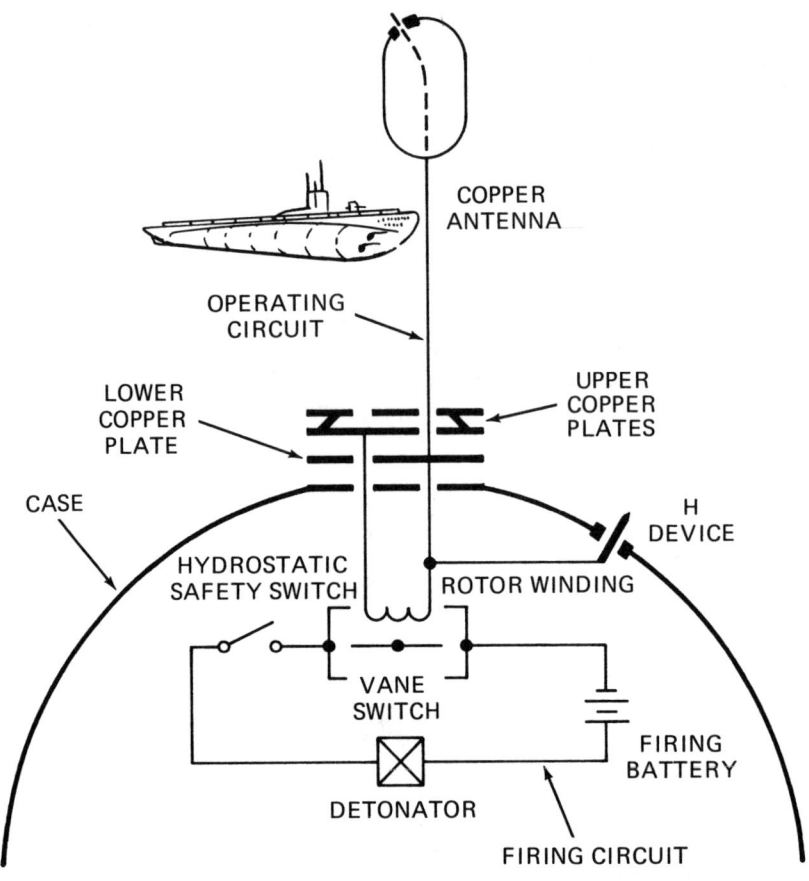

Figure 5. Circuits of the K-Device

The electrolytic firing influence called the *K-device* (why, I don't know—perhaps for the Knott Apparatus Company) worked as depicted in figure 5. The K-device was part of a proposal made by Mr. Ralph C. Browne of Salem, Mass., who worked for the L.E. Knott Apparatus Company of Cambridge, Mass. His proposal was for an underwater gun using a new electrolytic scheme for target detection. Commander Fullinwider of BuOrd saw the possibility of the electrolytic scheme for a mine-firing device and persuaded Browne to work on this application instead of the gun. It called for a sensitive relay, a self-contained battery which became a limitation on the life of the mine, an electric detonator, and a hydrostatic safety switch. There was also a hydrostat to release the antenna float when the case reached its planned depth. The copper wire was carefully insulated from the steel float and the steel case. Because the wire was connected only to insulated copper plates, there was no flow of current when the mine was immersed in salt water, as all of the copper was at the same potential. However, when the steel hull of a ship made contact with the copper antenna wire, the steel became one electrode of an electrolytic cell connected by the wire to the copper plates which became the other electrode. A current then flowed through the antenna wire with the return path being the seawater itself. This small current closed the sensitive relay which connected the firing battery to the electric detonator.

It shall be left as an exercise for the reader to decide whether the electrolytic firing device resulted in a contact or an influence-type mine. The fact is that although it was necessary to touch the copper float or wire, it was not necessary to touch or impact the mine itself. This gave a larger area of influence to the mine, better matching its damage radius, and decreasing the number of mines needed for barrier coverage.

It was this same Commander Fullinwider who played such an important role in the whole development of mine warfare during both world wars. A review of his career and a tribute to his accomplishments was appropriately and generously given by L. W. McKeehan at the dinner meeting of the first conference on the Naval Minefield in 1958. His remarks are quoted here for the record.

"THE CONTRIBUTIONS OF CAPTAIN SIMON PETER FULLINWIDER, USN, TO NAVAL MINING AND MINE DEFENSE".

Dr. L. W. McKeehan
Laboratory of Marine Physics
Yale University

It is proper at this time to honor a man who is no longer with us, but should be recognized for his vast contributions and years of devotion to the defense of this country in the field of U.S. naval mine warfare and mine countermeasures. I speak of Simon Peter Fullinwider, born 29 August 1871, graduated with U.S. Naval Academy class of 1894, and retired 30 June 1914 as a commander of the U.S. Navy. He was recalled in 1917 to head Desk N in the Bureau of Ordnance where he was in charge of design, development, and procurement of the Mark VI Mine [now Mk 6], the antenna-type moored mine used for the United States in part of the North Sea Mine Barrage of World War I, and for design, development, and procurement of depth charges at these same times. Early in 1918 he started a small laboratory for these projects and for others concerned with proposed acoustic and magnetic moored mines. After the armistice of November 1918, this work was continued in a new mine building in the Washington Navy Yard which later became the Naval Ordnance Laboratory. He also started the Mine Depot at Yorktown during this tour of duty, and was personally responsible for its first stages.

In the spring of 1940, when nearly 70 years of age he was called back again to head the same Desk N in the Bureau of Ordnance to be responsible for all underwater weapons, except torpedoes. He had additionally the responsibility for degaussing. Though relieved as Head of Desk N later in the summer of 1940, he remained on duty in BuOrd, and later in the Office of Naval Operations. He was largely instrumental in the establishment of the Mine Warfare School at Yorktown, the Mine Disposal and Bomb Disposal Schools in the Washington area, and the Mine Warfare Test Station at Solomons, Maryland, where the Naval Ordnance Laboratory had previously done experimental testing. He fostered the Mine Warfare Operational Research Group, first in BuOrd and later in OpNav. These enterprises now exist as other installations.

Commander Fullinwider was promoted to captain on 25 February 1942. By this time one of his three sons, all of whom had graduated from the Naval Academy by 1926, was already of that rank, and there were by 1 August 1943 four Captains Fullinwider on active duty. All remained on duty throughout the war, and the eldest survived it by ten years, dying on 19 November 1955.

It is clear from all this that Simon Peter Fullinwider, Captain, USN, can justly be considered the patron saint of American Naval Mining and Mine Defense.

Let us all honor him with a standing ovation.

The execution of the North Sea Barrage plan, once it had been settled on, was a miracle of enterprise and effort. In order to meet all the requirements, it was necessary to manufacture, load, and ship these mines and lay them at the rate of 1,000 per day! To meet the explosive loading requirements, a new plant was built at St. Julian's Creek, Virginia. Ground was broken in October 1917, and it was ready for operation in March 1918. This plant loaded over 73,000 mines at the rate of 300,000 pounds of TNT in 1,000 mine cases per day without an accident! Twenty freighters were assigned as transports for the various mine parts being manufactured all over the country. The mines were assembled at two depots in Scotland at the rate of up to 6,000 per week and were first available when the first minelayers arrived in June 1918. Since the Navy initially had only two minelayers which were converted cruisers, eight additional merchant vessels had to be modified to serve as minelayers. The location and extent of the Barrage is shown in figure 6. Laying was done in a total of fourteen excursions up through 30 October 1918. The United States planted 56,611 mines of American manufacture, and the British planted 16,300 mines of British manufacture. Reference 10 states 56,033 American; 15,093 British including 1,360 swept up in the western section because of shallow depth setting. The war ended before the whole plan was carried out.

The effect of the Barrage has been much debated. The Germans have said that it had no effect at all—they simply skirted the Barrage along the Norwegian coast where it was completed only at the end of the war. Ruge [43] says the antenna floats could be seen on the surface and hence avoided. Hezlet [19] says there were six German submarines lost. Some American estimates go much higher. Patterson [33] says that the cost of the minefield was roughly equivalent to the cost of conducting the war for one day and surely the Barrage shortened the war by one day! It has been estimated [47] that the cost of the minefield to the United States was about 80 million dollars including 41 million for minelayers, carriers, and bases, and 39 million for 125,000 mines! If the war lasted twenty months for the United States this would give a total war cost to the United States (and Allies?) of 48 billion dollars. Whether this is reasonable or any of the money arguments are reasonable, is not clear. In the light of later astronomical magnitudes for later wars, the numbers do not seem excessive.

Admiral Hezlet in his book *The Submarine and Sea Power* [19] makes some telling observations about both the Dover minefield and the later North Sea Barrage. The efficiency of the former was due to the combination of patrols at night aided by searchlights on shore which forced the U-boats to dive into the minefields. In all, thirteen U-boats were thus destroyed in 1918, and the U-boats stopped using the straits, preferring instead to add six days' transit time to reach the Atlantic by going north around Scotland. This, of course, was the motivation for the effort to close the North Sea and bottle up the entire U-boat fleet. Even with more time, it is unlikely, says Hezlet, that the northern minefields would have prevented the U-boats from coming out. This is presumably because sufficient surface patrols could not be maintained to force the submarines to submerge. The distance from Scotland to Norway is about 250 nautical miles. The water depth varies from about 300 feet over most of the distance to about 600 feet at the eastern end. The middle section, figure 6 (marked A, 134 miles long), was mined with U.S. mines from 10 feet down to 300 feet, for a width of 50 miles. There were no plans for

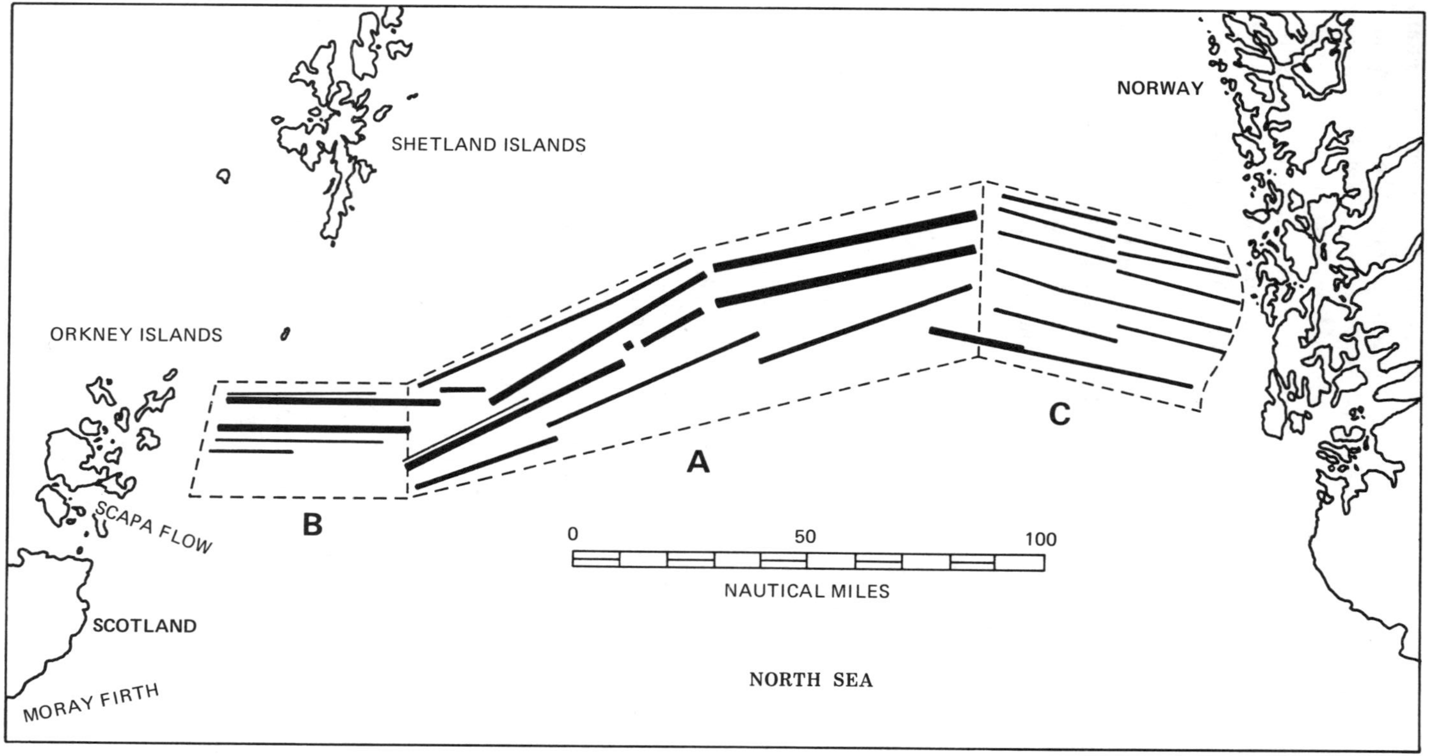

Figure 6. The North Sea Barrage

surface traffic here. The section near the Orkney Islands was mined with British mines from 65 feet to 100 feet deep—the section being 50 miles long and 20 miles wide. Finally, the Norway sector was mined again by British mines from 65 feet to 200 feet deep, the sector being 60 miles long and 50 miles wide. Surface traffic was thus possible without risk, assuming that the mines stayed moored. The mines were also kept out of the Orkneys proper so that passage between islands for British units was safe.

A different situation existed in the Straits of Otranto at the southern end of the Adriatic. This passage was too deep to mine, and says Hezlet, even though the U-boats were often detected by a screen of trawlers equipped with hydrophones, only two were sunk during 1918. The U-boats had little difficulty in passing these straits in spite of the large concentration of 280 ships in the barrier force. "The success of a barrier strategy depended entirely on whether it was practicable to impose a sufficiently high casualty rate to force the U-boats to abandon the attempt to pass as too dangerous. If it could, then the passage was as good as sealed; if not, then the very great effort was largely wasted." [19, P. 100]

It appears that mines would make the difference if properly designed, properly reliable, and properly supplemented by other forces. Either without the other, however, could be largely a waste of effort.

Mines come of age

After World War I, both the United States and Britain were left with large stores of mines which simply could not be ignored. The problems of providing better minelayers and of practicing with this materiel received some attention. The improvement of mine materiel and the means for countering mines were high on the agenda of the mine establishments. Particularly was this true in Britain where the question arose of what to do in a possible future war if the enemy were to discover or steal the use of magnetic influence mines. The British intensified their study of the magnetic fields of steel ships and how these might affect a mine mechanism. Before the next war (World War II) had started they were experimenting with reducing the magnetic moment of a ship, so that its net magnetic field would be less and it could pass close to a magnetic influence mine without triggering it. The reduction in the ship's magnetic field intensity, H, was called degaussing—although owing to a change in scientific nomenclature in 1932 it should perhaps more correctly but less euphoniously have been called deoersteding. It was doubtless more fitting to degauss a ship against a German mine. The reduction of the ship's magnetization, B, induced by its location in the earth's field was indeed degaussing, but the measurement of its effectiveness was made at a distance in oersteds.

In the area of minelaying the British commissioned the *Porpoise*-class submarine in the 1930s which was able to lay fifty moored contact mines from a casing outside the pressure hull, as well as cylindrical bottom mines from torpedo tubes. This was an interesting advance. At the same time aircraft were fitted to deliver bottom mines, but not moored mines. These latter mines, of course, were quite heavy and had not been designed with the aircraft option in mind.

Although in the period between world wars the Soviets chose to continue work on mines and set up a small facility in Leningrad to develop, test, and manufacture them, their products did not play a significant role in World War II. This was doubtless due to the manner in which hostilities developed and the location of the main battle fronts in the heart of the country. Nevertheless, the Soviets did make some significant advances in mine technology during this period. They developed an antisweep device which would cut the sweeping cable which was trying to cut the mine mooring cable—perhaps the first counter countermeasure. They developed an antenna mine after the K-device principle. Some of these mines, incidentally, were encountered in Korean waters in 1950–53. The Soviets have also specialized in riverine mines, small mines sometimes used with nets for harbor defense, and a magnetic bottom mine of hemispheric shape for possible use in strong currents. This latter sounds like the British M-sinker, but had an anti-countermine feature. It required two magnetic looks separated by two seconds in order to be satisfied that a target was present. This prevented the mine from being triggered by a single pulse which could come from a nearby explosion. A shock from such an event could move the mine in the earth's field—or otherwise, through vibration, produce a magnetic signature. The Soviets also had devised a moored-mine inertial firing device which required that a weight be displaced by the movement of the mine when it was struck by the target. Patterson [33, P. 41] tells us that this device was used before World War I in the Russian M-12 moored contact mine, in the M-16 developed during World War I, and in their M-26 of 1926. This also suggests a new departure in mine nomenclature—naming the mine after the year of its origin. At least such a practice would keep the chronology straight. He also says that "it is thought that the AMG-1 moored contact mine with chemical horns is laid by aircraft without parachute and can be planted through ice." If so, this is very remarkable indeed.

In the United States between the wars an effort was made to improve mine development, mine laying, and mine sweeping. Frequently all three of these efforts found a focus in the Mine Building set up at the Washington Navy Yard in 1919. Lieutenant L. W. McKeehan, USNR, was the first commanding officer. Dr. R. C. Duncan was the first chief scientist and remained such right into World War II. The USS *Cormorant*, a "bird"-class minesweeper built from World War I funds, was assigned to the Mine Building for testing mines and accessories. The new home of U.S. mine development was a two-story building 120 feet long and 60 feet wide, giving space to a small organization of physicists, engineers, and draftsmen dependent for logistic support and shop work on their host organization, the Washington Navy Yard.

The secrecy of mine work did not increase the understanding between tenant and host. The Bureau of Ordnance itself, sometimes called the "Gun Club," was reluctant to designate this new organization as anything more than a Building. Ten years later the small Experimental Ammunition Unit which had come to share the Mine Building was formally included in the organization, which was then renamed the Naval Ordnance Laboratory (NOL). Thus for the first time, and possibly quite by accident, the functions of material design, reliability, and explosive safety were put together under a single head. The new laboratory was given at least a nominal measure of control over its own projects. The tasks of the laboratory were to improve the Mk 6 mine; to improve the K-type firing device; to investigate magnetic and acoustic ship influences; to design a magnetic firing device; and to design a 21-inch cylindrical mine (also moored), to be laid by discharging it from a standard submarine torpedo tube and to be fired by Hertz horn. This latter assignment later became the Mine Mk 10 and saw a limited but successful use in the Pacific in World War II. These assignments, however, in general far exceeded the abilities of the laboratory to perform, because of limitations in staff and money.

In 1929, before this reorganization, the Senior Inspector of the Gun Factory had recommended to the Chief of the Bureau of Ordnance that a board be appointed to study the work of the Mine Building and recommend how this work could be carried on by other sections of the Gun Factory. The final report of the board was entirely contrary to the inspector's ideas. It recommended that the newly named laboratory be authorized to work on any ordnance development or research problem assigned to it and that its staff be increased as required. The chief approved the recommendations of the board except that there was to be no increase in the staff! What progress might have occurred was very much inhibited by economy. As Duncan says,

> Promotion of personnel was almost unheard of, new employees were out of the question, machine tools were largely those discarded by other shops in the Yard, and the purchases of experimental equipment were nil. Experimental work on all projects was strictly limited.
>
> ... At one time work on depth charges was limited to $25 per month or one machinist for two days. All purchases above $10 had to be submitted as formal requisitions on which bids would be requested, and then the article was purchased from the lowest bidder. One urgent job required a small DC motor that could be bought from any motor manufacturer for about $15. In order to save time, the Laboratory obtained one day a purchase order of less than $10 for the motor frame, and the next day an order, also for less than $10, for the motor armature. The supplier was told that the Laboratory would accept the two parts assembled. [12, P. 86]

It must be remembered that a lot of this went on in the doldrums of the great depression. However, attitude and understanding could have gone a long way toward mitigating the effects of such petty economy. It appears that for a few years after World War I, the Bureau of Ordnance had a research fund for mine work. This fund, at the Bureau's own request, was added to the general appropriation and speedily became lost for mine work thereafter.

But new projects were assigned from time to time. The Navy finally obtained authorization from Congress (Act of Congress 28 May 1924) to build a submarine specifically designed to lay moored mines. The Bureau of Construction and Repair (later the Bureau of Ships) designed and built the boat, called the *Argonaut*, at the Portsmouth Navy Yard (at Kittery, Me.). The boat was commissioned 2 April 1928. In Duncan's account, this design included the mine-handling equipment. The Mine Building was directed to design the mine assembly to fit the mine-handling equipment designed by the Bureau of Ships. This was done in due course and a model of the new mine, based entirely on blueprints of the submarine, was sent to Portsmouth Navy Yard and tried out in the vessel before commissioning took place. The mine and the mine layer passed all tests at the time and so several hundred complete mines known as the Mine Mk 11 were built by the Naval Gun Factory. Storage was provided for 60 Mk 11 anchored mines between frames 161 and 200. The storage consisted of

Uncle Sam's largest submarine arrives at Washington Navy Yard, 20 November 1928. The fleet submarine V-4 was the only mine-laying submarine of the United States Navy at that time.

The USS *Argonaut* (V-4) in New York, March 1929. (United Press International)

This remarkable and unique submarine, launched 10 November 1927 as V-4 at Portsmouth Navy Yard and commissioned in 1928, had a remarkable history. Although designed as a minelayer she had two 6" deck guns which can be clearly seen in the pictures. She was renamed the *Argonaut* (SS 166), 19 February 1931, and joined SubDiv at Pearl Harbor in 1932. On 7 December 1941 she was at sea, patrolling off Midway Island. She returned to Mare Island Navy Yard for conversion to a transport submarine January to August 1942. She joined the *Nautilus* (SS 168) in carrying the 2nd Marine Raider Battalion for the Makin Island Raid, 17–19 August 1942. She carried 121 combat-loaded Marines, stood by, and brought them back.

Argonaut's third war patrol took her into the Southwest Pacific. On 10 January 1943 she was sunk by a Japanese destroyer with loss of all hands while attacking a Japanese convoy off New Britain.

three groups of open racks, two rotating cages, each of which could hold six mines, and one pressure-proof "mine compensating" tube, holding eight mines, which extended the full length of the mine stowage section. This information comes from General Information book USS V-4 (SMI) Fleet Submarine (Mine Laying Type) in the Navy and Old Army Section of the National Archives, Washington, D.C. The round tubes were large enough to accommodate the square anchor boxes of the Mk 6 mines, and so there was room for a larger moored section. The Mk 11 carried 500 pounds of cast TNT instead of the 300-pound load of the Mk 6. No tests of mine loading or laying were ever carried out in the presence of both the mine designers and the ship constructors.

To Duncan's chagrin, when the *Argonaut* was finally fit for sea it was ordered in great haste from Portsmouth to Hawaii and stopped in Washington only long enough to take on a load of mines. When the ship finally started to practice plants in Hawaii, a number of mine modifications were suggested by the ship's officers. The captain requested that a mine designer be sent to Hawaii, but the chief of the Bureau ruled that no travel funds were available. Duncan does not say which Bureau! The modifications which were made were exceedingly slow and were not as satisfactory as if the designers and their designs were in close contact. The *Argonaut* made several successful test plants but none in war. Immediately after Pearl Harbor she was redesigned to be a submarine supply ship, and sometime thereafter (10 Jan 1943) she was unfortunately sunk by the Japanese. Thus ended the only U.S. effort to build and outfit a submarine minelayer. Perhaps the whole effort was poorly conceived or at least poorly timed. The Navy, however, did have a number of Mk 10 moored mines, the result of one of the Mine Building's projects, that were for launching from submarine torpedo tubes (82 Mk 10 mines planted in World War II). Further, a much larger number (576) of Mk 12 ground mines, in aluminum cases and with copied German magnetic needle mechanisms, were launched from U.S. submarine torpedo tubes in the early years of the Pacific War. Hence, the conversion of the *Argonaut* did not destroy submarine mine laying, but only the chance to lay large moored mines clandestinely in waters say over 200 feet deep. It is not known whether such a combination of circumstances and necessities did in fact exist. Perhaps we can say that the *Argonaut* and the Mk 11 mine were never missed.

We can, however, fairly say that in 1939 the United States was more or less where it had been twenty years before with respect to mine warfare, but now with a terrible new mine threat in the hands of the Germans and with some degree of preparation available on the part of the British. Up until the war started, the United States got very little information concerning mine R&D activity in Great Britain or Germany. The British had, however, improved their moored mine, had studied magnetic needle and magnetic inductive mines, and had considered methods to reduce a ship's magnetic field. Germany had developed a magnetic needle type mine deliverable by aircraft (and submarines). They were unaware of British countermeasures work, and assumed that the British would be helpless against their magnetic mines for at least a year or two. Of such miscalculations are victories and defeats made.

The new idea unleashed on the world at that time by the Germans was that the magnetic-field change due to a passing ship could be detected by a stationary mine lying on the bottom. Up to that time all offensive mines had been moored. Since the mine did not float, it could consist entirely of mechanism and explosive. There was no longer any need to transport a great inert anchor mechanism. The greater explosive charge made the mine effective in fairly deep water —say up to 200 feet, although most harbors and their approaches were much shallower—about 40 feet for dredged channels. But the chief advance was the possibility of laying these mines from aircraft, replenishing the field by aircraft with no danger to the German mine layers from their own mines. It was, of course, necessary to retard the fall of the mines by automatic parachutes so they would survive the impact on water. Very early in the war, in November 1939, the British were fortunate enough to find a mine which had been laid on land in error (on the tidal flats at Shoeburyness in the Thames estuary), to disarm it without blowing themselves up, and hence to establish what the vaunted secret weapon was. This feat was accomplished in the main by Commander John Ouvry, RN, and Dr. A. B. Wood, Chief Scientist at HMS Vernon, later Mine Design Department. In a memorial number of the Journal of the Royal Naval Scientific Service dedicated to Dr. Wood, Commander Ouvry says:

> It was my duty during the Second World War to take charge of the Enemy Mining Section of the Mining Department, HMS Vernon, Portsmouth, and provide specimens of German and Italian underwater weapons for Dr. Wood (Mine Design Department) and his assistants to examine and to probe into their secrets. The Germans never failed to provide plenty of problems of this sort, and we were pleased to keep Dr. Wood busy.
>
> We had a good start with the arrival of Hitler's first "Secret Weapon," the Magnetic Mine. On this occasion Dr. Wood arrived on the scene at Shoeburyness when we were stripping the weapon

of its external fittings. After it had been conveyed to HMS Vernon the following day a personal message was received from the First Lord of the Admiralty, Mr. Winston Churchill (as he then was), to state that investigation was to proceed without ceasing until the answer was produced. This Dr. Wood was able to give in detail wthin twelve hours of the reception of the mine in HMS Vernon. [52, P. 14]

Dr. Wood's own account is also of interest:

On the morning of Thursday 23 November 1939 I was attending a meeting on M. sweeps with D.S.R. [Director Scientific Research] (C. S. Wright) and Sir Frank E. Smith at Westminster when I was requested by phone to go to Southend where a German parachute aircraft mine was exposed on the beach at low tide. Rear Admiral Minelaying arranged for a car to take me to Southend and thence to Shoeburyness where I met Commander Maton, Lt. Commander Ouvry, Lt. Commander Lewis, Warrant Officer Baldwin, and [A. B.] Vearncombe. I was equipped with waders, and a bull terrier followed me out to the mine around which Ouvry, Baldwin, and Vearncombe were standing. I met Lewis on my way out and he held a small aluminum object in his hand which he thought was the magnetic mechanism of the mine and proposed opening it to see. I advised him to wrap it in cotton wool and send it by car to Woolwich Arsenal to be x-rayed—it turned out to be a bomb fuse! On arrival at the mine the bull terrier began 'ratting' under the horned (anti-roll) nose of the mine which had a diaphragm about four inches diameter fitted in it. This diaphragm made me suspect the possibility of an acoustic firing unit, although the aluminum body of the mine suggested a magnetic pistol. I sent the bull terrier home! The process of recovery of this mine and another one like it about two hundred yards away, and the subsequent examination of its contents in M.D.D. Vernon, is described in a report which I dictated on my return to Portsmouth. In the conclusion of this report it was stated: "It is clear that the German Mine is of the pivoted magnetic type operating on changes of vertical magnetic field—in the direction of N-pole downwards—a feature common to all steel ships in N latitudes. Assuming all the German mines possess this characteristic (although it certainly does not apply to British M mines) then a simple method of making ships immune at once presents itself, i.e., to magnetise the ships so as to have no north pole downwards, or if any polarity is shown it should be south downwards—a method obviously simpler than that of demagnetising." [52, P. 84]

Wood goes on to say that in time the enemy magnetic mines were made bipolar, operating on either N or S poles, and more sensitive. "This made demagnetising more difficult and minesweeping easier. The first N-pole mine required about 50 milligauss to fire it; in about a year the M mines were N or S pole operating on ±20 milligauss. Later mines were filled with automatic latitude adjustment to compensate for ambient magnetic field and sensitivities were increased for actuation by only a few, 4 or 5 milligauss N or S change of field."

A German mine recovered by the British was sent to NOL in 1940, and this led to a decision to make a copy of the mechanism. The firing device consisted of a magnetic needle mounted in gimbals and released on a knife edge after planting and after springs had been automatically adjusted by clockwork so that the needle was in equilibrium with the vertical ambient magnetic field. In this way, the adjustment to the local field, which was, of course, a function of latitude and other variables, was handled. The mechanism was expensive, delicate, and required careful machining. The mine required a nonmagnetic case for which aluminum was used. Compared with the later induction mine, it was a clumsy device—but it did work to the point of practicality and was a great improvement over the M-sinker gadget which involved two magnetic needles and which behaved so capriciously that it had never been seriously pursued. There was a limit to the sensitivity of this needle mechanism because of friction, the necessity for damping, and the limited magnetic moment of the needle.

As we have seen, a way to deal with the magnetic mine was to remove the magnetic field from the ship, a procedure called degaussing. The Naval Ordnance Laboratory became involved in this, and designed both the coils to be installed on board ships and ship degaussing ranges. The Laboratory designed the magnetometers to measure the net fields of ships both before and after degaussing. During the war nearly 13,000 ships were fitted with degaussing equipment in the United States at a cost of approximately $300 million.

In measuring the field of a ship, it was found more practical to let the ship pass over a small coil in which wire was wrapped around a permalloy core, than to use the large loops the British started with. Such a coil could be mounted in a copper pipe and driven into the bottom of the range where it would be held still and leakage problems would be minimized. These induction coils became the sensitive elements for the next improvement in mine design. The great advantage of this is that no mechanical adjustment is required for the ambient earth's field. A steel case can be used. Any change of any component of the field can be detected, and the signal can be required to correspond to a ship's signature; that is, signal processing can begin to enter the mine developer's bag of tricks. Figure 7 illustrates the principle of operation of the magnetic induction firing device.

It is significant that the induction mine could not be moored because it would fire itself moving about in the earth's field, whereas up to this point all mines had been moored. This led to an effort to find a way to use a magnetic firing influence in a mine which itself was free to move, or in a depth charge which was sinking and perhaps tumbling freely. The answer was to use three pickup coils mutually at right angles and to add their squared outputs so that the result was a measure of the total magnetic field. This was then independent of the orientation of the body in it. It was very convenient to have a magnetic material for the core in which the magnetic induction was proportional to the square of the magnetic field with minimal hysteresis. This led to the development at NOL of the material called Parabonol, because the B, H curve was a parabola. The suffix, "nol," became a sort of trademark for NOL-developed materials.

The fact that the new ground mines could not be "swept" by towing wires to sever their mooring cables, (because they didn't have any) led to a curious bureaucratic impasse. The Bureau of Ships in charge of minesweeping maintained that since minesweeping was a countermeasure and degaussing was a countermeasure for the new bottom mines, the Bureau of Ships should be in charge of degaussing. The Bureau of Ordnance, on the other hand, maintained that degaussing was the same sort of thing as providing armor to ships to withstand the impact of gun projectiles. And they, after all, were in charge of armor. Meanwhile, the specification of degaussing coils and generators and the measurement of their results on ships was pro-

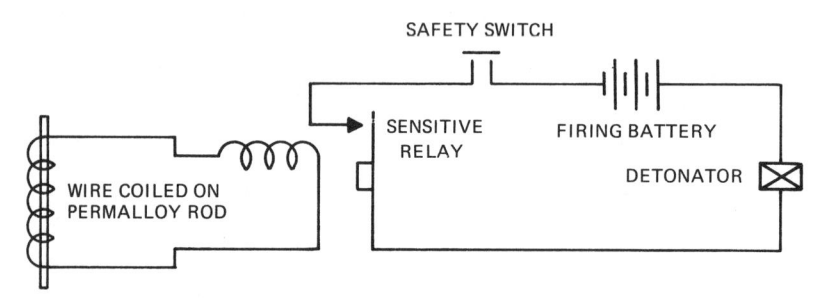

Figure 7. Functional diagram of the Induction Type Magnetic Mine

ceeding under extreme urgency—mostly at NOL, under direction of BuOrd. The debate went to CNO and SecNav who decided that BuOrd would design the degaussing coils and measure the fields and BuShips would install the coils and furnish the power to operate them. This decision was largely predetermined by the fact that BuOrd had already made much progress in the problem and "a change of cognizance in the middle of a project would only result in duplication of effort"—a precept which is sometimes honored and sometimes not.

A second cognizance battle was settled entirely in favor of BuShips. In the first days of the war, the British had found that a magnetic field could be formed to simulate the field from a large ship by passing current in the sea from one electrode to another. As Francis Bitter described in his book on magnets [3] when he arrived in Liverpool in 1940 on an information-gathering mission (first NOL representative in the U.K.), "Ships broken in half in the middle lay around the harbor. They were victims of the new ground mines. We also saw the first minesweeping going on. At first we could not understand it. Tugboats were steaming about the harbor dragging enormous black snakes behind them. These were large cables covered with buoyant insulation so they would float. On the tugboats there were powerful motor generators to produce a current. One electrode was at a relatively short distance behind the tug, [at the end of one cable], the other at the far end of the other cable. A current was made to pass from the distant electrode through the sea water in a rather wide circling path to the forward electrode. This current produced a magnetic field behind the tugboat. The tugboats, therefore, had to be very carefully degaussed because they had to travel over the magnetic mines without exploding them. Then, behind the tugboat were the magnetic fields produced by currents in the sea to imitate the field of a ship.

This seemed at first a rather odd way of producing a magnetic field. The standard way of doing it is to make a coil. One could make much stronger fields with smaller currents by means of coils of wire, but the problem of making a coil that could be conveniently towed behind a tugboat was extremely difficult. This was actually a most brilliant solution which made it possible to tow much less wire than would be needed in a coil and to have the wire in an extremely simple form for towing." [3, P. 121] And, I would add, much more resistant, or perhaps totally resistant, to the dreaded mine explosion when it came. The Bureau of Ordnance recommended that the design of influence minesweeping gear be assigned to it in connection with its mine-firing design problems. The Bureau of Ships, however, demanded that all sweeping remain under its cognizance, and after much discussion and many conferences the Chief of Naval Operations allowed it to so remain. After all, the towing of electrical cables for magnetic "sweeping" purposes was from a ship-handling point of view similar to towing mechanical cables for mechanical sweeping purposes.

There is one small irony in connection with this, as Duncan says, "Since the U.S. general policy was opposed to using a mine unless the United States could sweep it, the initial use of several mines was delayed to give the Bureau of Ships the opportunity to develop satisfactory sweeping techniques. Finally, this policy was dropped and at the end of the war, there were some mines planted in Japanese ports which even the United States could not sweep or destroy." [12, P. 92]

Although the United States duplicated and manufactured replicas of the German magnetic needle mine very early in the war (used in the Mk 12 mine), it was at about the same time that the Laboratory initiated the design of several new mines responding to various ship influences—namely, magnetic, acoustic, and pressure separately and in combination. Several other possible influences were rejected as less practical, namely gravitational, optical, cosmic ray, and electric potential. These new mines were designed to lie on the bottom; to be launched from aircraft; to contain 500 pounds of high explosive (Mk 36), and another series to contain 1,000 pounds of high explosive (Mk 25); and to contain many features like a variety of sensitivity settings, delayed arming, clock and electrolytic sterilizers (so the mine would become a dud after a definite time and would not have to be swept); and ship counts which would foil what enemy influence sweeping was done. They also contained vacuum-tube amplifiers, electrical circuits for differentiating the ship's signals with respect to time so that the maximum could be detected, or indeed a minimum between two maxima or a second maximum.

There were many new problems to be solved, not the least of which were the effects of 500 or 1,000 pounds of explosive on the sea bottome on a ship of what size and construction at what position, in what water depth. Another problem was what sensitivity to build into the mine so that it would not be triggered prematurely or require the ship to be unnecessarily close. Either of these situations would reduce the effectiveness of the mine. Finally, how many mines should be laid and in what arrangement, and what is the so-called threat of the resulting minefield as a function of number of mines, and the ratio of sweeping to actual traffic. It should be pointed out that the calculated threat is simply an estimate of the fraction of losses that will be incurred if the enemy chooses to traverse the field. If he chooses not to, then, of course, he has agreed to a 100-percent blockade of his traffic. Since he presumably cares about his ships only as items to use—to carry cargo, to move strategically or tactically—the effect of a high-threat minefield is to deprive the enemy of the use of his ships, even though he may not lose any. He might as well not have them if he can't use them. All of these points and questions gave rise in the early days of World War II to an evening discussion group at NOL who played "mine warfare games." This was the beginning of "operations research" in this country. The group soon was moved to the CNO, augmented, and named Mine Warfare Operations Research Group (called the "Morgue" for short). George Shortley, in his retiring address as president of the Operations Research Society of America in 1966, recounts some of this early work. [44]

The ground mines, large and small, were started at NOL under the direction of BuOrd in 1941 and early 1942 and began to be delivered in quantity in 1944 and 1945. It was these mines which starved the Japanese economy in the last four months of the war. The pressure-detecting mechanism was susceptible to wave actuation, but when combined with a magnetic detector so that the mine would ignore either influence except when accompanied by the other, it became virtually unsweepable. The use of the pressure-detecting mechanism was not permitted until very late in the war, because it was feared that the enemy might learn its secrets and make use of it against the United States and Britain. It turned out, however, that the Germans had a similar device, the famous "Oyster" mine, and used it to delay the Allied invasion of Europe in 1944. Our pressure-magnetic mines were then used in large quantities in the final mine attack on Japan in 1945. As Cowie [10, P. 163] says, the remarkable thing about the Oyster mine was its extreme simplicity in the method of detecting the pressure drop under a moving ship. A rubber bag exposed to the sea covered a space which was divided into two parts by means of a diaphragm. When the mine is laid, the pressure on either side of the diaphragm is equalized through a leak-hole, and any slow variations of pressure due to swell or tide are taken care of in the same way. "When a ship passes over the mine, however, the sudden reduction in pressure corresponding to a head of about two inches of water cannot be dealt with quickly enough by the leak-hole. The pressure on either side of the diaphragm therefore becomes unbalanced, and it is sucked towards the rubber bag and so completes the firing circuit to the detonator." The mechanism is shown in all its simplicity in figure 8.

Although mining was carried on by both sides in the European theatre of the second Great War, the efforts were, let us say, steady

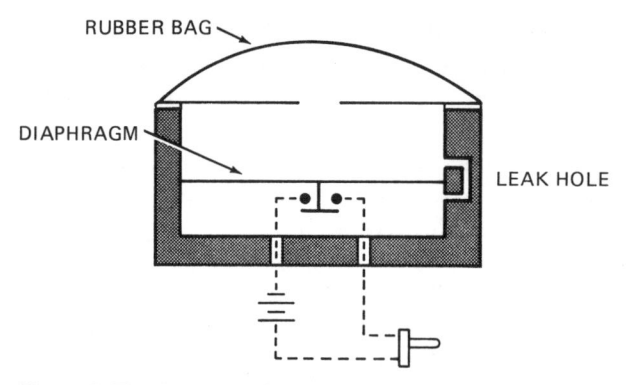

Figure 8. The Oyster Mechanism.

rather than spectacular. There was, of course, a great deal of worry over what might happen, particularly if the waters around prospective amphibious landings were mined. Fortunately, the Germans were unable to exploit this defense as much as they might have, partly because of uncertainty in command, partly because of an eventual lack of mine materiel, and partly because of uncertainty as to where in fact the invasions would occur.

In the Battle of the Atlantic there were no fleet actions, and the major problems, as in World War I, were the defense of shipping and hunting and killing submarines. In the antisubmarine war, the ability to deal with submarines grew through the development of better sonar and better weapons. The first officer in charge of the Mine Building in 1919, L. W. McKeehan, returned to active duty in approximately 1940 from an academic career at Yale to be the naval officer in charge of mines and depth charges in BuOrd. One of the new weapons which came out of this effort was the first antisubmarine acoustic homing torpedo, called the Mk 24 mine. Some thought that this designation was an astute move on the part of security to confuse the enemy, but others knew that it was because the Torpedo Section of the Bureau was too busy to fool with dubious newfangled ideas, and the McKeehan staff had no license to develop anything but mines. And so a mine it was. The Mk 24 mine, also called "Fido," and launched from both ships and aircraft, was credited at war's end, in its very short operational deployment, with the loss of at least one German submarine.

The United States laid a large minefield of 3,460 Mk 6 mines in April 1942 on the Gulf side of Key West. The channel through this was difficult to navigate and three U.S. merchantmen were sunk by these mines. In addition, the USS *Sturtevant* (DD 240) was a victim of the field, having failed to read the warning notice before departure from Key West. Much of the feeling against mines among some naval officers, it is said, can be traced to this incident and others like it. This information and more about the mine war in the Atlantic is contained in Patterson's "History of Mine Warfare," Nimrod Report, Chapter 2 [33]. There were also defensive Mk 6 minefields at Trinidad, Hatteras, and Chesapeake Bay, and both ground and moored control mines in various harbors.

The U.S. use of defensive minefields in World War II along the Atlantic coast (and elsewhere) is being surveyed by J. M. Martin and G. B. Ramsey. My own recollection of these minefields goes back to a secret report during the war listing U.S. ships lost to our own mines. As I recall it, there were seventeen ships which had holes blown in them by Mk 6 mines. Because the mines were planted at known depths below the surface, these data gave strong and quantitative support to the ordnance contention that a noncontact, close, under-bottom explosion was more damaging to a ship than a charge of the same weight would be if exploded in contact and at the side. The Bureau of Ships in wartime was very doubtful of this proposition and was making a great effort to protect large ships, battleships, and aircraft carriers from torpedo explosions occurring against the side. The ordnance view was that torpedoes should explode under the bottom of the ship where protection of any kind was hard to provide. The BuShips people insisted on protecting the sides, because they insisted that these were a ship's most vulnerable parts and the most likely to be hit. A final blow in the debate, of course, was that exploders designed to go off under the bottom didn't work consistently and so it was better to try for a direct impact. This debate, incidentally, seems to be still going on, although now one substitutes "missile" for "torpedo."

The U.S. defensive minefields and in fact most defensive minefields seem to be better at sinking their own ships than those of the enemy, although they can perhaps be credited with keeping the enemy out. In spite of these fields, however, German submarines managed to lay up to 338 mines in 1942, '43, and '44 off the U.S. Atlantic coast in the vicinity of various ports. They succeeded in sinking or damaging twelve ships for a score of twenty-eight mines laid per ship attacked. In Reference 33, eighty of these mines are listed as "undiscovered," meaning that their existence became known from enemy sources at the end of the war. Table 1 from Reference 13 summarizes the results, differing slightly from Reference 33; i.e., there were 317 mines laid and eleven ships sunk or damaged. The German mine-laying operation also resulted in the closing of several ports from one to sixteen days. I would say their offensive mine-laying operation was well

TABLE 1.
GERMAN SUBMARINE MINE LAYING,
WESTERN ATLANTIC, WORLD WAR II

	Days Port Closed	Mines Laid	Ships Sunk (or Damaged)
St. Johns	0	81	2
Halifax	1	60	1
Boston	0	10	0
New York	2	10	0
Baltimore	0	10	1
Norfolk	3	47	5
Wilmington, N.C.	6	0	0
Charleston, S.C.	16	36	0
Savannah	3	0	0
Jacksonville	3	12	0
San Juan	0	17	0
St. Lucia	6	6	2
Trinidad	0	14	0
Panama	0	15	0
Mississippi River Area	0	9	0
	40	327	11

~ 29 mines/ship "sunk or damaged"

* From Reference 13.

worth it, since they lost no submarines doing it. Our defensive mine laying, on the other hand, cost us several ships lost or damaged, and the advantage gained from it is much more debatable—although it provided anchorage areas free from enemy submarines for the assembly of convoys.

The invasions of Africa at Casablanca were unopposed by mines as were the operations at Algiers, Iran, Tunisia, and Sicily. The Germans used mines in the Bay of Salerno and at Anzio, causing more confusion and delay than casualties. With regard to the Normandy invasion, mines played only the part of an annoyance even though the Germans first introduced the pressure mine here. The situation is well related in the following paragraph from Patterson:

> It is remarkable that up to this point during the entire course of the war in the European and African theatres, no influence mines had been encountered in amphibious operations, although magnetic mines were used in the Thames Estuary in 1939. The Germans certainly had them, with the pressure mine a great potential menace. If the Luftwaffe would have had its way their use might have completely frustrated the invasion, but bungling at staff level saved the Allies from such fate. The pressure mine had been devised early in the war, but the Naval Staff would not produce or lay it lest a similar mine design be employed against them in the Baltic. Hitler intervened to order 4,000 units for use against an invasion. One

half this number was sent to Le Mans Airport and stored for use when needed, but Goering, fearing an invasion through Brittany would overrun Le Mans, removed the stock to Magdeburg. The transfer was finished on 4 June. By 7 June the high command had ordered them to be laid immediately, but it took four days to get them to airports in reach of the Normandy coast. Over 4,000 were laid by the end of July, but in one of those fateful bits of good fortune the British got one of the mines on dry land at Lac-sur-Mer and had developed countermeasures within hours: simply slow down to four knots. [33, P. 50]

We note that this is an easy countermeasure that will be useful if one is approaching an anchorage or a dock, but might be harder to manage if one is in transit somewhere. After the taking of Cherbourg Harbor around July 1, it was found that the Germans had left it an utter wreck with mines thickly laid in the inner and outer harbors, many of influence types with ship counters. Eight magnetic and acoustic sweeps were conducted every day for eighty-five days before it was felt that the area was reasonably safe.

The British-German score for the war is given by Patterson, who is quoting from a Soviet article: "The best technical cadres in many countries, including those in England and Germany, worked on the solution to the problem of eliminating the mine menace. Nevertheless, some 280 British combatant ships and 296 merchant ships . . . were blown up and sunk by German mines, while English mines accounted for some 250 combatant and 800 merchant ships. Mines blew up 22 percent of the landing craft lost by the U.S. and British navies in the 1944–45 landing operations." [33, P. 55]

The mining campaign against Japan which began in 1942 and ended in August 1945 was in two major parts. In the first part, nearly 13,000 mines were laid in over 150 enemy harbors and channels all over the newly extended Japanese empire. The first mines were necessarily laid by submarines, because ships were scarce and vulnerable, aircraft were also scarce and the minefields were out of their range. The second part or inner zone campaign started in March of 1945 and laid 12,000 mines by air in most harbors and passages of the Japanese main islands. The geography and the situation made Japan an ideal mining target since she was vitally dependent on imports for raw materials such as oil, iron ore, coal, and even food—whereas her overseas operations were equally dependent on her naval support and on supplies from home.

The United States had available in 1941 the Mk 10 mine (a moored sub-launched mine with Hertz horns, developed before the war) and the Mk 12, an aluminum-cased sub-launched ground mine with the copied German magnetic-needle mechanism. Although there were 1,200 of the Mk 10s in the stockpile in 1941, only 82 of them were planted. Five hundred seventy-six of the 600 Mk 12s on hand were used, all in the Pacific. The Mk 10s which were used were modified and called Mk 10 Mod 3s. They were equipped with the M-5 magnetic-needle mechanism. To solve the case motion problem the needle array was in a pair of gimbals and the needle was mechanically compensated by a counterpoise having the same moment of inertia. This too was a copy of the German moored magnetic mechanism.

The first minefields were laid in October 1942 by submarines operating out of Perth, Australia. A total of 160 Mk 12 mines were laid in approaches to Bangkok, Haiphong, and in the Hainan Strait. These mines immediately sank six ships totaling 22,000 tons and damaged six more, another 18,000 tons. A total of 421 mines in twenty-one fields sank twenty-seven ships and damaged twenty-seven more throughout the course of the war, or one ship sunk or damaged for every eight mines. If the remaining 237 sub-laid mines in fifteen other fields (from which no casualties are known) are included, the score is twelve mines laid for every ship attacked. This, as we shall see, is a very much better record than for air-laid mines, possibly because of the greater accuracy with which mines may be laid in channels by submarines. Of course, submarines take 50 to 100 times longer to traverse a given distance than aircraft and are likely, in the course of this time, to encounter many ships which they can attack with torpedoes (self-propelled, modern). If they are lucky, they can sink a ship with one or two torpedoes, whereas it takes eight or twelve mines, as we have seen, to do it more or less by chance. Conversely, aircraft being on station or in transit for shorter times are ideal as minelayers. They can go to the area to be mined at the least risky time, deliver their mines, and return home for the most part in safety and comfort. Therefore, it is fairly clear that submarines are useful as minelayers for special missions which are inaccessible to aircraft for one reason or another. No submarines were lost on mine-planting expeditions in the Pacific. Because experience showed that submarines in World War II generally returned from a period at sea with a few torpedoes unexpended, it became the practice to carry a few mines instead of the unexpended torpedoes and leave them planted so that the submarine could return empty.

By 1944 the growing availability of the newly developed influence ground mines, and the growing conviction that a mine blockade of Japan would be conclusive, led to concerted efforts on the part of the

Navy to find aircraft to be used for minelaying. Naval aircraft (TBFs from carriers, and PBYs had been used to bottle up lagoons full of Japanese ships, with great effect: viz., several atolls in the Marshall Islands and Palau. However, Navy planes were not large enough to handle payloads of mines over long distances, whereas the Army Air Forces B-29s could carry twelve 1,000-pound mines or seven 2,000-pound mines to destinations as much as 1,500 miles away (and get back). Continued planning and discussion at all levels, from Admiral Nimitz and General Arnold on down, resulted in the agreement to undertake the aerial mining campaign against Japan. The Navy would furnish the mines and the technical people to prepare them plus qualified officers to help plan the operations and brief the pilots. The Navy wanted a complete blockade of Japan's sea lanes and the use of enough B-29s to do the job. As Lott says, the clinching point was the Navy's statement: "The Air Force will get all the credit." [27, P. 223] On 22 December 1944, General Arnold issued orders for mining operations to begin the first of April 1945. Once the go-ahead had been given, the Navy moved fast. On 19 January 1945, a group of mine experts arrived in Tinian, and one month later they had a mine depot built and in operation for mine assembly. The first mines were actually planted in the Shimonoseki Strait on the night of 27 March 1945, just four days before the assault on Okinawa. It has been said that Operation Starvation was an Air Forces show with a Navy-prepared script. Had it commenced in January, as the Navy wanted, the attrition of enemy shipping would have considerably reduced Japanese resistance by the time of the Okinawa assault.

Almost certainly the loss of imports would have brought the Japanese to negotiations which might have prevented Hiroshima and Nagasaki. Lott continues, "Here should be emphasized one of the little-known but highly humanitarian aspects of mine warfare: a mine blockade enables the winner to win without killing. Enemy ships in a minefield enter it by their own choice; the enemy is free to keep his ships in port and save them if he wishes. But more important, mines never destroy homes, hospitals, or industrial facilities necessary to peacetime rehabilitation, nor do they wipe out noncombatant civilians." [27, P. 223]

Let us look briefly at the results of Operation Starvation. Late in 1944, Japan depended on sea traffic from China and Korea to bring in 80 percent of her oil, 88 percent of her iron ore, 24 percent of all coal, and 20 percent of her food. Within the islands, 75 percent of transportation was waterborne. The situation is shown in figure 9, Japanese Traffic in March 1945, where the width of the lines is proportional to average daily shipping in March 1945. After four and

Practicing off Corregidor. A mine rolls off the stern gate of the USS *Epping Forest*.

Launching a Mk 6 mine off a small boat, Mine Warfare School at Yorktown, Virginia (late fifties).

A Navy boat laying mines near Mainz, Germany, during "Exercise Harvest."

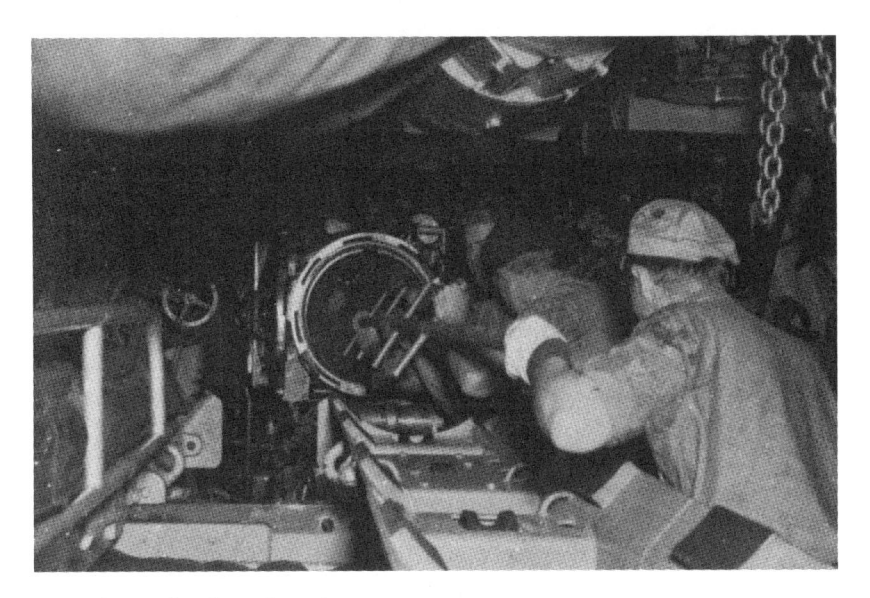

From the inside of a submarine.

After surface launch (and momentary broaching).

Mine-laying Navy Avenger loaded with mines, prior to the raid on Palau harbor in WW II. Thirty-two Japanese ships were bottled up by this operation and later sunk by aerial bombing.

A Mk 26 mine is photographed during an actual drop at Chichi Jima.

Actual photo of mine laying from B29 Superfortress over Japanese home waters at Shimonoseki Straits.

High-speed mine laying reduces risk to our planes.

FARMINGTON PUBLIC LIBRARY
100 WEST BROADWAY
FARMINGTON, NM 87401

one-half months, the situation in August 1945 is shown in figure 10, Japanese Traffic in August 1945. During this time, 12,000 mines had been laid by 80 to 100 bombers of the 21st Bomber Command under General LeMay. Aircraft made 1,528 trips and delivered 4,900 magnetic, 3,500 acoustic, 2,900 pressure-magnetic, and 700 low-frequency acoustic mines. The mine-laying effort represented 5.7 percent of the 21st Bomber Command's total effort. (The rest of the effort, about 1,500 planes, was spent in bombing.) Further data are given in figure 11, Starvation Campaign, April-August 1945. A detailed history of these events is given in References 23 and 49.

After World War II, some operation researchers on both sides of the Atlantic decided that all major powers would work out their air defenses with modern technology so that no aircraft could fly within reach of harbors or other mineable waters. After all, such aircraft might be carrying atom bombs which would be far too dangerous to let approach. The researchers at that time did not consider that such bombs would be deliverable by missiles. They concluded there was less use developing air-delivered mines because they could not be delivered. It was judged that the days of mine warfare were over. In the two succeeding U.S. wars, this conclusion was shown to be premature. In Korea, the United States was the recipient of a mining effort which held up our operations at Wonsan for weeks and created a great deal of tension in the process. True, these Russian mines, many of 1904 vintage, were laid by sampan and junk, not by air. In Vietnam, however, more mines were laid by air than had been laid in all previous wars! Part of the difficulty in sweeping the mines in Korea, apart from the lack of minesweepers and the large number of mines (3,000), was that they were laid in a combination of moored (old) and ground mines (new) which presented a new menace to the minesweeper.

The difficulties in Korea arose from many sources. First, the possibility of encountering numerous mines was not clearly visualized far enough in advance, and mine countermeasures had only ten days allotted for the job with too few sweepers. Second, intelligence information on how many mines and of what sorts was lacking. Third, the maps and charts of the area were inadequate. A message was sent to the Pentagon by Rear Admiral A. E. Smith, to whom Captain Spofford, in charge of mine countermeasures, reported. The message said that the U.S. Navy had lost command of the sea in Korean waters; Admiral Joy said later that the main lesson of the Wonsan operation was that "no so-called subsidiary branch of the naval service, such as mine warfare, should ever be neglected or relegated to a minor role in the future." The Navy decided that it would place mine

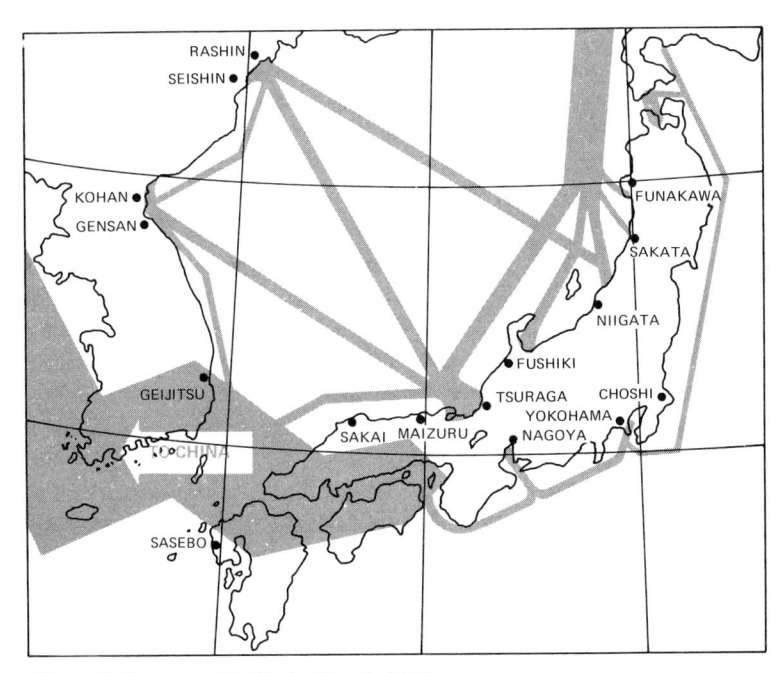

Figure 9. Japanese Traffic in March 1945.

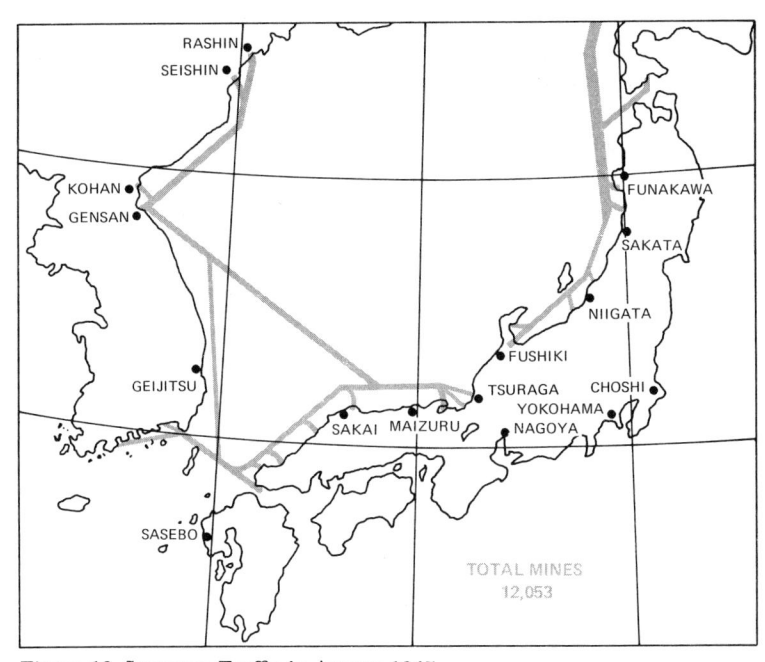

Figure 10. Japanese Traffic in August 1945.

countermeasures in the highest priority, or at least give it priority over mine development, which was not quite the same thing. The Mine Countermeasures Station at Panama City, Florida, was expanded, and a new building was erected. The new emphasis resulted in a new name: The Mine Defense Laboratory. Its present name, under a somewhat expanded mission which includes swimmers and their vehicles, is the Naval Coastal Systems Laboratory.

It is believed that in Korea the Soviets had shipped in about 4,000 moored and magnetic ground mines and that about 3,000 of these were laid at Wonsan. Of these only 225 were swept or destroyed in the swept channel. The rest lay outside the swept channel. The U.S. casualties were four minesweepers and one fleet tug sunk, and five destroyer types severely damaged. Several South Korean vessels were also sunk or damaged. Thus, the yield to the enemy was quite good for the expenditure of old mine materiel and three weeks of unskilled labor devoted to mine laying from towed barges. It was learned after the war that Soviet military personnel had supervised the planning and assembly operations, had made the adjustments on the magnetic mines, and participated in laying them. These induction mines were laid nearer the shore and so were not discovered until the sweepers had become somewhat comfortable cutting the cables of the old moored contact mines. The new combination of ground influence mines sowed among moored contact mines made sweeping extremely slow, complex, and hazardous. If the sweepers had been under attack from shore batteries at the same time, as was the situation thirty-six

MINES LAID	SORTIES FLOWN	PLANES LOST (FEW BY ENEMY ACTION)	SHIPS SUNK	DAMAGED BEYOND REPAIR	DAMAGED AND REPAIRED
12,026	1,528	15	294	137	239

8	MINES LAID PER SORTIE
431	SHIPS LOST OR 900,000 TONS
9,000	TONS OF MINE PAYLOAD
100	TONS OF SHIPPING LOST FOR EACH TON OF MINES LAID
29	SHIPS LOST FOR EACH PLANE LOST
3½	SORTIES FOR EACH SHIP LOST
28	MINES LAID FOR EACH SHIP LOST

Figure 11. Starvation Campaign, April–August 1945.

years before in the Dardanelles, the mines could not have been cleared and the landing at Wonsan could not have occurred.

The Korean War, as we have seen, was a rude shock to the Navy, emphasizing a lesson learned in the Civil War, which had been neglected for nearly a century. In the wars between, the United States had carried out mine warfare against the Germans and against the Japanese but had not been forced to protect herself from mine attacks, except for a sporadic German effort on the East Coast in World War II. Just because something has not happened does not mean that it won't. As a result of this restated lesson, a more balanced approach to mine warfare was pursued in the Navy in the fifties. Renewed scientific advice was brought into the mine community by means of the Mine Advisory Committee (MAC) established in 1951, at first by an ONR contract with Catholic University under the chairmanship of Karl Herzfeld. In 1955 the contract was shifted to the Physical Sciences Division of the National Academy of Sciences. The MAC has conducted a series of studies including Monte, 1957, (on countermeasures), Bowdoin, 1958, (on deep water mining), Masking Techniques, 1961, Monte Plus Five, 1962, Swimmer Countermeasures, 1962–65, Pebble, 1965, Precise Navigation, 1966, Shallow Water Naval Warfare, 1966, Power Sources, 1967, High Resolution Sonar, 1968, and Nimrod, 1967–70.

During these years the Mine Defense Laboratory at Panama City, the Edwards Street Laboratory at Yale, the Naval Ordnance Laboratory at White Oak, and others carried on promising research and engineering studies in mine sweeping, mine hunting, mine classification, and mine neutralization. Helicopter sweeping was developed at MDL and the 50-series of influence mines was developed at NOL. These mines were the product of plans made in the late 1940s to provide a series of antisubmarine mines with advanced and modular features. They included the air-laid, Mk 52, 1,000-pound bottom mine with six different influence combinations; the air-laid, Mk 55, 2,000-pound bottom mine with the same influence combinations; the first air-laid moored mine, the Mk 56, with a total field magnetic influence mechanism; and the Mk 57, a submarine-laid, moored magnetic mine with a plastic mine case and a total field mechanism.

During the years from 1958 thru 1973 the NOL sponsored and conducted a series of sixteen technical annual mine conferences at which a total of 602 papers was given. These conferences were attended by government and contractor personnel, by naval personnel from both afloat and ashore, and by representatives of the mine community from research through production to readiness. The

The YMS 28, the first naval vessel to enter the harbor of Marseilles, France, has detonated a mine, September 1944.

A floating mine in the Kii Suido Straits has been detonated by rifle fire, 1946.

Mines exploding in water off the coast of Korea as seen from the side hatch of an airborne PBM 5, 1950.

The minesweeper 512 blows up from a contact mine in Wonsan Harbor, Korea, December 1950.

83

proceedings of these conferences are on file at White Oak. A cumulative index of titles and authors dated October 1973 is also available there.

After Korea, the next use of sea mines, apart from the mining of the Suez Canal (!), was in the war against North Vietnam.

The operations research conclusions about control of the air did not apply to Vietnam. It was perfectly possible to fly over enemy territory and to drop bombs in an attempt to interdict trails, destroy supply dumps, and cut off supplies. It was not until the mining of Haiphong Harbor in May 1972, however, that the movement of supplies into North Vietnam by sea was stopped. Virtually all the ships in the harbor remained there immovable for months, and no new ships came in. The story of the mining of Haiphong, its materiel, tactics, effectiveness, and how the mine blockade was cleaned up has yet to be told. The thin film magnetometer developed at NOL in the sixties widened the variety of munitions which could be used as mines. The magnetometer consisted of a high-frequency oscillator whose frequency shifted measurably when the magnetic field was changed by the presence of a target. This device permitted munitions to be used without parachutes, and reduced the size and power requirements of magnetic detection by a factor of at least ten. It is indicative of the sagacity of Le Duc Tho in agreeing to a so-called Protocol with Dr. Kissinger on 27 January 1973 in Paris that "The United States agreed to clear all the mines it has placed in the territorial waters, ports, harbors, and waterways of the Democratic Republic of Vietnam. This mine-clearing operation shall be accomplished by rendering the mines harmless through removal, permanent deactivation, or destruction." The Vietnamese would render assistance, but as part of the "Agreement of Ending the War and Restoring Peace in Vietnam" the United States had to agree to clean up the mines! Of course most of the mines had been equipped with sterilizers set for six months or less.

Important lessons and questions

From this brief history one can see that the development of mine warfare is coincident with the development of naval power and depends as that did on the growing technology which also had led to the Industrial Revolution. We have seen that mines depended on the new discoveries in electricity, chemistry, and materials. Before there were any mines it would have been difficult to imagine the uses to which they could be put—and no one did. Perhaps the first lesson then is that the evidently growing role of mine warfare in world events should not be neglected or forgotten. It would indeed be prudent for leaders, aggressors, and defenders to acquire and retain a good understanding of the subject. The second lesson is that there is a continuing need for readiness. A balanced capability in being means a working stockpile with trained men at hand. It means a mobile, available countermeasure force which is in a high state of effectiveness. A third lesson is that the dependence of mine warfare development on advances in technology suggests that in the future there may indeed be some surprises yet to come. It seems therefore essential that the Navy keep a careful and continuous eye on the uses of new technology for mine warfare, both pro and con. The time-tested method for doing this is to retain a continuing in-house excellence in the conduct of research and development.

These are perhaps the most general conclusions one can draw. The details as to how one reaches these desirable ends are less clear. How do we motivate and instruct? How should the mine-warfare interests of the Navy be organized for greatest efficiency? What is the proper balance in funds and effort between delivery vehicles and sweeping vehicles, mine research and development, and procurement? How can the whole process between setting requirements and satisfying them be arranged and settled? Who is to be in charge of what? Some of these questions, of course, are not peculiar to mine warfare, but their solution in a cost-effective manner may be. Solutions tend to go with large funding, and large funding goes with the large lobby. Since mining is now and probably always will be a small fraction of military spending, it will always have to conserve its funds and argue carefully and responsibly for enough to do the job it feels is necessary and possible.

These issues and others may find better answers after we have increased our knowledge of some of the specific matters to be discussed in the next chapters.

3 Mine Targets and Types

Influence fields

Anything that floats can be a mine target. Of course, some targets are easier to detect, or more vulnerable to attack, than others. Targets may be detected and localized by using various signals emanating from them. The crudest and most obvious signal of the presence of a target is its actual impact or collision with the detector. As we have seen, the contact mine was the only thing available in mine-warfare operations from their beginnings up to World War II, when a ship's magnetic field was used as an influence to activate a magnetic detector at a distance from the ship. The early devices for detecting impact were triggerable springs (like mousetraps), inertial devices, frangible glass tubes, or plasticly deformable horns. The so-called K device which found extensive use in 1918 in the North Sea Barrage was almost an influence device but not quite. It was necessary for the target to contact the mine antenna but not the mine. The action occurred at a distance from the mine so that the full damage range of the mine could be exploited, but the separation was achieved by a physical extension of the mine in the form of an independently floated copper wire, and not by any influence transmitted through the medium.

In these modern days when everyone takes miracles for granted, the detection of objects automatically at a distance has become quite commonplace. One needs only think of invisible optical beams that open doors when people approach, various burglar alarms, even smoke alarms now being required in people's homes. For the salesman and the consumer the question How does it work? means is it reliable, does it really detect when and only when something is there, or how often do you need to replace or repair it. For the technically minded person, however, the question means On what set of physical principles is the operation of the device based? We are so accustomed to being surrounded with various technical wonders that appear to be inscrutable, that many have given up asking how they work. You push a button and the thing turns on and works very well or else you call for help. Fortunately, a discussion of the influence fields of ships need not be frighteningly complex.

Most ships are made of iron and steel. Even wooden ships have iron parts such as nails and machinery. This magnetic material assembled and immersed in the magnetic field of the earth acquires a net magnetization of its own which is made up of two parts—a permanent magnetization and an induced magnetization. The first part is acquired when the ship is built. If a ship, for example, were built on the earth's magnetic equator in a north-south orientation, then its permanent magnetization would be horizontal and the ship could be represented by a single bar magnet lying fore and aft. If the ship were constructed at the north magnetic pole, the permanent magnetization would be vertical in the ship and there would be no longitudinal or athwartship components. Evidently the permanent magnetization of a ship depends not only on its size and types of material but also on where it was built and the orientation of the keel in the shipyard. It is clear that it will be necessary to measure the magnetic field around several ships in order to determine the variations in magnitude which will be caused by all these variables. In fact the so-called permanent field is also subject to variation depending on the ambient field, the prevailing bearing of the ship, and the amount of vibration which can reorient the little magnets whose net sum makes up the total permanent magnetization.

The second aspect of the ship's magnetism makes for greater complications. When a magnetic material is placed in a magnetic field it acquires an induced magnetization, B, equal to the field strength H, multiplied by the magnetic permeability of the material, μ. The induced magnetization of the ship in its three components (fore and aft, vertical, and athwartships) is then dependent on the ambient field of the earth which depends in direction and magnitude on the location of the ship on the earth.

The changes in magnetization are also constrained by the hysteresis loops for the magnetic materials of which the ship is made. The most important input is the ship's heading. When a ship heads north, a north magnetic pole is induced towards its bow and a south magnetic pole is induced at its stern. In order to reduce and control the magnetization of ships, two things have to be done. The first is to remove or reduce the permanent magnetization by imposing successive decreasing magnetic fields on the ship in alternating directions. This is done by wrapping coils around the ship and sending through pulses of current reversing the direction each time and using essentially trial and error to reduce the permanent magnetization to an acceptable level. This technique was first developed in France before Hitler took over in 1940 [3]. Once the perm has been removed, the induced and variable part of the magnetization has to be taken care of by degaussing. This, as everyone knows, consists of using coils around the ship in a horizontal plane like a belt to control vertical magnetization, or using vertical coils enveloping parts of the ship to control longitudinal magnetization. The flow of current in these coils is changed depending on the heading and location of the ship to reduce the net induced magnetization. It is apparent that all of this can get to be a major project. In order to measure the effectiveness of a degaussing installation it was necessary to have ranges in various parts of the world where ships could be checked before and after installation and periodically thereafter.

In the first days of World War II the British simply installed horizontal loops on the bottom at a harbor entrance and as ships came in and out they passed over these coils. Their magnetic fields swept through the coils creating an induced voltage proportional (as Faraday found out in 1831) to the rate of change of magnetic flux through the coil. The resulting currents were recorded on paper tape on shore. These records were called ships' signatures and formed the chief basis for judging the effectiveness of deperming and degaussing. The signature was used by the mine designer to determine the sensitivity of his magnetic mine influence mechanism—the device by which he detected the presence of a ship and decided that it was close enough to be damaged if the mine exploded. Although such signatures gave an integrated effect and were similar to what a coil in a magnetic induction mine would see, they were a direct function of the ship's speed, and the accuracy of the result depended on knowledge of the ship's location while underway. It was also difficult to forecast what the ship's signature would be like at one depth if it were measured at another depth. Natural fluctuations in the magnetic background of the earth itself introduced some noise to the measurements, and any shifting in position of the coils due to waves or currents further diluted the accuracy of the results.

The coil ranges were gradually replaced by ranges consisting of hundreds of magnetometers fixed at regular points in the bottom. The ship to be measured could be anchored in position and its magnetic field at each point could be separately recorded. Using such data it was possible to extrapolate the measured field at one depth to predict the field to be expected in a horizontal plane at another depth. There were three rooms at the White Oak Laboratory which were given over to a "magnetic extrapolator" in which the ship's field was represented by an array of forty vertical electromagnets. The current was set in each electromagnet so that the vertical field at that point was equal to that measured on the range. Then the total field beneath the ship at any point at a deeper depth could be determined by reading a magnetometer placed at that point. Years later we retrieved the space occupied by this electromagnetic simulator when we acquired an electronic computer which among other things could solve Laplace's equation with arbitrary boundary conditions. Duncan [12] says that by 1942 the degaussing technique had become so perfected that the amount of field reduction beneath a ship was determined by economic not scientific considerations. The situation was a bit curious when it came to be realized that it was impossible to protect a large capital ship against a sensitive mine in shallow water, i.e., in the worst cases. After the war the policy evolved of protecting minesweepers and submarines by degaussing but not major ships. Design criteria for the degaussing of sweepers and submarines continued to be the product of model experiments conducted in the so-called nonmagnetic area at the White Oak Laboratory. During the war, however, nearly 13,000 ships were fitted with degaussing equipment in the United States at a cost of approximately $300 million.

A typical magnetic signature with and without degaussing is shown in figure 12a. If this undegaussed field sweeps over an induction coil in a mine at the bottom, then the maximum voltage will appear where the rate of change of magnetic flux is at a maximum. At the maximum of the field there will be zero induced voltage because the rate of change of flux is zero. Hence, the induced current in the mine induction coil from an undegaussed ship will look like figure 12b.

In designing a mine to work on this signal, it would be natural to require an induced current of a certain level in order to trigger a sequence of events such as the closing of a relay, which could then cause other events to occur such as imposing a delay, causing immediate firing, or registering one ship count. Mine circuits were quickly developed which required a second actuation but in the opposite direction and only within a certain time slot to help discriminate against sweeping pulses or against unwanted targets. The effect of degaussing is not only to reduce the signal but also to introduce

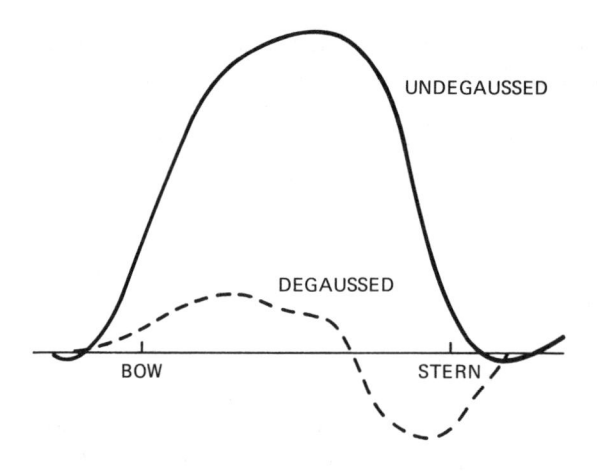

A. SHIP'S MAGNETIC SIGNATURE WITH AND WITHOUT DEGAUSSING

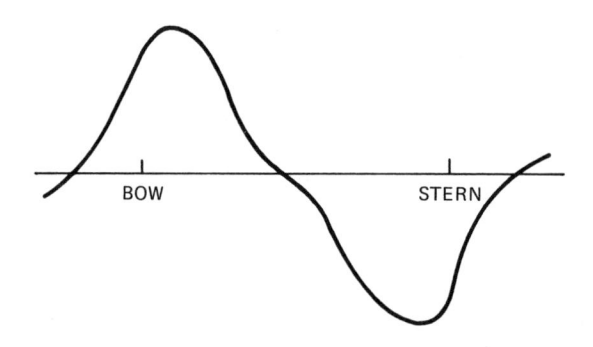

B. CURRENT INDUCED BY THE PASSAGE OF AN UNDEGAUSSED SHIP

Figure 12. Typical Ship's Magnetic Signatures (a) and Induced Current (b).

more wiggles in the induced voltage signal and make it more difficult to tell the mine what to do.

Although the British, as we have seen in our brief history, were the first to investigate and develop a magnetic influence-mine firing mechanism (using a magnetic needle), the Germans were the first to use such a mine in actual operations (1939). The Germans also were the first to use acoustic mines (1940) and pressure mines (1944) as well. The United States and the British were working on pressure and combination influences during the war and were prepared to use them once the Germans had taken the first step. Although acoustic and pressure fields can be lumped together under the designation hydrodynamic as opposed to electromagnetic, I like to keep them separate. The pressure field of a moving ship is not just an acoustic field of near zero frequency, in my opinion, because it is not caused by an acoustic source. It is rather a consequence of Bernoulli's theorem relating the pressure and velocity in a channel flow. It cannot be controlled by damping the vibrations of moving parts in a ship or by other acoustic techniques. The pressure field of a moving ship is quite limited to the vicinity of the ship and is also almost impossible to reproduce without having either a ship or an object that looks like a ship. The pressure influence, in short, is an ideal way to actuate a mine.

During the war there was a very small group at White Oak working on the idea of using the pressure effect for a mine. One of the group was John Bardeen, who later was to become the only man ever to win two Nobel Prizes in physics. His first award, incidentally, with Brattain and Shockley, was for work which led directly to the transistor and the whole development of solid-state electronics. These developments, making possible information processing with very low power and in a very small space, have had an enormous effect on mine development as well as on all other electronics applications. They were of course not available in the early 1940s when Bardeen was at White Oak.

When a ship traveling at more than 4 knots (to get measurable effects) passes over a point in 50 feet of water, then the pressure on the bottom at that point rises slightly as the bow passes over, and then drops below the ambient level until the stern comes by. The pressure then rises slightly above ambient and then returns to normal as the ship moves away. These pressure changes were measured in inches of water and were of a duration determined by the transit time of the ship. These effects were very real but hard to measure because they also were very small. The originally used German oyster mine (Invasion of Normandy) was designed so that a pressure drop of two inches or more for about seven seconds would fire the mine. This would catch a 120-foot ship traveling at 10 knots. Now it so happens that there are occasional ocean waves which have periods and amplitudes which yield reduced pressure or "suction" times of the order of seven or more seconds with an amplitude exceeding two inches of water. Consequently, the pressure influence if used alone in open water or water subject to the influence of ocean swells, is apt to be prematurely swept. In confined waters it is another story. An effort to sweep the pressure mechanism in quiet water was undertaken by J. V. Atanasoff at NOL during the war. Atanasoff, in charge of the acoustics research division, had many ideas, one of which led many years later—in fact in approximately 1972—to a court decision that he was among the first to apply the binary coding system now used in all electronic computers. One of his ideas which paid off in an unexpected way was the construction at the new White Oak Laboratory of a gigantic anechoic room for acoustic calibrations and experiments. Its later use was quite marginal (in fact, it was used only once for some turbulence measurements), and so it was converted to a three-floor facility for the installation of a computing machine, an indirect descendant of J. V.'s earlier work. The mathematicians who were installed with the machine did not like their windowless formerly anechoic environment and migrated en masse to another facility which could promise better amenities and also a better and larger machine. Sic transit gloria.

On the matter of minesweeping, J. V. had the idea of towing a large underwater balloon made of rubber and looking like a gigantic hot dog. This bag, filled with water, would produce a pressure signature like that of a ship. If the mine went up, the bag was expendable, and there was a good chance that it would not be damaged at all. After all, the shock wave produced by the explosion should pass right through the bag because there were no air spaces in it. The matter of rigging this bag so that it would fill properly and so that it could be towed off to the side in order that the tug would not have to pass over the mine was a mechanical detail to be worked out. Many trials were made with various designs of this device, which came to be called the Loch Ness Monster. The hydrodynamics of this bag filled with water were somewhat horrendous. Who could tell that it would leap up and down like a giant porpoise when towed, or that so many would end up a palpitating mass of shreds? The swept path width was very narrow, and the mechanical and material problems were not adequately solved. When the responsibility for countermeasures was solemnly turned over to the Bureau of Ships, they were only too glad to shelve this project and start one of their own not involving any Loch Ness monsters. However, it is conceivable that there is some merit in the idea and that a redesign with different shapes and materials, and with helicopter towing instead of tugs, could result in a practical pressure sweep, and a portable one at that!

Another possibility which was investigated experimentally was to make waves by large underwater explosions. Such a tactic could produce the necessary suction phase in restricted waters where it would not be practical to maneuver a ship at high speed to produce similar effects. There is always the risk also that the ship used to sweep one mine will be sunk by another. An example of a ship's pressure signature is shown in figure 13. Considering that 1 inch of water is less than 2 mm of mercury, and that a change in atmospheric pressure of $\frac{1}{4}$ of 1 percent is equivalent to this, it is clear that a fairly sensitive device with a long-term stability is needed. Since tidal changes can amount to about six feet in six hours (or much more in some places) we can have a two-inch change in ten minutes. The time for a ship to pass a given point will be between one-tenth of a minute and a minute, depending on its speed and length. Hence, the mine mechanism must respond to this but adjust itself to much larger variations in pressure occurring in a longer interval of time. It does this by slowly equalizing pressure through a tidal leak which is too slow to respond to ship variations.

One ingenious scheme for detecting pressure changes was developed by David Muzzey, late Associate Technical Director for Engineering at NOL. It consisted of an ionic solution of potassium iodide and iodine which was polarized by a special low-current battery. When a change in pressure shifted the fluid, its conductivity changed and the resulting current change could be detected. The device, at one time secret, called the solion for ions in solution was abandoned because it developed corrosion problems in storage. These problems were reduced by storing the solion cells in cans containing nitrogen only, but its shelf life was limited and it was consequently set aside in favor of solid-state devices which were competitive with solion in power consumption.

The third major ship's influence, of course, was acoustic. The sounds coming from a ship underway are extremely complex—coming from vibrating machinery, from the ship's structure and plates serving as diaphragms, from the propellers churning the water causing cavitation noise from oscillating and collapsing bubbles (caused

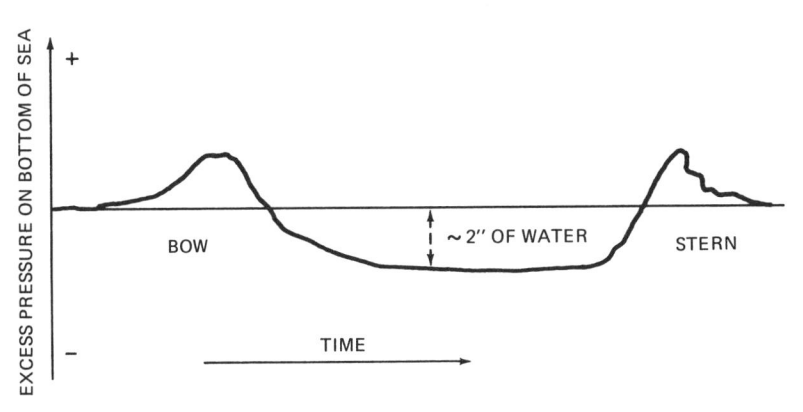

Figure 13. Typical Ship's Pressure Signature.

by our friend Bernoulli), and even from flow noise caused by the transition of laminar to turbulent boundary layers in the water along the ship. All of these noises have different frequency bands, amplitudes, and phases, and vary with speed and, in the case of submarines, with depth. Ship's noise is a continuous band with line components superimposed. Some of the noise can be mitigated by various quieting techniques used in the ship's construction. However, the sounds which appear at a given point on the bottom of the sea are not only those that come straight from the ship. They are the sum of sounds which have traveled through the water and have been reflected one or more times at surface or bottom with appropriate phase changes which depend on the angle of incidence of the sound on the reflecting layer. All these sounds are subject to a differential absorption, increasing as the square of the frequency. Additional selective absorption is caused by ionic dissociation of $MgSO_4$ dissolved in sea water. To the ship's sounds must be added ambient sounds originating from sea state, marine life, and other shipping, all processed and modified by propagation through a nonhomogeneous medium consisting of warm water, cool water, muddy water, moving water, of nonuniform depth and covering a bottom which can vary from rocks to deep mud.

It would appear that the acoustic signature from a ship is the most capricious of all the influences used to fire a mine. The characteristic low-frequency thumping of the shaft travels a long way and hence is not good for localization. In the relatively shallow water of most ports and channels the multipath interferences and reinforcements of all the net sounds can lead to loud signals when the target is still far off. The sweeping of acoustic mines called for some sort of noisemaker to simulate the ship's sounds, and the more noise the better. This, of course, led to the setting of some mines at low acoustic sensitivity so that they would respond only to the noise sweeps at close range and thus were specifically antisweeper mines. It became quite apparent that the use of acoustic influence mechanisms alone and not in conjunction with the magnetic influence was rather ineffective. This applies to mechanisms which passively listen for noises in shallow water and attempt to process whatever noise comes in to reach a decision to fire. In deep water the propagation of signals of relatively high frequency is a much cleaner proposition: there are fewer multipath reflections, fewer ambient sources. If the mine adds an active high-frequency source to its kit of diagnostic tools, it can illuminate a suspected target briefly with its acoustic flashlight or search beam, can confirm its suspicions obtained passively, and then take whatever action has been programmed into its data processor. Figure 14 il-

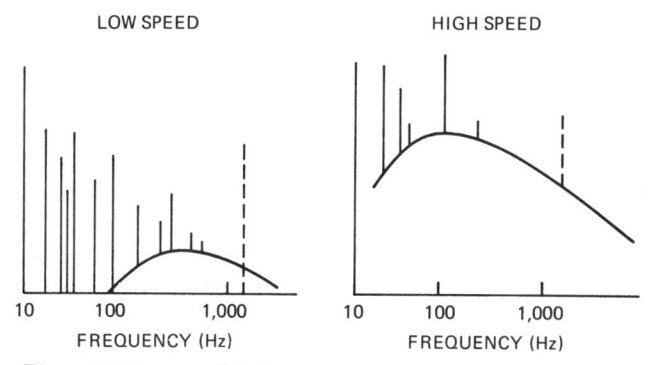

Figure 14. Spectra of Submarine Noise at Two Speeds.

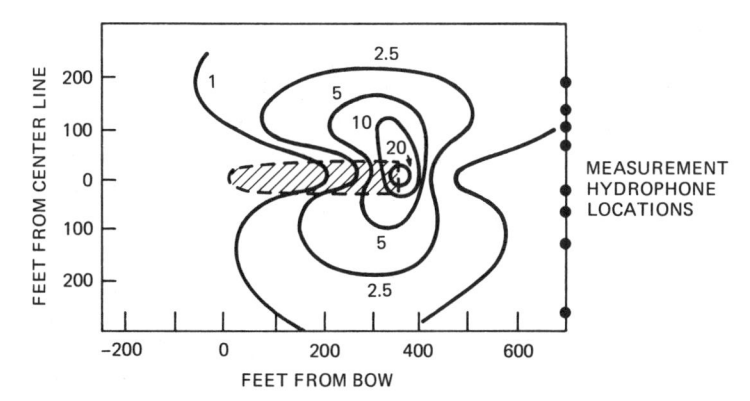

Figure 15. Sound Pressure Contours.

lustrates the frequency content of the acoustic signature of a submarine at two speeds. Figure 15 shows sound-pressure contours (caused by a freighter passing at a speed of 8 knots) in dynes per square centimeter in a 7-Hz band on the bottom. The water depth was 40 feet. Reference 47, from which these figures were taken, contains detailed discussions of the acoustic fields of ships.

As we have just seen, there are many influences emanating from a ship. These have various attenuation laws. For example, low-frequency acoustic energy in shallow water, i.e., water depth less than the sound wavelength, will decrease inversely as the distance. Acoustic energy in shorter wavelengths from "point" sources will drop off inversely as the square of the distance and faster due to absorption. The magnetic field from a dipole source will decrease inversely as the cube of the distance. If one makes a magnetic detection mechanism which measures at two points and detects the difference, i.e., a gradient mechanism, then the effect attenuates inversely as the fourth power of the distance. If one tried to detect a ship by actively inducing eddy currents in its metallic structure and detecting their radiations, then the attenuation law would be the inverse sixth power of the distance. This latter scheme is, of course, extremely short range and has been used for hand-held detectors of land mines or metal, but not as a mine mechanism itself. There are many other influences which could give notice of the presence of a ship. For example, a ship is opaque. If it is daylight there should be a detectable shadow beneath a ship provided the water is not too muddy or the detector is not too deep. Is there a radiation window in water such that some electromagnetic wavelength other than yellow sunlight could get through more easily? If this is not achievable, what about the penetration available to cosmic rays?

According to Archimedes, the weight of a ship is equal to the weight of the water it displaces. Consequently, the average mass per square foot of area above a mine on the bottom is the same whether a ship is present or not. However, the distribution of mass is altered, some parts of the area having low mass and other parts a high mass. Also the material itself is drastically altered, a ship being mostly air with some metal. Therefore, the passage of cosmic rays through a ship should in principle be detectable. Such an influence detector would localize the explosion under the ship and thereby increase the number of sinkings compared with the number of merely damaged ships. It would also be extremely difficult to sweep. The cosmic ray mine has been "looked at" hastily in the past and then hastily discarded on various grounds, some of which were technical—perhaps the detectors were too fragile or took too much power; some of which were operational—perhaps the system would work poorly on submerged submarines, and these were the priority target at the time; and some of which were just plain practical—we have a good functioning pressure magnetic mine and so why look further?

Another rather esoteric but omnipresent influence to consider is the distortion in the gravitational field caused by the presence of a ship. This too has been examined and discarded on similar grounds. It is possible, however, that new ideas for gravimeters could turn the conclusion around. Other influences such as the electric potential field surrounding a ship have been in more lively contention. This field is due to the fact that a ship normally consists of dissimilar metals—steel hull, bronze propellers, and so forth—and being immersed in an electrolyte (salt water), there is always some sort of electrolytic action taking place. For example: if a battery is formed by the bronze propeller and the steel ship in seawater, the current will flow only when a conductor connects the propeller and the ship. Such conducting contact as occurs in the propeller shaft bearings is generally varying with time as the shaft turns. And so there is a varying resistance in the battery circuit which modulates the current flow at some frequency related to the shaft rpm, and to the number of contact locations in the bearings. The magnitude of current depends on the contact resistance, which is influenced by the amount of lubrication, the condition of the packing glands, the speed of rotation, and so on. The current flow produces a slight magnetic field, which can be detected by the fact that it is modulated, normally called the AM (alternating magnetic) field associated with the UEP (underwater electric potential) field. Further, the magnitudes of these fields vary enormously between otherwise similar ships because of paint condition, marine growth, age, and corrosion. Many mine designers decided long ago that this galvanic influence would not be a reliable one on which to base a mine design, whereas those responsible for sweeping felt that provisions for sweeping such mines should be on hand. Such differences of opinion are the sources of bureaucratic tension, especially at budget time. Still other possibilities for mine mechanisms involve the velocity flow around moving ships (related to the pressure signature), and even nuclear emissions from nuclear propulsion plants in submarines.

There are a number of new developments which may pose new problems for mine design. For example, the supertanker presents to the mine a bigger ship than it has ever seen before—a deeper draft, a longer length, and perhaps a new vulnerability. This should be an easy problem to overcome. There will not be any difficulty detecting the target; it will be a matter of adjusting the mine logic so that it will respond properly. Almost the opposite problem will occur, however, with the new fast ships now in an evolutionary state. These include hydrofoils, surface effect ships, and those of catamaran design called SWATH ships (Small Waterplane Area Twin Hull). Some of these ships can travel at about 80 knots and have a small displacement and a shallow draft. Since most of these are in an experimental stage, the amount of data on the signatures of the various sorts is meager. It is quite clear, however, that the mine which is going to fire under one of these ships will have to make up its mind many times more quickly than it has with slower targets in the past. With the use of helicopters for mine countermeasure work, the fields and the vulnerability of helicopters themselves also become of interest to the mine designer.

These matters are also, let us say, of considerable interest to the helicopter pilot!

Vulnerability

It is obvious that a mine should not explode at too large a distance from its intended target. It must wait until the victim is within damage range. But how far away is that? It is necessary to have an idea of the relation between charge size and position, and the strength or weakness of ships.

A high explosive detonating underwater produces almost instantaneously a mass of gas (equal to the mass of the explosive) at a pressure of about 100,000 atmospheres. This gas expands against the water and causes a shock wave to be formed and to spread ahead of the gas-water interface. This shock wave, traveling initially with slightly supersonic velocity in water, carries outward a pressure pulse which is roughly exponential in shape and has a peak pressure which decreases approximately linearly with distance from the explosion (actually as $r^{-1.13}$). It is this shock wave which carries explosive energy to a distance, and impinging on an air-backed metal plate such as a ship hull will set it in motion in response to the pressure acting on it. If the acquired velocity is fast enough, the resulting kinetic energy of the plate will exceed its ability to absorb this energy through plastic deformation, and the result will be rupture of the plate. Whether rupture occurs or not there will be a transmission of particle displacement throughout the structure holding the plate. In other words there will be a transmission of shock and vibration throughout the structure of the ship. This can have a variety of serious effects such as the breaking of tubing and the misalignment of bearings, or of minor effects such as throwing circuit breakers, or breaking crockery and light bulbs. A severe shock can immobilize a ship, especially one that is not protected through design against such effects.

A detailed calculation of plate deformation and damage by an impinging shock is a straightforward but complex thing to do. The plastic properties of the material at high rates of loading must be known. The plate moves away from the water in many cases before the shock wave has decreased very much, so that there is a short period of cavitation between the plate and the water, followed by a reloading phase as the plate slows down and residual shock pressure is still there forcing the water against the plate. It is necessary to consider the reflection of the shock from any rigid boundaries to which the plate is attached. Such calculations confirmed by controlled experiments have shown that this damage mechanism is well understood. It is, however, far easier to assume for approximate purposes that just the shock-wave energy impinging on a plate or diaphragm is used to create plastic deformation or to create shock.

In the early days of World War II, our destroyers carried depth charges which were tossed over the fantail and went off at different depths. The shallow ones were much closer to the ship when they went off than the deep ones because the sinking time was less, but nevertheless the deep ones gave the ship more shock. Fleet complaints about excessive shock when depth charges went off over a relatively close hard and rocky bottom gave further clues to what was happening. The amount of shock was postulated by Al Focke and me in the Bureau of Ordnance to be proportional to the amount of energy which impinged on the bottom of the ship when the charge went off. The ship looked bigger to the charge when it was deep even though it was farther away. The effect of the hard bottom was to provide, in effect, an image of the explosive source so that the ship received two shocks, one direct and the second reflected from the bottom. It was possible to construct curves below the ship such that a charge situated anywhere on a given curve would communicate the same energy to the ship and hence the same shock. We called these curves "iso-damage contours" and established them in such a way that the ship's bottom subtended the same solid angle at any point on the contour. These curves looked like circles around the ship, with the closer circles corresponding to the greater damage. If the charge weight were W pounds of TNT or equivalent and the solid angle subtended by the ship at a given point were ω steradians, then if the product of W and ω were less than 0.01 it would be safe to explode the charge at that point without shock damage in deep water. We also found that if the energy criterion were drastically increased, say four hundredfold, so that $W\omega > 4$, then rupture of the ship was likely to occur. These relations were very qualitative but useful for making estimates in practical situations. The circular contours around the ship did serve to delineate damage zones. Of course, because of the variations among ships and all the other variables, the zones were themselves rather imprecise and got names like critical damage, moderate damage, or light damage.

There is an entirely different phenomenon connected with the underwater explosion which becomes of great importance for producing damage when the ship is directly over the explosion. As we have seen, the high-pressure explosion gases expand against the water producing the shock wave which travels faster than the gas-water interface and very soon leaves it far behind. The gas globe continues to expand, however, carried on by the momentum imparted to the water—far beyond the radius where the explosion gas pressure has decreased to the ambient pressure at the depth of the explosion. In fact, in a typical case the gas globe expands until the pressure inside it is as low as a tenth of an atmosphere or less. Then everything comes to rest momentarily, and all the energy which remained behind after the shock wave departed has been converted to potential energy. The large gas globe now starts to collapse, and the bottom of it comes up faster than the top goes down.

The result is a new collapsed and condensed gas globe at a point typically many feet (twenty or thirty for a depth charge) above the location of the original explosion. This new explosion at a point much closer to the target is a new source of damage which is caused not only by a newly radiated pressure pulse but perhaps more importantly by a jet or flow of water which is associated with the massive migration upward of the bubble (i.e., of the water surrounding it.) This now has the effect of lifting a ship at the middle and causing it to flex violently as a free-free beam. Such whipping action can break a ship in two or cause other serious structural damage at distant points on the ship where high stress or high displacement can occur. The almost endless complications resulting from all these effects and their causes were extensively investigated during and after the Second World War. Perhaps the best single technical reference on underwater explosions is a book of that name by Robert H. Cole [8]. The *Compendium on Underwater Explosion Research* published by ONR in 1950 is a collection of the key scientific and technical papers from both the United Kingdom and the United States in three volumes entitled *The Shock Wave, The Gas Globe,* and *The Damage Process* [18].

As a result of this work, and much subsequent work by H. G. Snay and coworkers, it has been possible to construct equal damage contours around a ship which take into account the effects of gas-globe damage as well as shock-wave damage. These contours, of course, are considerably farther away from the ship at the bottom than at the side. If a ship is harder to damage (because of thicker plates, higher strength steel, or reinforced construction to prevent whipping and so forth), then the damage contours for severe damage are drawn closer to the ship. Figure 16 illustrates damage zones for a typical vessel. The figure pertains to a definite TNT equivalent charge weight exploded on the sea bottom at any chosen depth and athwartship distance, but in the median plane of the ship. It can be seen at once that curves of this sort are merely guides to enable the best average match to be made between the damage potential of a selected mine and the damage expected for a selected class of targets. The zones show the

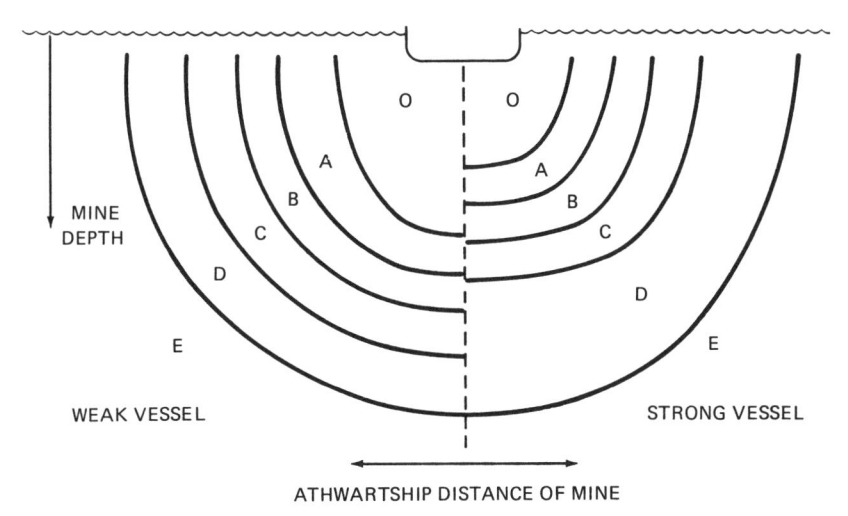

Figure 16. Damage Zones of a Typical Vessel.

various damage levels which can be selected. For example, a mine in zone O will produce immediate sinking, whereas a mine in zone A will cause hull rupture, machinery destruction, and eventual loss of the vessel. Zone B, called moderate damage, will not necessarily split the hull but will cause small leaks in stuffing boxes, submarine hatches, or inlets. Shock damage may lead to immobilization. Zone C is still in the major shock-damage category and may result in loss of combat effectiveness for submarines or warships. Zone D creates shock damage which tends to be unpredictable, varying from partial immobilization of a submarine to negligible damage. In Zone E a well-designed submarine or ship with shock-mounted equipment will have negligible damage if any.

Now if one placed a given mine in a certain depth of water and passed a ship over it at various athwartship distances, there would be a high probability, perhaps even a certainty, of its firing if the ship went directly over it, and a decreasing probability of firing if the ship passed at increasing distances. One can construct a curve showing the probability of actuation for one mechanism or another as a function of the athwartship distance of passage. Figure 17 is such a curve.

Note that not all influences will necessarily have a shape of this sort. For example, if one had a heavily magnetized ship which could actuate a magnetic mechanism at a large athwartship distance, then the probability curve could have a shape like the broken line. This could be used to fire mines at a distance and was tried by the Germans with their Sperrbrecher sweeper. The mine planner's response to this, of course, is to reduce sensitivity until the sweeper actuates a

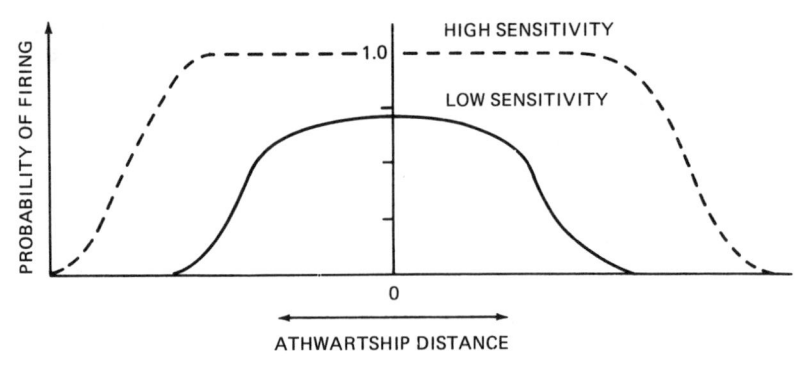

Figure 17. Probability of Actuation for Influence Mines.

mine under its own hull, and not at a harmless distance. There are probability-of-actuation curves in principle for each mine mechanism and setting for each water depth and class of target. Probability is introduced because the ships of a given class are not identical, nor are mine mechanisms of a given setting, nor indeed are environmental and background conditions. Nevertheless, if a minefield planner wishes to immobilize the maximum number of ships, rather than sink a smaller number, he will choose his damage zone accordingly; thereby, in the known depth of his proposed minefield, he will find the athwartship distance beyond which the damage produced is undesirably small. He would like to have a mechanism whose probability of firing beyond that distance was zero for that class of targets at that water depth. By selecting mechanism and sensitivities he can in effect make the mine respond to his wishes. The area under a probability-of-firing curve can be set equal to a width, w, multiplied by unity. The quantity w is then called the average firing width for that particular mine setting, and is the width associated with a probability unity. In other words, one can think of all mines as firing within that width, i.e., $\pm \dfrac{w}{2}$ of the centerline of the ship, and not firing outside these limits. Although this is not strictly the case, the concept of average firing width will give the same number of ships damaged for random passage as the use of the probability curve will.

In order to plan an influence minefield, a great deal of information must be brought to bear. One must select the location; describe the expected traffic; assess the vulnerabilities of this traffic; decide what kind of damage is to be sought and what the minefield threat is to be; select the mine mix; decide on what countermeasures are to be encountered and what to do about them; select the numbers and settings of mines; determine the laying means and schedule, including replenishment; and, one would hope, provide for some sort of surveillance to assess the accumulating results. All of this is a nontrivial operation requiring a high degree of knowledge, skill, and cooperation among technicians, planners, and operators, and in many cases between services.

The kinds of mines

It is obvious from the foregoing that many different mine capabilities are necessary to be able to conduct efficient naval mine warfare in various situations. For example, if the water depth is doubled from one minefield location to another, say from 50 feet to 100 feet, then for the same target the charge weight should be four times as large. Charge weights between about 300 pounds and 1,200 pounds are convenient for bottom mines. In fact, bottom mines run out of usefulness against surface ships in water over a couple of hundred feet deep because their weight will become excessive. Most harbors and their approaches, however, are shallower than this, and so there is wide scope for their use. Mines may be classified in several ways, for example as offensive or defensive, bottom or moored, contact or influence, as well as by size, type of influence, intended target, method of delivery and intended water depth. If each of these possible combinations were represented by a separate and unique mine design, the number of mine types would be very large indeed.

There is a further classification that can be introduced, and that is whether the mine is self-propelled at some stage of its operation. We have carefully explained the essential characteristic of mines to be their habit of lying in wait for the target. Nevertheless, it is a simple extension of this idea to have the waiting mine identify a target at a distance which is beyond the damage zone of any practical explosive charge. The mine could then release itself and be propelled toward the target either using buoyancy forces or an engine of some sort. The mine in effect would act like a homing torpedo of short range after its mine mechanism had detected and confirmed the target. Such mines, for example Captor, have been developed for use in deep water against submarines. In these cases, special precautions are taken to prevent attack against surface ships. It should be noted that the problem of target acquisition, which is a considerable problem for an air-launched torpedo, has been taken care of by the mine mechanism. The homing phase is not entered into until the target is affirmed as within range. Buoyancy mines and self-propelled mines are merely a logical extension of the damage width of the mine and make it possible to deal with occasional traffic using a relatively small number of mines.

Another example of mobility in mines occurs in the so-called mobile mine which is launched out of a torpedo tube in a submarine. The mine is carried by a quiet propulsion system (battery-powered) to a distant point where it is deposited on the bottom. The laying submarine thereby does not have to approach a possibly dangerous harbor defense system as closely as it would otherwise. Such missions, or the use of mobile mines for the replenishment of offensive minefields in enemy waters, are clearly very dangerous, and the popularity of the mobile mine among submariners is understandably not great. Nevertheless, this represents a capability that is otherwise not available, especially if mine laying by air is denied, as in heavily defended areas.

Table 2 gives examples of various real combinations. The control mine for harbor defense, the descendant of Colt's defensive initiative, was developed to a high state of proficiency by NOL during and after World War II, when harbor protection was transferred from the

TABLE 2.
U.S. MINE TYPES

Use	Target	Planted by	Postion in water	Fired by	Nomenclature
Defense	Ship/Sub	Ship	Bottom	Manual	Control Mine Mk 2
Defense or Offense	Ship/Sub	Ship	Moored	Galvanic Antenna	Mk 6
Offense	Ship/Sub	Swimmer	Contact	Time Delay	Limpet
Offense	Ship/Sub	Aircraft	Bottom	Influences	Mk 52
Offense	Ship/Sub	Aircraft	Deeper Bottom	Influences	Mk 55
Offense	Ship/Sub	Aircraft	Moored	Total Mag Field	Mk 56
Offense	Ship/Sub	Submarine	Moored	Total Mag Field	Mk 57
Offense	Ship	Aircraft	Bottom Riverine	Influence	"Destructors" Free Fall, Ready Round
Offense	Sub	A/C, Ship or Sub	Deep Moored	Influence	Captor
Offense	Ship/Sub	Submarine	Bottom	Influence	Mobile Mine
Anti Sweep	Sweep Wire	Aircraft	Moored	Contact	Sweep Obstructor

Army Coast Artillery to the Navy. It has now been undoubtedly rendered obsolete by the extended defense systems represented by air patrol, antisubmarine measures, and offshore submarine detection systems. If these systems should ever fail and if the United States should ever be in real danger of invasion, then such defensive mine installations might be useful to deploy if they could be recovered from the mothballs. Similarly, the Mk 6 moored mines are obsolete in design and are no longer in the mine stockpile. The availability of moored mines to be laid by surface ship for either defensive or offensive purposes needs to be considered carefully. The present readiness in moored mines relies on the aircraft and submarine-laid moored mines Mk 56 and Mk 57 respectively, which were developed in the fifties as part of the so-called Long Range Mine Program. The principal thrust of the U.S. mine program since World War II has been against the submarine threat rather than against the merchant ship, except for the riverine adventure of Vietnam. Whether this is a reasonable position for the long-term future will depend on the likelihood of possible future antagonists having or depending on an extensive merchant marine or surface fleet.

4 Technology and Mine Design

Definitions and discussion

The influence of technology on mine design can best be shown by historical examples. Modern examples are more difficult because of security limitations and also because the technology itself is more abstruse. This is particularly so in the area of modern solid-state electronics. Almost everybody has an electronic hand computer and knows how to use it, but hardly anyone knows how it works. In fact, there are so many devices in our society which are sold and used in modular format (which tends to preclude repair), that the question How does it work? is losing its original meaning. When I asked a local salesman how his smoke alarm worked, he said, "Oh, very well." This was before I learned that there were two kinds of detectors, namely those using photoelectric and those using ionization principles. The question How does it work? comes to mean How well does it work? rather than By what physical principles does it work? There is, then, a resulting emphasis on alleged capabilities rather than on operating principles which, properly engineered, can lead to the safe, reliable, sensitive, economic, long-lasting, and foolproof devices which everybody wants or everybody advertises. It seems to me that everyone who uses a device in any sort of professional manner should have not only an empirical knowledge of it based on actual use, but also an understanding of the principles of its operation. Clearly, no advances in a device can come without an appreciation of the way a new material, a new process, or a new idea can fit in with what is already there. All improvements and almost all new applications start from a previous base, no matter how primitive. The user can make more meaningful suggestions for improvements if he understands the principles involved.

There is, of course, the possibility that the more developed nations have reached the plateau of the brave new world, or of Utopia 14, in which everything is as small and as efficient as possible, where the amount of work to be done has been automated and mass-produced to the point where no further improvement is needed or imagined, and where no understanding on the part of the user is necessary to the enjoyment of that use. It may indeed be no longer necessary to have some understanding of technology remain with those who use that technology.

When I first went to Caltech where Robert A. Millikan was president, I decided to study physics because I thought it would give me a precise understanding of the world—(perhaps I should say the physical world). My mother was always embarrassed by this, because she did not know what physics was and could say "physicist" only with difficulty. Mrs. Millikan, it is alleged, had a definition for a physicist which everyone could understand. A physicist, she said, is a man who knows how to fix the vacuum cleaner but won't. Now, the vacuum cleaner in 1930 represented a technology based on certain aspects of electricity and magnetism, aerodynamics, and mechanical engineering. These are subjects not entirely without relation to physics. Her definition does indicate that it is possible to understand certain principles but not be able to use them quickly in some practical situation. It does not follow, however, that the repairman should be ignorant of the principles of his trade. If he understands the principles correctly, he will be a better repairman than if he does not. I suppose the danger in this is that if he has that much interest in the principles he is likely to stop being a repairman and become a physicist.

Our conclusions applied to mine warfare are that those officers who are involved in mine warfare should have an understanding of the physics behind their subject even if they could not actually design a mine; that minemen who can assemble, service, and test mines should also have an enlightened view about what they are doing. This view is valid, even when the technologies used in mines become so specialized and abstruse that it is difficult for anyone to understand them unless he spends full time at them. In other words, good minemen have to become specialists.

The technologies we are talking about include, for example, solid-state electronics replacing an older technology using vacuum tubes for rectification of electric current, for amplification, and for modulation. The computer technology for information-processing, the analysis and comparison of signals, and for decision-making, is of growing importance. The technology of chemical and other sources of power, batteries and the like, is another area of great importance. Even in such a well-worked area as signal detection, for example in detecting magnetic changes, there have been many new schemes for magnetometers. These have advantages and disadvantages that have to be compared and judged if one of them is to be selected for mine use. In any choice of an influence field it is necessary to have a complete knowledge of the competing background field, its fluctuations, and the causes of its variations. Thus, the search for an understanding of the magnetic fluctuations which could on occasion be severe enough to trigger sensitive mines, has transcended this earth and taken us into investigations of sun spots and the solar wind. It appears that streams of protons fired toward the earth by solar flares become deflected in the earth's magnetic field and cause fairly rapid changes in the ambient magnetic field. The changes are rapid because the flares are quite directional and sweep over the earth quite rapidly. Also, they can point at the earth every twenty-seven days as they turn with the sun's rotation. And so we can expect some background fluctuations to occur on a periodic basis. This and much more information on the magnetic background may be found in the book [4] *Natural Electromagnetic Phenomena Below 30kc/s*, the subject of which is perhaps not a technology primarily associated with mine design, but it is a related body of knowledge, which is highly relevant and indispensable to the design of sensitive mines.

A general remark may be in order about *technology*, a word which has been used so much lately that it has lost a precise meaning. Webster says that the first meaning is "industrial science or systematic knowledge of the industrial arts." It is also a synonym for applied science. We are immediately confronted by a question. Is technology merely a knowledge of industrial processes or does it require an understanding of why the processes work. There are many processes which are not fully understood and which sometimes fail because of slight changes in impurities or in conditions not previously thought to be important. A trade or commercial secret may produce fine and unique products for a customer over a period of years, then unaccountably fail to work at some crucial later time. If the customer is the U.S. Government and if the failure is critical, an understanding of the process is essential and to be valuable should have been worked out beforehand. There are many examples of this sort of thing—such as failures of plastic parts because of improper controls in heat treating, or because of a gradual slip in instrument calibration, or a shift to a new source of supply for some component. Thin films for photo detectors, magnetic films, and semiconductors in general are very sensitive to process control. To find out why a process works requires first finding out what the process is. Many times this is called a trade secret, and so, as if things were not difficult enough, we have economics and the law to contend with.

It is desirable in a technical development to know not only what will work but what will not work and why. This is in the nature of an insurance policy. Companies may say, if a process works, why look for trouble. As Mr. Bert Lance is reputed to have said, "If it ain't broke, don't fix it." The government, or a competitor, on the other hand, may want to know in advance what could happen and what to do about it if it did happen. The technology business is thus a blend of science, process control, and testing which can take the form of industrial art and trade secrets on the one hand, and of a sophisticated experimental and theoretical understanding on the other. The tensions of the military budget process aggravate the situation. For a while the Mansfield amendment insisted that any research work done in an agency be relevant to the mission of that agency. Of course, some work was more relevant than other work. The amendment did not specify in whose opinion the relevance was to be determined, or what the unit of relevance was.

The result of this was to eliminate "basic" work in mission-oriented agencies, because it could not be shown to be relevant. It might turn out to be relevant to the mission of some other agency or worse still, not relevant at all! Instead, the budget items moved toward "technologies" which turned out to be groupings of subjects clearly related to hardware which was in turn clearly related to mission. These technologies, then, had a very utilitarian aspect which no one could deny. Another idea came along in Congress from Representative Deddario, which was also based on some sort of economics. The idea was that if a new program or a new agency arose which needed help in research and development, it would make more sense to go to an existing government laboratory for such help, if it happened to have experience in the required technologies, than to start a new government laboratory. This proposition is self-evident as a short-term solution. In the long term there are questions of interference, responsiveness, direction and probability of expertise, and specialized facilities which make each agency eventually feel the need for its own laboratory or in some cases its own captive contractor.

As a corollary, however, the idea developed that a given "technology" could be "transferred" to another agency by applying it to a project of that agency. In this way the knowledge and experience

TABLE 3.
MINE WARFARE NEEDS RELATED TO NAVY BASIC RESEARCH

Basic Navy List (As of 1959)	Naval Mine Warfare Needs (Continuing)
Physics (Nuclear Solid State, and General)	Isotope power supplies, cosmic ray detectors, X-ray inspection, influence fields & detectors.
Astronautics and Aeronautical Engineering	Mine/Aircraft carriage and separation. Parachutes of all types, free fall, stability.
Material Sciences	Plastics, adhesives, metals, alloys.
Electronics	Semiconductors, solid-state circuitry, etc.
Mechanics	Shock, impact, hydrodynamics, damage.
Medical Sciences	Shock treatment, protection, explosion damage.
Biology and Biological Sciences	Sea growth, fish noise, ...
Oceanography	Currents, environment, waves, acoustic propagation, bottom topography.
Chemistry	Corrosion, batteries, explosives.
Geography	Climate, environment, trade routes, channels, islands.
Psychology	Minefield planning, training.
Operations Research	Mine warfare was an initial field for these techniques and continues to use them.
Meteorology	Acoustic and wave background, extremes predictions.
Astronomy and Astrophysics	Background magnetism.
Mathematics	Computer simulation of minefields.
Combustion	Initiation, detonation.
Earth Physics	Magnetic and other background, minefield surveillance.

developed in one agency by adherence to the relevance criteria for that agency could, perchance, be used for solving another problem in another agency. Whether this same solution would have developed in the other agency, given time but under its own particular relevance criteria, is not known. It is clear that technology, however much of it is known, can be transferred or borrowed in a properly permissive atmosphere. It may also on occasion be sold or stolen. It also can be developed. There is a slight resistance in the private sector to the idea of stressing technology in government laboratories as a safeguard against industrial failures. In order to build technological knowledge (as opposed to scientific knowledge), it is necessary to have some experience in actual industrial processes. The acquisition of such experience is often condemned by industry as encroaching on the domain of free enterprise and as a form of government competition with profitable industry. Most claims of this sort are invalid partly because the actions of government are mainly undertaken to fill a void and to protect the public interest. For the most rapid solution of technical problems, it is essential that technical information and experience be available to would-be problem solvers across all branches of technology. It is impossible to predict with any precision just where the solution to a technical problem will come from. In this sense a free technology is full of pleasant surprises. There is little doubt that the solution will arise in some "technology," but uncertainty remains as to which one. A rigid adherence to mission orientation as a guide for research is shortsighted and is contrary to free enterprise of the intellect.

At the Naval Ordnance Laboratory, there was an interesting example of the pleasant surprise not immediately related to mine warfare. The magnetic materials laboratory with its experience in alloy processing was working on a Polaris-inspired project to make alloys expected to have very high melting points and low conductivity. The best of a series of alloys under investigation would then be a candidate for a heat-shield material to project missile nose cones upon reentry into the atmosphere. At a project review meeting in 1961 a small piece of wire was brought in which had been drawn and was made of an alloy of nickel and titanium. Dr. Dave Muzzey, who had once been chief of the Magnetic Materials Division, was attending the project review. Although the NiTi alloy did not have an extremely high melting point, it was noteworthy that it was malleable at room temperature and also was extremely resistant to corrosion. As this was being described, Muzzey wound the wire around a pencil to see how malleable it was, slid it off, and then to test corrosion or perhaps to enliven the meeting, lit a match under the coil. To his astonishment the wire instantly straightened out to its original length with no sign of having been bent around the pencil.

This event caused a lively curiosity among the researchers, who eventually were able to explain the whole thing in terms of a transition from a face-centered to a body-centered crystal structure when the temperature changed from one side of a transition point to the other. Most of the theoretical work was done by Wang and the metallurgical work by Buehler. It was learned how to change the transition temperature and other properties by changes in composition. This material, which was called Nitinol in accordance with our policy for naming NOL-developed metals, was unsuitable for its originally hoped for use. It has, however, found application in several unrelated fields—for example: as a self-erecting antenna for space satellites; as an automatic device for detecting brake temperatures in railroad freight cars (and preventing hot boxes, the major cause of train wrecks); and as a replacement for stainless steel in orthodontal bridges, where it has been found that Nitinol wires can provide a steady force through a longer displacement. This means that the bridge does not have to be adjusted and the tooth straightening can actually be done in half the time. I suppose that according to the strict relevance criteria these benefits should not have been pursued at the Laboratory because they were not part of mine or reentry technology.

Examples of the influences of technology on mine warfare

1. The influence of shock-wave research.

There are many historical examples of the interaction of technology with the design and development of specific devices. In 1959 the Naval Research Advisory Committee sponsored a study on Basic Research in the Navy [35]. The study was conducted by Arthur D. Little & Co. and was submitted to Mr. Gates, the Secretary of the Navy, on 24 April 1959 in an attempt to show him that money spent on basic research was in the long run well spent toward furthering the Navy's mission. The argument, illustrated with examples, was that "the success of the Navy in accomplishing its mission in competition with other world powers depends largely on a continuous flow of new and better weapons and techniques. This, in turn, requires the continuous development of new technologies which have their roots in the results of basic research." To make the argument more watertight, one can show that the gap in time between a basic discovery and its application is rapidly decreasing in the modern world; therefore, we cannot afford to be dilatory with our basic research efforts.

On 24 June of the same year Mr. Franke, who had just succeeded Mr. Gates as Secretary of the Navy, wrote to the committee congratulating them for their thorough and constructive analysis of the problem of basic research in the Navy. He said that the recommendations contained in the report would receive very serious consideration and would be invaluable in the budgetary deliberations. "I am sure, however, that the committee is aware of the dangers which would attend fixing any part of the budget at an arbitrary percentage. I appreciate the opinions that basic research should be favored at this stage in our national affairs. At the same time, we must realize that our extensive national commitments require great care in maintaining a balance between the various portions of the total budget." As a result of this great care or in spite of it, the research budget continued to decline. The committee had recommended that the budget be doubled in order to restore the former relationship between basic research and the total R&D effort. The declining budget no doubt led to a new description of some of the work to show that it was really of an applied nature after all and hence eligible for "technology" funding if not research funding. This illustrates the difficulty of administering a program whose description is more a state of mind than an objective reality.

As an example of the interaction between basic research and the application of knowledge, the authors of the report chose the subject of shock waves to show how a new understanding of something can affect the development of technologies even before they exist. Stokes, in 1848, puzzled by gas-flow problems, proposed a surface of discontinuity in a gas or fluid across which there was an abrupt change in velocity and density. The conservation of mass, momentum, and energy across this surface was maintained. Thus was born the shock wave, which turned out to be the key to success in many new fields.

If, for example, one understood the behavior of shock waves, one could decide how to shape a bullet's nose or a rocket nozzle, or decide at what height to explode a bomb for maximum damage, or how to achieve supersonic flight or reenter the atmosphere from space flight. The applications of shock-wave research to mine warfare, though not specifically identified in the referenced report, are many. They include, for example, the generation of an underwater shock wave by various explosives, and the consideration of these generated waves in producing damage—all related to mine warfare. These efforts have led to many advances in instrumentation to measure these waves, to determine their pressures and decay rates with time and their attenuation with distance, and especially their reflection from targets or absorption by targets. The shock-tube technology has been used for underwater gauge calibration. The propagation of shock from one mine to another has affected mine design. The shock of water entry is another area affecting design. Even the emergence of a solar flare causing a jet and a shock wave has affected the frequency response of background fluctuations in the earth's magnetic field, as we have previously seen, and places a limitation on the range of magnetic detection or mine sensitivity. All of this tends to illustrate the uncertainty of identifying the relevant technology much in advance of its application.

2. Technology and the development of magnetic firing mechanisms.

As an example of an evolving device and its dependence on technology, let us review briefly the magnetic mine mechanism. The first device was only slightly removed magnetically from the original Chinese lodestone. It was, of course, a compass needle—a small, permanent magnet mounted on gimbals and later balanced with a spring controlled by a clock so that it could be unlocked after the mine had settled in whatever ambient magnetic field existed at the location. The next device was not a magnetometer at all but rather an induction coil for responding to the rate of change of magnetic field rather than the field itself. The coil could be made sufficiently sensitive to actuate a relay by placing inside it a rod of permalloy. This was a magnetic alloy of 24 percent iron and 76 percent nickel which had a magnetic permeability of 8,000 as compared with 200 or 300 for iron. It was learned that iron was not a uniform substance but a mass of crystals and magnetic domains which made one piece of iron very different from another. By adding amounts of molybdenum, zinc, copper, cobalt, or other metals or by changing the heat-treatment schedule, the magnetic properties of the resulting alloys could be drastically changed. Their permeability, coercive force, retentivity, and the shape of the hysteresis loop could all be changed or controlled. The reluctance of Bell Laboratories to continue supplying permalloy and mu metal to the Navy after World War II led to the establishment at NOL of a magnetic materials laboratory. This laboratory enjoyed the continuing advice of Dr. Gustaf Elmen, who had discovered and developed these materials at Bell Laboratories before the war. The Bureau of Ordnance was further interested in the development of new magnetic materials for magnetic amplifiers and for detectors and magnetometers. With their support and much continuing technical interest, the new laboratory developed a very successful record not only in magnetic materials but also in other metallurgical discoveries. An account of this is now available in a monograph by D. F. Bleil, "History of Research in Magnetic and Metallic Materials," published in May 1977 as NSWC/WOL TN 77–55.[4] In the course of the work, a nonmagnetic material would show up once in a while. Sometimes nonmagnetic components yielded a magnetic alloy (as, for example, in the case of the bismuth/manganese alloy called Bismanol, which had a higher coercive force than any other known material).

At an early stage of the evolution of mine mechanisms, a good deal of effort went into the design of relays—to make them more sensitive, corrosion-resistant, shockproof, and so on. In the next stage, concern about relays disappeared, because the relay itself as a component of the design also disappeared and was replaced with vacuum-tube amplifiers, and still later by transistors. A further step in firing-mechanism development was taken in the use of the thin-film magnetometer. The operation of this depended on detecting a frequency shift in a high-frequency oscillator whose tank circuit contained an inductance which was sensitive to the local magnetic field. This inductance consisted of a coil of wire around a small glass plate on which had been evaporated a film of permalloy so thin that eddy currents were inhibited. These thin films were laid down by evaporation in vacuum and relied on a technology which was far removed from the earlier metallurgical skills.

The advantages of this new device were great, of course. Sensitivity and frequency response were high, and power consumption and physical size were low. The thin film magnetometer with transistor oscillator, solid-state information processor, and new battery power supply became a very important application in the Vietnam War. The device was small enough so that ordinary 500- and 1,000-pound bombs could be converted to mines simply by replacing the bomb fuze with the magnetic detector and associated mechanisms. Thus, it turned out that the number of offensive mines used (mainly in rivers) in this war became of the same order as had been used before in all locations in all wars by all nations. The total number made was in excess of one half million!

3. Limits of magnetic detection.

The horizontal component of the earth's magnetic field in temperate latitudes is between .15 and .30 oersted. At this field intensity we can take the permeability μ of magnetic iron to be almost 300. (See graph p. 5–214 of Reference 1). Assuming a submarine to be made of this material and to be placed along the field, its magnetic induction will be no larger than B = 300 × .3 or 90 gauss. If this submarine has a 25-foot (7.5-m) diameter and a pressure hull which is 1 inch (2.5 cm) thick, then the strength of its induced magnetic pole is $90 \times 750\pi \times 2.5$ or 53×10^4 gauss cm^2, and its magnetic moment is this multiplied by its length. If the submarine is 300 feet (90 m) long, then its magnetic moment is M = 48×10^8 gauss cm^3. The field of the submarine directly over or under it (treated as a horizontal dipole) is M/r^3. Treating the submarine as a sequence of vertical dipoles in the vertical component of the earth's field is more complex. The simpler calculation illustrates the order of magnitude to be expected. Suppose the submarine is traveling at 3 knots (which is about as slow as it can go and be manageable). Then it will take one minute to pass by a fixed point. A look at a graph showing the Geomagnetic Frequency Spectrum [4, P. 3] reveals that the natural background of magnetic fluctuations having a period of one minute is about 0.3 γ (one gamma is 10^{-5} oersteds). Assuming that a signal-to-noise ratio of unity is a good measure of the limit of detectability short of special correlation techniques, we see that the maximum distance at which the submarine can be detected under these circumstances is given by

$$\frac{48 \times 10^8}{r_d^3} = .3 \times 10^{-5}$$

(taking mu for water equal to unity) or $r_d = 1170$ meters.

The MAD or Magnetic Airborne Detector, developed at NOL in the early years of World War II, is limited by self-noise due to its own motion through spatial background variations. Geomagnetic temporal fluctuations would rarely interfere with its detections which occur in the time required for an aircraft to fly over a submarine. This might be of the order of a second rather than a minute. The background for this period is about $.005\ \gamma$. If the self-noise of the airborne detection system from all causes could be eliminated entirely, or brought down to this level, then the limit of detection would be

$$\left(\frac{48 \times 10^8}{.005 \times 10^{-5}} \right)^{1/3}$$

or 4700 meters. The practical range is of course much less. Note that these estimates apply to an undegaussed submarine.

Fields of science related to mine warfare

In the Naval Research Advisory Committee Report, Volume I, [35] already referred to, there is given a list of the fields of science which are related to the missions of the Navy. The list is given in order of then current (1959) Navy basic research expenditure. As is there stated, this cannot be viewed as a priority list because the costs of performing basic research vary with the field of science. For example, something might be of the highest priority but be relatively inexpensive, in which case it would appear low down on the list.

It may be of interest to consider to what extent the basic informational needs of naval mine warfare depend on this same list. As shown in table 3, all of these seventeen fields have made direct contributions to one or more of the technical needs of mine warfare.

Mine warfare as a system

We may think of technology as a growing tree with one discovery branching out from another and with developments hanging like ornaments from the branches. Such analogies are not accurate because technology grows in a more disorderly way than a tree does, sometimes with new branches growing out of a coalition of old branches.

Hence, it may not be possible to construct a meaningful technology tree for future mine development, using experience as a guide. In one sense it can be said that everything depends on everything else, or that any given development depends in the end on everything else.

TABLE 4.
LINKAGE OF SYSTEM COMPONENTS TO TECHNOLOGIES AND NEEDS

Component(s) of the "Mine Warfare System"	Technologies and Needs of Mine Warfare
R&D, Eval, Prod.	Energy materials Gunpowder, gun cotton, explosive trains, high explosives, power, safety.
R&D, Eval, Prod.	Initiation Spark, percussion cap, heat, hot wire, exploding wire, safety.
R&D, Eval, Prod.	Detection Impact Electrochemical Influence (magnetism, acoustic, pressure).
R&D, Eval, Prod.	Information Processing Circuitry, analogue, relays, amplifiers, modulation, vacuum tubes, solid state, logic.
R&D, Eval, Prod.	Power Supplies Dry cells, storage, wet cells, NiCd, Hg, ZnC, Radioactive Isotopes.
R&D, Eval, Prod.	Materials Impact strength, corrosion, nonmagnetic, weight, cost. Metals—stainless steel, plastics, tempered glass(?).
R&D, Eval, Prod., Delivery vehicles	Delivery devices Flight gear—parachutes, torpedo tube launching, safety.
R&D	Dynamics Drag, stability in air and water, water entry, vortex shedding, bottom stability.
R&D, Eval, Prod.	Anchors Shallow, deep, automatic.
R&D, Eval, Prod.	Propulsion and Control Buoyancy, propeller & batts, U/W rockets; homing.
R&D, Eval.	Countermining Shock resistance, damage, actuation.
R&D, Eval, Prod.	Auxiliary devices Clocks, delayed arming, sterilizers, ship count, command and control. Self destruct.
R&D, Eval, Plans and Requirements	Target Damage Target vs. charge weight/shape/type. Position and distance. Shock, plastic deformation, rupture.
R&D	Target influences vs. size, speed, location, depth.
R&D	Influence background Magnetic storms, anomalies, ambient noise, natural, manmade, waves, tides.
War Plans	Effectiveness Analysis, threat definition, evidence, economic effect on enemy.
War Plans, Training, Logistics, Operations	Minefield planning What mines and how many for what threat at given location. Should be done in conjunction with strategic and tactical plans.
Operations, Logistics, Vehicles	Planting (deliver) Replenishment, surveillance of minefields.

Component(s) of the "Mine Warfare System"	Technologies and Needs of Mine Warfare
Logistics, Stockpile	**Depots** Where are the mines, how long to assemble them, test and load? Are there enough of what kinds in the right places? Spare parts if any?
War Plans, R&D, Production, Vehicles	**Politics** (Propaganda) Information to maintain understanding and continuing support of mine warfare activities—i.e., new development as needed, R&D in support of it, production, training, recruitment.
Delivery Vehicles, Operations	**Delivery** Identity, suitability, availability (dedicated?), training status, understanding mining vs. competing missions.
War Plans, Training, Stockpile, Vehicles, Logistics	**Readiness** Encompasses materiel, vehicles, plans, logistics, state of training and understanding, stockpile surveillance. Depot functions. Budget?
R&D, Plans, Operations	**Operations Research** Computer aided simulation of minefields.
R&D, Production, Stockpile, Logistics	**Evaluation for reliability** Statistics vs. cost, sample size, overtesting in order to reduce sample size. What are the tradeoffs?
R&D, Evaluation	**Simulation of Physical effects** Dynamics: impact, water entry, vibration. Static: life, temperature, corrosion, etc.
R&D, Production, Operations, Logistics	**Evaluation for safety** Accident paths, energy materials, design criteria, arming for meeting operations and environment requirements.
Training and Operations, R&D, Logistics, Vehicles	Fleet exercises in mine planting and mine sweeping. What should the training round be? Dummy-loaded mines with an inflatable recovery float to be actuated by sweeping, or at end of exercise by time?
Production, R&D	**Manufacture** Producibility, design disclosure documentation, contract selection, management; prime contractor vs. special industries; complete mines vs. assembled components.
Training	**Education** Mine schools; R&D technologies, management.
R&D, Evaluation, Production, Stockpile, Logistics, Vehicles, Training, Plans, Operations	All aspects of countermeasures not specifically included above. Sweeping, sweepers, signal simulation, unsinkable ships, towed sweeps, helicopters, searching, hunting, destruction, navigation, channel marking, withholding, traffic control, alternate ports or beaches, stockpiling of materials.

This would not be a meaningful statement, nor a useful one, for example, in planning a laboratory structure or a program, because it makes no choices and does not emphasize or exclude anything. A tree is something that grows, that starts small, that develops many branches, some of which are discarded, and that in the end dies—often when full grown. The analogy, we hope, will stop short of this final event. Such a tree could no doubt be derived from a history of mine developments carried out over the years, and that is in one sense what this book is about.

But a broader view could be obtained of naval mine warfare by looking at it as a *system* which depends not only on technology but on the recognition and management of the components of that system. The components are somewhat arbitrary, but I have listed eight which together make up a system for the effective use of mines and mine countermeasures in naval warfare. It is clear that both mines and mine countermeasures participate in each of these components. The total mine warfare system could be shown as two concentric circles, of identical components, one labeled "mines" and the other labeled "mine countermeasures," but this is not necessary as long as it

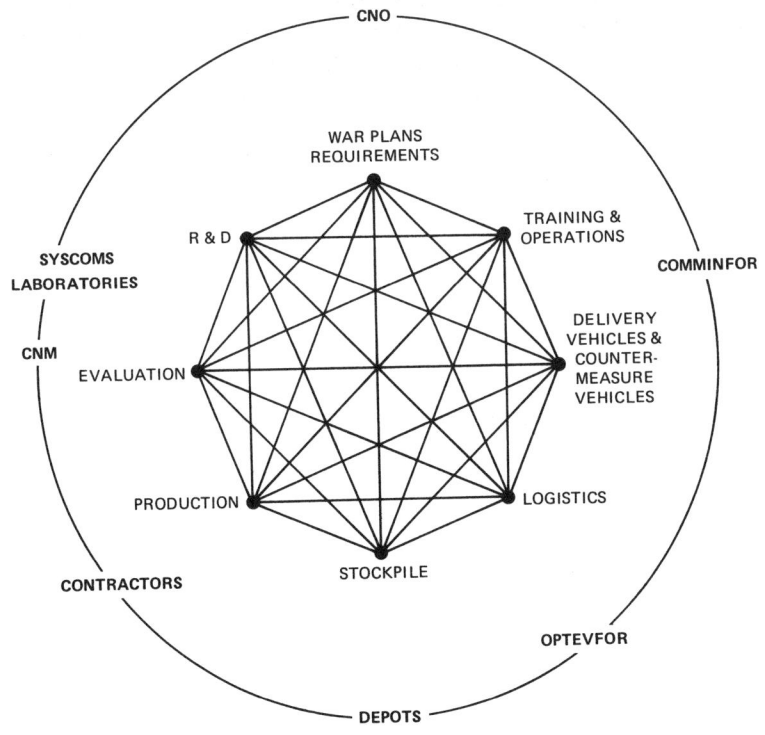

Figure 18. Mine Warfare as a System.

is understood that neither is to be omitted. Figure 18 shows the eight components, viewed as large areas of work to be done, in a sequential arrangement; so that if one starts with war plans, at the top, for example, one works counterclockwise through all the things to be done and finally arrives back at the war plans stage again. In actuality, each component must be in timely communication with every other one. This is indicated by lines connecting each to the other. This makes twenty-eight two-way channels of communication. Also note that as one goes around the circle the components shift from shore establishment execution to fleet execution. The management of all this would seem to require a benign and possibly extraterrestrial power looking down on the whole system and encouraging it to work.

In the real world the actual management of the whole system is difficult, partly because the Navy is not organized that way and partly because many of these components are only visitors in someone else's house. For example, the mine-delivery vehicles have many other functions to perform. Mines must compete with torpedoes for submarine space, and with bombs for aircraft space. For our purposes here it suffices to note that in the mine warfare system as shown each component depends more or less heavily on some aspect of technology. This is obvious in the R&D component, but is nevertheless true of all of the components. Table 4 attempts to link components of the mine warfare system to various needs in mine warfare and their related technologies.

5 Countermeasures

The problem

If an enemy uses mines against us what are we going to do? First of all, assume that the mines are known to be present. If their threat against ships is thought to be marginal and if there is an important need for the movement of our traffic, then the advantages of moving our ships may be thought to overcome the risks. In other words, we could ignore the mines and take whatever losses resulted. Incidentally, early in the history of mine warfare this is what Admiral Farragut did at the entrance to Mobile Bay and he got away with it, even though he had already lost one ship as the price of finding out that mines were there in the first place. There are other possible actions in the face of a mine threat—including inaction (i.e., stopping the traffic and waiting), or using an alternative route if one exists. But the direct approach is to attack the mines themselves and to actuate them harmlessly by simulating the ship's influences the mine is supposed to respond to. Another possibility is to find the mines somehow and nullify them by some means, including blowing them up with explosives. These two general schemes, called mine sweeping and mine hunting, constitute the world of mine countermeasures.

In the early days before World War II, all mines—except some of the control mines for harbor defense—were moored. It was therefore obvious that it would be possible to drag a cable horizontally through the water behind two ships or paravaned from one and thus to sweep up any moored mines in the swept path. Although it was not the safest operation in the world, this was done extensively to get rid of mines, and the U.S. Navy had a lot of experience with this sort of countermeasure in sweeping up the North Sea Barrage which had been laid down with so much effort and enthusiasm in the last months of World War I. I am indebted to Dr. Tom Phipps of the White Oak Laboratory for referring me to Rudyard Kipling's poems about mine warfare. One called "Mine Sweepers" 1914–18 is as follows:

> Dawn off the Foreland—the young flood making
>   Jumbled and short and steep—
> Black in the hollows and bright where it's breaking—
>   Awkward water to sweep.
>   "Mines reported in the fairway,
>   "Warn all traffic and detain.
> "Sent up Unity, Claribel, Assyrian, Stormcock, and
>   Golden Gain."
>
> Noon off the Foreland—the first ebb making
>   Lumpy and strong in the bight.
> Boom after boom, and the golf-hut shaking
>   And the jackdaws wild with fright!
>   "Mines located in the fairway,
>   "Boats now working up the chain,
> "Sweepers—Unity, Claribel, Assyrian, Stormcock, and
>   Golden Gain."
>
> Dusk off the Foreland—the last light going
>   And the traffic crowding through,
> And five damned trawlers with their syreens blowing
>   Heading the whole review!
>   "Sweep completed in the fairway.
>   "No more mines remain.
>
> "Sent back Unity, Claribel, Assyrian, Stormcock, and
>   Golden Gain."

It will be noted that there were no casualties and that after a hard day's work the task was successfully completed—"no more mines remain"—and traffic could move again. In our own sweeping operations in the North Sea, there was the added comfort that we were sweeping our own mines and therefore had complete knowledge as to what and how much was there. Our Vietnam minesweeping effort in 1973–74 was similar in that we were sweeping our own minefields, whereas, for example, the minesweeping problem in Cherbourg Har-

Mines washed up on beaches are often corroded and covered with sea growth, making them hard to identify.

Japanese mine washed up on Oregon coast, WWII. Lieutenant H. A. Ridenour digs sand away from mine detonator!

Disposing of Japanese mine floating in the Kii Suido Straits off Japan after the mine has been swept loose from its anchor. A total of 312 mines were disposed of in this manner in 1946 after the war.

A battered Liberty ship, the SS *Pratt Victory*, serving a postwar mission in Japanese waters. One of three of the "Guinea Pig Squadron" this ship, controlled from on deck, runs through channels to check whether U.S. pressure mines are still active. The crew members of the sacrifice ships stand on mattresses or raised wooden grates to lessen the shock if their "Guinea Pig" hits the explosive jackpot. December 1945.

U.S. minesweepers at Wonsan Harbor, Korea, clearing a channel through the minefield. 1950.

Well deck of LSD-16 at Wonsan, Korea. Minesweeping boats being launched from deck. April 1952.

Streaming magnetic minesweeping cable over the fantail of the USS *Gull*, AMS-16, at Chinnampo, Korea.

bor in 1944 and the Korean sweeping efforts in 1951 had the added dimension of the unknown.

Since the mine countermeasure business is directed primarily toward countering enemy mines, there is an added element of secrecy in it because it is important that an enemy not know how much we know about his mines. As is said, there are intelligence reasons for guarding this information or even this lack of information. It is, however, clear that without knowing whether a given potential adversary has a specific mine, there is a finite set of physical principles and technical devices which should make the group of possible enemy mines very similar to our mines deriving from our own state of mine research and development. In other words, secrecy does not ensure a monopoly on knowledge; it merely prevents an enemy from knowing whether we know what he knows.

We have noticed in chapter 2 that there has been a debate from time to time over the organization of mine warfare with respect to the location and prosecution of countermeasures. It was interesting that in World War II, when magnetic ground mines were used in great numbers, the countermeasures against them took two forms. One was to simulate the magnetic field of the ship by towing a cable which passed an electric current through the water. This made use of the former minesweeper, now outfitted with electric generators, and could still be called "sweeping." The other countermeasure was to reduce the magnetic field of the ship so that the ship could pass near the mine without activating it, or in some cases, as in the German Sperrbrechers, to increase the field and to project it ahead of the ship so that mines could be actuated far enough away to be harmless when exploded. As we have seen, there was a debate about this between the Bureaus of Ordnance and Ships as to which should be in charge of what. This was settled by the CNO with a touch of Solomon-like compromise: degaussing design, coil specification, and ship measurement would be done by Ordnance, whereas installation and the provision of power for the coils would be a Ships responsibility.

The Korean war dramatically highlighted the importance of readiness or lack of readiness in the mine countermeasure field and led to the establishment of the Mine Defense Laboratory at Panama City, Florida. This was built upon the ship-like remnant of the wartime Mine Warfare Test Station which had been set up at Solomons, Maryland. (The ordnance-like part of this station was then attached to NOL as a field station for countermine tests.) The CNO also declared that mine countermeasures enjoyed a higher priority than mine development, which is not to say that funding for either was adequate for smooth and rapid advance.

The cornerstone for the new Mine Defense Laboratory was dedicated in about 1956. In the late 1960s there was a move stimulated by the DOD toward consolidation of defense laboratories in order, as it was said, to increase their size and scope toward a "critical mass," which could then undertake responsibilities for system developments. The Naval Ordnance Laboratory, having long since achieved this desirable state, had only marginal interest in further consolidation. Nevertheless, a committee was established to make recommendations on how to implement the rather unlikely merger of NOL, White Oak; NWL, Dahlgren; and the newly established Mine Defense Laboratory a thousand miles away in Gulf Coast Florida. One of the management injunctions was that the merger was not to result in the establishment of another echelon of control over the already existing command structures in the three laboratories. The only way to do this appeared to be to abolish two of the existing commands. This resulted in at least two of the three laboratories developing a rather desperate opposition to the idea of consolidation, on an ingenious variety of reasonable grounds.

What is of interest here, however, is the set of arguments raised by Panama City and by their Ships Systems sponsors to retain a separate management structure for mine countermeasures work. Mine countermeasures work should be retained under Ship Systems management, because sweeping was inherently a ship-dependent activity requiring knowledge of seamanship and ship rigging and handling. It was important to have the countermeasures research and development independent of mine R&D, because if they were mixed together the countermeasures work—being more difficult and less glamorous—would suffer by comparison. It was important to keep the two activities separate because each tended to paralyze the other, and there could be no stability or progress in either development effort, since each new idea in one area would lead to another new idea in the other area which would neutralize the first. In contrast to these arguments it was suggested that it would be beneficial for the working people to have knowledge of work in the whole field, and not to develop positions based on ignorance. It was said that if the activities were joined, the work force and talents available to both types of work would exceed what is now available to either. Depending on priority and need, a much greater effort could then be mounted in the desired direction. There was seen to be an advantage in flexibility. It was pointed out that the British had made the join many years ago, but this argument was not particularly persuasive since they had also been unable to afford to do otherwise. Perhaps, then, economy would be a reason for consolidation?

Finally there was a point which I think to be of considerable importance and it is this. As mines and their mechanisms and their data-processing computers become more sophisticated—and they are virtually at this stage now—they will be able to distinguish between a real target and the countermeasure designed to simulate it. It is now virtually impossible to sweep a mine which requires magnetic, acoustic, and pressure influences properly sequenced in time without providing to the mine a simulator of these influences which is actually a ship! Therefore, the future of countermeasures techniques may depend more on ordnance-oriented devices than ship-oriented devices. For example, mine hunting and destruction can be carried out by remotely controlled self-mobile sensors of an acoustic, optical, or magnetic nature (depending on range), and these devices will look a lot like underwater guided or homing missiles. Therefore, the joining of these relevant technologies under one central roof should be in the long-range interest of better mine countermeasures. A vote was taken on the issue and the committee decided that it would be less disturbing (at least to itself) to attach the Mine Defense Laboratory to a BuShips-type laboratory than to an Ordnance-type laboratory, if indeed it had to be attached to something. This resulted in its becoming, from 1968 to 1972, a part of the Ships R&D Center, an innocent bystander headquartered at Carderock, Maryland. After that time for various political reasons it was returned to autonomy under a new title, U.S. Naval Coastal Systems Laboratory, retaining the separate mine countermeasures mission. As countermeasures move into the deeper sea it may be necessary to revise the title again, although it will be noted that the idea of coastal waters is being expanded to take in waters over the Continental Shelf, or waters out to two or three hundred miles from the coast, in some cases where there is no shelf to speak of.

The question of priority between measures and countermeasures was put into a state of balance by a CNO instruction as long ago as 1964, when fortunately there were no particular operational pressures for either side. In the instruction, mine warfare was defined as the strategic and tactical use of sea mines and their countermeasures. It was said to include all available offensive and defensive measures for laying mines and for protection against mines. It stated reasonably enough that an offensive mining potential for critical areas and an effective mine countermeasures force to counter anticipated enemy mining attacks are essential to peacetime readiness. The total offensive and defensive requirements of a balanced naval force must be considered. As if these policy statements might not serve to provide all the answers, the instruction ended by saying that if conflicts in assignment of funds, personnel, or forces arise which cannot be settled by the activities involved, the matter will be referred to the CNO.

The problem of priorities is always with us as long as there are more things to do than there is money to pay for them. In comments made to a recent Mine Development Conference held at the White Oak Laboratory in January 1977, Admiral Zumwalt, retired former CNO, spoke of the difficulty of shifting effort from one naval bureaucracy to another. This inertia, though not all bad, makes increases in mine warfare very difficult to achieve, largely because its constituency is not large or powerful, as are the bureaucracies (he called them unions) surrounding the major career structures inherent in carriers, planes, submarines, ships, or shipyards. He said, "It is in retrospect almost unbelievable for me to realize the great risk that we had to take in order to get more money available for doing something to improve our mining capabilities. Only a little of this came from the other unions; the major amount to give us our capability to do helicopter minesweeping came by wrapping up almost completely the surface mining ships and taking that great gamble that we could get the CH-53s in time and get them operational and ready for the minesweeping that it was perfectly clear would be coming along. We gave up what we had in the hand in order to reach out and get one in the bush, and fortunately we delivered in time."

The simulation of ships' fields

Clearly it would be advantageous to build ships (and submarines) which had such small influence fields that no mine could detect them. This, of course, is impossible—a ship being a large object that displaces water, moves through it, makes noise internally and externally by the flow itself, and has other effects like altering the magnetic field, the conductivity of the medium, the electric potential difference between selected points, the penetration of cosmic radiation, and so forth. After doing the best we can in reducing a ship's influence fields, there will still be plenty for mines to operate on. The mine countermeasure of reproducing the ship's signature initially worked well when the signature in question was the magnetic field which varies with time as the ship passes by, or the noise emitted by the ship, its engines, its propellers, or its internal machinery.

These successes gave rise to a variety of magnetic cable sweeps, and to a proliferation of hammer boxes, parallel pipes, and noisemakers actuated by the flow of water into a Venturi tube as the device was towed. There were also noise-makers based on the timed explosions of a gradation of small charges which could deploy themselves like an unfolding Christmas tree. These noisemakers could give way in modern times to sophisticated underwater speakers which could reproduce from tape recordings the exact frequency spectrum of a real target, in proper amplitude and varying with time in a manner consistent with the actual motion of a real target. However, very early in the development of influence mines it was learned that the use of acoustics alone did not provide proper or consistent localization of the target, and so this influence came to be used in combination with others. In fact the pressure-magnetic combination, without acoustics, proved to be extremely difficult—and, with no further sophistications added, it still is. The pressure influence is the result of a massive but slow flow of water. To simulate this requires essentially a massive object moving through the water. If ocean swells are present, the long-duration waves, if of sufficient amplitude, can simulate the pressure signatures of some ships some of the time. To prevent this kind of natural sweeping, the magnetic influence is tied in, so that sweeping in swell conditions cannot occur by itself. If swell conditions cannot occur, as inside harbors, the pressure-magnetic (p-m) mine becomes very difficult to sweep.

When the Germans first used the Oyster mine off Normandy and at Cherbourg in 1944, these mines were equipped only with pressure mechanisms, so that in principle the production of appropriate swells could trigger the mines. A considerable effort was undertaken at that time to determine what sort of waves could be made by appropriate sizes or groupings of explosive charges in various water depths and what would be the extent of the sweeping coverage. There was held out the possibility that a single properly arranged explosion could produce waves which would sweep all the mines in the harbor at one blow (except for those on ship counts). This effort is described in Reference 17.

If p-m mines are laid in sheltered waters, then the pressure signature has to be simulated by a large ship-like object. We have already mentioned the towed bag idea, or Loch Ness Monster as it was first named. Other projects flourished for a time in an effort to build an unsinkable ship which could provide the necessary influence fields, detonate the mine, and then survive the dire consequences of the explosion by virtue of its specialized construction. Thus there would be available a sweep which could deal with this mine. This is a good concept if it can be made to work. Many attempts were made to make highly compartmented barges—barges with liquid and air interfaces, with energy absorbing materials, deflecting plates, replaceable or flexible members, and so on.

The opportunities to try out these devices in a real situation were fortunately few. A British "Eggcrate," an "unsinkable" compartmented barge, was towed about outside Cherbourg Harbor for pressure minesweeping by the British after Normandy. But the weather was so rough that the pressure mines were actuated by the waves, making the pressure sweep unnecessary. Because of the high winds the thing apparently wound up on the beach. Another strictly experimental device (American), called XMAP, consisted of an unsinkable steel cylinder 100 feet long, 25 feet in diameter, and of unsymmetrical design so that the hull was 1 foot thick on the bottom and 6 inches thick on the top. This cylinder was supposed to be towed back and forth where mines were to be swept and to survive any explosions that might occur under it. The influence field of this beast was very narrow, but the main difficulty was that it could not be controlled when towed. On one occasion is refused to turn when its tug went into a harbor and wound up stubbornly towing the tug backwards out of the harbor entrance. The XMAP was eventually abandoned. There were at least two things wrong with these attempts: first, those unsinkable ships could not survive many explosions—perhaps not more than two or three; and second, they were extremely cumbersome, slow, and not at all readily available at some distant place where they might be needed. Their expense was such that it was not practical to pre-position large numbers of these things, and beyond that it was not known where they might be needed overseas—if ever.

Because of the cost, lack of promise, and immobility of these methods, emphasis began to shift toward mine hunting methods for dealing with the so-called unsweepable mines such as the pressure-magnetic combination. Further, it was found that a helicopter exerting a thousand horsepower could provide as much towing force as a seagoing tug of the same horsepower. Why not use helicopters for towing cables, sweep wires, noise-makers, even permanent magnets on small floats? If the helicopter is high enough up and far enough forward, it will escape any underwater explosion plume caused by mine actuation and it also presumably will not itself actuate a mine.

There were many questions on the influence fields of helicopters themselves and their vulnerability that had to be answered. Would it be possible for a helicopter by its downdraft to produce a shiplike pressure signature on a submerged mine? And what would happen to the helicopter if it did? This work did go on as Admiral Zumwalt indicated, and helicopters were used in some mine clearing operations at Hanoi in 1973-74. It must be noted that they were used against our own minefields and were thus spared any possible nasty surprises. They did, of course, demonstrate their obvious advantage of mobility. But the basic difficulty of simulating pressure fields remains. Pressure mines have to be dealt with primarily by mine-hunting techniques. This then regenerates the need for small surface vessels to serve as sonar platforms, navigation aids, buoy carriers, and bases for Explosive Ordnance Demolition (EOD) swimmers and for mine destruction weapons. Such small special-purpose vessels could then be called mine hunters rather than mine sweepers.

Mine hunting

It has been clear for a long time and certainly since the Korean War that the technology of mine hunting had to be developed and improved. The Mine Advisory Committee was established in 1951 under an ONR contract with Catholic University under the initial chairmanship of Professor Karl Herzfeld and was later transferred contractually in 1955 to the Physical Sciences Division of the National Academy of Sciences. It gave its first attention to the problems of mine countermeasures and completed many studies on relevant subjects—such as countermeasures in general (Project Monte 1957), Masking Techniques (1961), Monte Plus Five (1962), Swimmer Countermeasures (1962-65), Pebble (1965), Precise Navigation (1966), Shallow Water Naval Warfare (1966), and High Resolution Sonar (1968). The Committee also made significant contributions to mine development, particularly in Deep Sea Mining (Bowdoin 1958, which analyzed and favored the Captor option for this purpose); in a study on Power Sources (1967); and finally in its great swan song, Project Nimrod (1967-70), which undertook a complete look at mine warfare, its prospects and role here and abroad, its technology, and to some extent what ought to be done about it. Throughout all this time, and indeed previously, it is fair to say that L. W. McKeehan was a constant guiding light, participant, leader, questioner, and at times gadfly. His lifetime interest in mine warfare, starting from his initial command of the Mine Building in 1919, through his later scientific work in magnetism and his resumed naval career in charge of mine development in World War II, and subsequently in countermeasures at the Edwards Street Laboratory at Yale, lent a continuity and drive to the Committee efforts. He also inspired some excellently qualified people to participate in the work; as for example: Prof. Carl Menneken, Naval Post Graduate school; Prof. Alfred Focke of Harvey Mudd College; Dr. Chester McKinney, Applied Research Laboratories of the University of Texas at Austin; Prof. Andrew Patterson, Jr., of Yale University; Dr. J. S. Coles, President of Bowdoin College; and many others.

If mines are delivered by submarine, then the problem of finding them is difficult indeed, because they have no air trajectory and they make no noise on delivery. However, for aircraft delivery it is possible to see the mines, track them optically or by radar, and locate their position at water impact. Johnson says that such methods were employed during World War II and sometimes with reasonable success. He likes the story that during the mining of the Suez Canal by the Germans, the British employed native mine watchers each having a long stick and a bag of sand. "Since the canal is narrow, several watchers could see each mine as it splashed into the Canal. Each watcher then laid his stick on the ground, pointing it in the direction of the mine. He placed the bag of sand on the stick to hold it in place. A countermeasure officer then investigated each reported mine and determined as closely as practicable its location. It was then disposed of, usually mechanically." [23, P. 288] Under these specialized conditions this mine-spotting technique worked well. It would be impractical where mines fell far from shore, or during nighttime. The Japanese used mine watchers and illuminated the most important channels with searchlights to be able to observe splashes. Radar for splash location would be preferable so that the target area would not be illuminated to outline the critical areas for the launching aircraft. Other possibilities are to place an acoustic range in the area, a grid of hydrophones wired to a computer which can determine splash locations by processing the time differences between sound arrivals at different hydrophones. More simply, as Johnson says, "There is an excellent possibility that a row of microphones [hydrophones] along predetermined channels can locate with reasonable accuracy the position of mines dropped in those channels." [23, P. 289] There are, however, many ways in which such mine-spotting schemes can fail, or be deceived. It is therefore necessary to have a means, if possible, of finding mines after they have entered the water and have found their resting place on or in the bottom.

The key problem in the mine-hunting business in this case is, of course, the fact that light does not travel well in water and particularly in water that is muddy or turbid. During World War II, I can recall that a well-known artist turned naval officer (Captain Charles Bittinger, Chief of the Section of Camouflage Design) offered his services in developing a camouflage for mines, and he was given some mine cases to experiment with. I was detailed to take him and his camouflaged cases to Solomons, Maryland, to lower the cases in the waters of the Patuxent River, and to test the efficacy of the disguises. The experiment was a failure because the water was so muddy that we couldn't see anything at all after the case was lowered a couple of feet. In fact, I believe we lost one entirely—it was so well disguised. Needless to say, the mine designers were not particularly impressed by the need for camouflage. We can take it for granted that if one is going to look for mines, he will need to use underwater sound rather than underwater light. Don't forget, however, that this was well before the days of television. It should be possible to search at close range in many environments using underwater lighting and underwater television. This sort of equipment is a standard part of the search and recovery system which has been used for many years by the White Oak Laboratory Station at Ft. Lauderdale, Florida, where mines and other objects are placed, found, and recovered in both shallow and deep water on a routine basis.

Nevertheless, the principal hope for penetrating sea water and searching for mines, especially in turbid waters, was pursued in the form of active sonar with a wavelength a small fraction of the mine dimension. The sonar took one of two forms—either a frequency-modulated sonar, or a short-pulse sonar. According to A. Patterson in his published account on Nimrod, these two developments during World War II were directed at mine detection. "At the Navy Electronics Laboratory [San Diego, California] the novel FM sonar system was quickly manufactured for installation on fleet submarines with which it was hoped to penetrate Japanese moored minefields. Successful transits of live minefields in the war zone were made by submarines so equipped. At the Harvard Underwater Sound Laboratory a short-pulse, rapid, electronically scanning sonar was developed, too late for actual use during the war. Techniques arising from both these designs found their way into the more advanced mine-classification sonars developed in the 1950s and 1960s, but the principal source of the design of a mine-hunting sonar was the underwater object locator system devised during the war by the General Electric Company. By the end of World War II this had evolved into the UOL Mk 4, a short-pulse, high-frequency sonic imaging device; successive versions of this device had become the AN/UQS-1 mine locator by 1950. The Korean War brought about a production run of some 150 of these units which were fitted in minecraft from 1951 through 1954. Experience with these units indicated their sensitivity to thermoclines and their lack of classification ability, leading to the design of the AN/SQS-15, essentially a variable depth (towed) version of the AN/UQS-1. Classification studies were started at a number of laboratories, including the Defense Research Laboratory (now Applied

Research Laboratories) of the University of Texas. A Chinese copy of the British ASDIC 193 high-resolution sonar was denoted the CXRP and fitted as a classification attachment for the AN/UQS-1. The AN/SQS-14, one sonar combining the AN/UQS-1 and the CXRP with refinements, reached operational evaluation in the early 1960s. The AN/PQS-1 hand-held sonar for underwater swimmers evolved from Navy Electronics Laboratory designs originated in the 1950s."

"Using the increased sophistication of available sonars and the abilities of swimmers, several organizations have conducted studies showing the relative advantages of mine hunting vs. mine sweeping." Patterson concludes this summary by saying "On the whole, one has the feeling that mine countermeasures have reached a position of considerable maturity." [33, P. 71]

There are two functions to be performed in the mine-hunting type of countermeasure. The first is to find the mine, and the second is to neutralize it or avoid it. There were several early efforts in both directions. In the Fiftieth Anniversary pamphlet published in 1969 by the Naval Ordnance Laboratory, the first listed development project was an item completed in 1942 called King Kong, the first ordnance locator. This device was developed at NOL in order to locate torpedoes which sometimes strayed from their true paths on test runs at Solomons and Piney Point, Maryland. Sometimes these torpedoes would bury themselves in the mud and become even more difficult to find. The success of the King Kong was confirmed in its subsequent use at Newport by finding at least one torpedo that had been lost since World War I! The device consisted of a large copper and brass cylinder about eighteen inches in diameter and six feet long, towed vertically five to eight feet above the bottom. There were two horizontal coils with air cores mounted one at the top and the other at the bottom of the cylinder, connected so that their magnetic fields were in opposition and driven at about 60 Hertz. Everything was in balance unless a ferromagnetic object intersected the field of one coil more than that of the other. A large object like a torpedo, consisting of a fair amount of iron, could be detected at a working range of eighteen feet. Because the magnetic field of the (dipole) object detected decreased as the inverse third power of the distance, the range at which detections could be made was relatively limited.

A variation using this principle was called the SMSD, standing either for Submarine Mine Surface Detector or for Ship Mounted Submarine Detector. It used two vertical air-core coils mounted fore and aft on a small launch. This avoided the towing problem but introduced more noise due to the presence of the launch.

The Queen gear was the next step—the result of a contract with

Bell Laboratories to develop a better ordnance locator. This gear, later called Ordnance Locator Mk 2 and Mk 3, was a powered magnetic gradiometer. That is to say that the ambient magnetic field was measured simultaneously at two different points, and the difference was detected. Each magnetometer was a mu metal core fluxgate type driven at 1,000 Hertz. The presence of an ambient field created a second harmonic in the output coil of the magnetometer. A difference in these two outputs indicated the presence of a magnetic material, e.g., mine case, oil drum, or other object causing a gradient in the otherwise uniform ambient magnetic field. The detection range for this device varied as the inverse fourth power of the distance.

The Queen magnetometers were mounted in a six-foot-long copper cylinder, about one-third the diameter of the King Kong. It was towed off the bottom in some applications and mounted in front of a launch in others. In one set of experiments the six-foot Queen gradiometer was fitted with plywood lifting surfaces so that the lift was increased with the speed of towing. There was a fathometer (active acoustic) built in so that the height of the gradiometer over the bottom could be measured directly, and the towing speed adjusted accordingly. This was to prevent the thing from being towed too fast and consequently being too high off the bottom to detect anything that might be there. In addition to this, there was a marking buoy which was towed along about twenty feet behind the detector. There was just time for an alert operator, on noticing a mine detection, to release the buoy and its anchor. This would then accurately mark the mine location for further consideration. One of the minor difficulties was that sometimes the buoy would have too short a line on it and would disappear after it was launched. This trouble was overcome when the gradiometers were mounted over the bow of a launch. In this case the buoy anchor, sometimes a cinder block, would be quickly lowered over the side manually by the operator, a newly graduated electrical engineer who could estimate where the detected object had to be and who could pay out an adequate length of buoy cable. This system was not pursued with great zeal because of lack of priority. The ordnance locating problem decreased as the torpedoes became more reliable, and the issue of mine countermeasures was not an immediate concern. These ideas, however, did lead to systems developed later at Panama City, some of them jointly with NOL.

To fill in the record on magnetic detectors, it is appropriate to mention some other developments that led to interesting applications. One of these was called the Electromagnetic Discontinuity Discriminator which for obvious reasons was always referred to as the EDD. This was a scheme for detecting any metal, not necessarily ferromagnetic, and to a lesser degree any plastic. It basically depended on the difference, positive or negative, between induced currents in the object detected and currents induced in the background material. The EDD consisted of three horizontal air-core coils, each several feet in diameter, which were towed on a horizontal "sled" near the bottom. The central coil was supplied with a 60-Hertz alternating current. The induced currents in the other two coils were balanced against each other so that there was no signal unless one of the coils ran close to an electric discontinuity such as a mine case. The detection signal for this scheme decreases as the sixth power of the distance, and the detection probability is therefore a very sensitive function of range. It is, however, a possible scheme for finding plastic or aluminum mine cases.

A much smaller EDD device called the Detector Mk 9 was developed by NOL to be hand-held by swimmers and used also for beach clearance. A further development with Ft. Belvoir led to the Mk 10, the hand-held detector of buried antitank land mines. Another even smaller magnetic detector called the Ordnance Locator Mk 15 was developed in 1967 using the new ring-core magnetometer having a ferrite core and operating at a higher frequency. The new locator was designed to search for small arms or weapons which might be hidden in the cargoes of small craft in Vietnam. This new detector was a magnetic gradiometer, not an EDD, mounted in a slender plastic pole so that it could be used as a hand probe. It was insensitive to the relatively small gradients in the various small craft, but was well able to detect the large gradient encountered near a small object like a hand gun hidden in a bin of fish.

Thus, the detector developments at the Laboratory went from the largest to the smallest because of different requirements in a twenty-five-year period. The Mk 15 Ordnance Locator was actually a technical descendent of the early Queen gear—both being gradiometers using second harmonic detection. Parallel with the Queen development during the Second World War was the development at the Laboratory before 1944 of the MAD which was short for Magnetic Airborne Detector or Magnetic Anomaly Detector. Which designation was used possibly depended on whether it was being used for detection of submarines or for geophysical prospecting. In either case it was airborne, towed behind the plane in its own stable vehicle, and was a flux-gate, single-component, second harmonic detection magnetometer driven at 1,000 Hertz. This device is of prime importance for the localization of a submarine target by an airplane, and has also been used extensively in mapping the magnetic field of the earth.

In spite of these efforts at magnetic detection of underwater objects, there arose a new nonmagnetic project called "Turtle" which was worked on jointly by the newly founded Mine Defense Laboratory and NOL. This system, deployed by helicopter, consisted of an underwater body which housed a modified PQS-1 (hand-held) sonar for detection, a television camera for classification, and an explosive charge for neutralizing any bottom mine found. If a mine were detected on the bottom, the charge would be dropped off and would later explode, destroying or detonating the mine. The Ordnance Laboratory took care of the explosive side of this device—the charge effectiveness, safing, arming, and delayed firing. The countermeasure laboratory took care of the detection and classification part and the overall system, which included such items as the hydrodynamic stability and the handling of the Turtle. A prototype was finished in the mid sixties but never progressed further because of the handling problems and the primary difficulty that poor visibility in turbid water made the television of reduced value. As a result of these evaluations, the Turtle system for mine hunting was not recommended.

A more promising and safer system evolved from the improvements in mine-hunting sonars which allowed an area of bottom to be scanned without having to pass directly over it. If the longer wavelength sonar received an unusual return, it could be shifted to a higher-resolution and shorter-range mode having about one-fifth the wavelength which would allow an identification or classification of the object to be made. For example, in the AN/SQQ-14 Mine Detecting/Classifying Set the wavelengths are 1.8 and 0.41 cm, respectively. The next part of the system would be a launcher by which a marking buoy could be projected to the vicinity of the identified mine. This buoy would be provided with an active acoustic signal which would be periodically transmitted for as long as seventy-two hours. The marker then would serve as a means for relocating the mine either in order to destroy or otherwise neutralize it, or possibly to mark it with a surface-visible navigation buoy if the location was to be avoided.

Neutralization could be carried out either by an EOD swimmer or by an underwater guided missile. One such missile, called Sea Nettle, consists of a self-propelled, wire-guided expendable torpedo carrying a rather sizable explosive charge of 75 pounds HBX. This torpedo is steered by commands until its sonar image, range, and bearing coincide with the sonar image of the target mine on the same display. The torpedo charge is then command-fired all at a safe range from the sonar command vessel. *Voilá!* In principle an enemy mine has been found and done away with! In principle the mine threat has been dealt with—except that not always is the mine found, of course, and not always is the destroying charge located within lethal distance. In fact, with everything working perfectly, the process requires time and incurs costs that inevitably become part of the inefficiency of warfare. The inevitable delay in traffic costs us freedom of action and shortages of supplies or forces. The effort mounted to carry out the countermeasures detracts from effort that would otherwise be directed toward offense.

The other method of dealing with mines, once they are located, is to send a diver to place a charge or otherwise deal with the mine. These brave men require a variety of equipments including hand-held sonars which clearly must be magnetically and acoustically quiet. A diver will approach to the vicinity of a known mine in a small self-powered boat and then complete his approach on his own.

Field evaluations were carried out on the Shadowgraph classification sonar system in 1962 and 1963 on several fleet minesweepers. The system has great flexibility since it is portable and can be easily installed on any ship. The system provides a permanent record of all bottom objects, and direct measurements of distance, height, and size of such bottom features can be made directly from the chart records. It is especially useful as a classifier in conjunction with other ships equipped with AN/UQS-1 sonar. The probability of detection for unburied mines on a smooth bottom is virtually unity. The Shadowgraph was found to be a significant advance.

All of these schemes and variations of them have been developed in the last two or three decades and presumably would have a good chance against most conventional mines in so called mineable waters, i.e., up to about 180 feet deep. There are, of course, many natural environments where such countermeasures would be very difficult to carry out—for example, where there are high winds, intense cold, high currents, or very soft muddy bottoms. It should also be pointed out that up 'til now the mine designers have paid little attention to the problems of mine concealment, or to mines which could provide booby traps for approaching swimmers. Of course, mines could be made in shapes which would be hard to identify, of plastic instead of metal cases, or perhaps of plastic explosives which form their own case. Or they could be provided with special rubberized anechoic coatings for disguise. It is therefore true in this case, as in most others, that warfare progresses from measure to countermeasure in repeated cycles which are paced in time by the rate of technological progress and the rate at which man strives for self-destruction. It would be encouraging to hope that this process will converge to an equilibrium where potential enemies will see that hostilities lead to mutual disadvantage. Then comes the millennium.

Vessels for mine countermeasures

Vessels for mine countermeasure work in the U.S. Navy have had various designations, many of which have become obsolete as the ships themselves have disappeared from Navy lists. Many of these ships, however, have been transferred to other navies of allied or NATO countries, so that the present roster of U.S. mine countermeasure vessels is a poor measure of past strength in these categories. For example, the old class of Minesweepers Fleet (MSF), steel-hulled, 220 (and some 180) feet long serve in the navies of South Korea, Mexico, Norway, Peru, Philippines, Taiwan, Uruguay, Burma, and the Dominican Republic. The last twenty-nine MSFs of the *Auk* and *Admirable* classes were stricken by the U.S. Navy on 1 July 1972, and in September 1972 and February 1973 twenty-one of these ships were transferred to Mexico. It is not clear whether the transferred ships in other navies are dedicated to minesweeping or are serving for general naval purposes.

Additional information gleaned from *Jane's Fighting Ships*, 1973–74 and 1976–77:

MSO "Acme" Class. 780 tons full displacement, 173 feet overall. Ocean minesweepers commissioned in 1957 and 1958. Two ships remain of this class, assigned to Naval Reserve training. Two ships were stricken on 15 May 1976.

MSO "Agile" Class, Ocean minesweepers. Built between 1952 and 1956. 750 tons full displacement, 172 feet o.a. Wooden hulls, non-magnetic equipment and Packard engines made of stainless steel alloys. SSQ-14 sonar with mine classification. Fifty-eight of these ships were built for U.S. service, 35 for NATO navies. Originally designated as minesweepers (AM) with UQS-1 mine-detecting sonar, they were reclassified as ocean minesweepers (MSO) in 1955. The present U.S. roster is:

Three ships in active commission, to provide support to research and development in mine countermeasures at the Naval Coastal Systems Laboratory in Panama City, Fla.; twenty assigned to the Naval Reserve Force; ten laid up in reserve (moth balls).

MSC "Bluebird" class, Coastal Minesweepers. All 13 ships of this class have been stricken. 370 tons full displacement, 144 feet o.a. Wood throughout, UQS-1 sonar. Previously 167 ships of this class were built in U.S. shipyards for NATO and other allied navies, viz.: Belgium, Denmark, Fiji, France, Greece, Indonesia, Japan, South

141

Ocean minesweeper, USS *Excel*, MSO 439, 172 ft. long. Crew of 6 officers and 65 enlisted men.

Coastal minesweeper, USS *Albatross*, MSC 289, 144 ft. long. Crew of 4 officers and 35 enlisted men.

Minesweeping boat, MSB 47, 57 ft. long. Crew of seven enlisted men.

Minesweeping launch, MSL 28, 36 ft. long. Crew of five enlisted men.

143

Korea, Netherlands, Norway, Pakistan, Philippines, Portugal, Singapore, Spain, Taiwan, Thailand, Tunisia, and Turkey.

Special Minesweeper [MMS-1] Ex-Harry L. Gluckman. A "Liberty" ship converted to explode pressure mines. Specially modified to withstand mine explosions and remain afloat and underway. Work completed in 1969 on ship of 1943 design. MSS-1 placed in reserve on 15 Mar 1973; stricken 10 Feb 1975.

NOTE: Ten "Liberty" ships were partially modified in 1952–53 to explode pressure mines. Only one ship placed in service, ex-*John L. Sullivan* as YAG-37. Fitted with four T-34 turboprop aircraft engines on deck and stuffed with buoyancy material, she was employed in mine countermeasures experiments until reduced to a floating wreck; scrapped in 1958.

MSI Minesweeper Inshore. 120 tons light displacement. 105 feet.

Cove (MSI 1) transferred to JHUAPL 31 July 1970, technically remains on the Navy list.

Cape (MSI 2) operated by NUC, San Diego. Neither in service nor in commission. Seventeen other ships were built for other countries under various programs.

MSB Minesweeping Boats. Eight boats; 39 tons full load, 57 feet, wooden hull. Of 49 boats only 8 now remain in active service, all based at Charleston, S.C. Used for riverine operations in Vietnam; built between 1952 and 1956.

MSL Minesweeping Launches. Sixteen launches remained in 1974, all based at Charleston, S.C. Not mentioned in Jane's '77 and presumed inoperative. Displacement 23,100 pounds full load. Length 36-feet overall.

Versatile minesweeping craft intended to sweep for acoustic, magnetic, and moored mines in inshore waters and in advance of landing craft. Complement four to six enlisted men. They are carried by large amphibious ships to assault areas. Completed in 1966 (plastic hull).

It is important to mention that the declining list of minesweeping vessels is accompanied by a growing strength in minesweeping helicopters, which can be deployed from bases or from larger amphibious ships such as assault ships (LPH) or transport docks (LPD). The only large mine-countermeasure ship remaining on the Navy list, *Ozark* (MCS-2), dates from 1942. She is currently transferred to the Maritime Administration Reserve and cannot operate the large CH-53 and RH-53 helicopters required for effective "airborne" minesweeping. These large helicopters can be deployed rapidly in Air Force C-5 long-range transports. This leaves the question of mine-hunting craft somewhat in abeyance, while planning continues for the development of future mine hunting/sweeping concepts. Future developments could include surface effect ships or hydrofoil craft which would be rapidly deployable, require reduced staffing and could hunt and destroy advanced technology mines.

Jane's Fighting Ships 1973–74 (page 471), comments as follows:

> The mining of Haiphong Harbour and other North Vietnamese ports in April of 1972 has caused a revival of interest in mine warfare in the United States. The mining operation, conducted by Navy carrier-based aircraft, is considered the beginning of that phase of U.S. military operations against North Vietnam which culminated in the cease-fire agreement of January 1973. Subsequently, U.S. minesweeping operations have been conducted to clear North Vietnamese waters. These operations were carried out by Navy- and Marine-piloted CH-53 Sea Stallion helicopters, and the few ocean minesweepers (MSO) remaining in active service.
>
> Both the mining and minesweeping operations provided opportunities for the Navy to test and practice mine warfare techniques that had been developed during the two decades since the Korean War of 1950–53. However, it appears unlikely that the Vietnam War will cause any change in the downward trend of mine warfare *ship* strength. During 1973–74 the active Navy minesweeping fleet consists of ten active ocean minesweepers plus 26 ocean and coastal "sweeps" manned by combined active and reserve crews and assigned to the Naval Reserve Force. This compares to 63 ocean minesweepers in active service during the mid-1960s. Similarly, the U.S. Navy has no surface warships capable of laying mines and the number of submarines capable of planting mines is declining, (according to *Jane's* 1972–73) . . . as the older diesel-electric submarines are replaced by nuclear-powered submarines with a more limited selection of tube-launched weapons. (The SSN-688 class will be able to fire only SUBROC and the Mk 48 torpedoes.) Navy carrier-based attack aircraft and land-based patrol aircraft have a limited aerial mine-laying capability; however, the Strategic Air Command operates several hundred B-52 Stratofortress bombers that are fitted for aerial mine laying.

Jane's 1973–74 prediction is well substantiated in their 1976–77 general remarks (page 624) on Mine Warfare Ships as follows:

> Currently the Navy operates three active and twenty-two Naval Reserve Force (NRF) minesweepers. The active ships provide support to mine research and development activities at the Naval Coastal Systems Laboratory in Panama City, Florida, the NRF ships are manned by composite active-reserve crews. In addition

145

the Navy flies 21 specially equipped RH-53D Sea Stallion helicopters. These helicopters, which tow mine countermeasure devices, are readily deployable to aircraft carriers or amphibious ships in overseas areas. They can counter mines laid in shallow waters but have no capability against deep-water mines. In 1976 the Navy's mine countermeasure capability was officially estimated at about one-third that of the late 1960s.

To end this review on an optimistic note, there is a plan for the new construction of ten mine countermeasure ships (MCM), the characteristics of which are not yet developed (or revealed). These will provide a deep-ocean capability to counter advanced Soviet mines. One ship is planned for the Fiscal Year 1979 program, three for 1980 and six for 1981.

Other current construction includes four coastal minesweepers (MSCs) for transfer to Saudi Arabia with planned completion March–September 1978. Two MSCs are also for transfer to South Korea.

In summary, the present total of mine countermeasure craft in the U.S. Navy has shrunk to three MSOs on active duty at Panama City, Florida, twenty-two on Naval Reserve training, ten in mothballs, all built in the mid-fifties; eight MSBs at Charleston, S.C.; and twenty-one RH-53D Sea Stallion helicopters. All other ships have been stricken.

The deep-water countermeasure

When Admiral Elmo Zumwalt was named CNO in April 1970 he immediately established "Project 60" to signify his determination to have recommendations to put before SecNav and SecDef by the time he had been in office no more than sixty days. In his book *On Watch* [53], he makes some interesting remarks relating to mine warfare. First, he points out some of the imbalance which in his opinion had developed in the Navy since World War II, namely, the emphasis on aviation at the expense of surface ships. He says that there have been three powerful "unions" in the Navy for the last quarter century: the aviators, the submariners, and the surface sailors. Their rivalry has played a large part in the way the Navy has been directed. He goes on to say that the union system has a curious side effect. Certain crucial activities are outside the jurisdiction of all the unions and therefore tend not to concern them very deeply. He says, "No union has a vested interest in mines, which have no bridges for captains to pace." Going back to Project 60, there were two projects which were emphasized (in addition to many others) in the area of mine warfare.

One of these was the emphasis on helicopter sweeping of moored mines and also of magnetic mines. "Developing the equipment and techniques took time and money, and meant going almost entirely without minesweeping capability for more than two years, which was a pretty big risk. Fortunately we got away with it. The new system was in operation by 1973, when the Navy was called upon to sweep the mines out of Haiphong Harbor as part of the Vietnam ceasefire agreement, and the force that did that job proceeded almost immediately thereafter to repeat its performance in the Suez Canal. In these operations the ability of the helicopters to sweep areas much faster than surface ships and with less manpower demonstrated that this concept was a winner." [53, P. 81]

After discussing certain other weapons Zumwalt says, "One other weapon whose development we proposed to accelerate was Captor, a rather spooky mine that, when it detects an approaching submarine, releases a Mk 46 torpedo to make a run against it. This was one of the cases of a program proceeding slowly for no other reason than that no union was pushing it; it clearly was a weapon that could be of great importance in fulfilling the mission of denying straits to the Soviets". [53, P. 82]

At the present time we have the Captor mine which has been devised to attack only submerged submarines and to be useful in water which is, say, between 300 and 3,000 feet deep. Such a span of depths would allow mines to be placed in critical straits such as, for example, the broad passage between Iceland and the United Kingdom. In the early days of the Captor development we were always referring to the waters between Greenland, Iceland, and the United Kingdom as the GIUK Passage, which on one occasion caused an earnest briefing officer to try to locate this passage somewhere off the coast of China! Such a mine barrier would make it very difficult for hostile submarines to pass from the Norwegian Sea into the broad Atlantic. This would be true only if we retained control of the surface or air so that the submarines would be forced to remain submerged. Under these circumstances, what could an enemy do to clear a path through the minefield? One could imagine launching from the submarine a torpedo-like body which would go forth trying to look for and respond to the Captor detectors and classifiers like a submarine. This would launch the Captor torpedo and cause it to become confused or lost, because there would in fact be no target within range. This sort of countermeasure would be expensive, would displace offensive torpedoes or missiles which are the reason for the enemy submarine's transit, and would have difficulty simulating the dimensions and other characteristics of a submarine. Such a countermeasure would reduce risk to the submarine but not ensure safety. There would be many more mines in the field than countermeasures which the submarine could carry.

There is another possibility which perhaps is equally fanciful, and that is to launch a small manned mini-submarine which would carefully avoid looking like a target. The man-sub combination could conceivably search its planned path for Captor and thus serve as a guide for its mother submarine. If it found a Captor mine, it could proceed further to approach it carefully and plant or launch a destructive charge which would clear it away. These schemes are, as has been said, very iffy, and can be thought practical or impractical depending on the assumptions made about costs, risks, search probabilities, timing, and so forth. In the end it comes down to compromises between fact and fiction which can give estimates that vary between rather wide limits. Considerations of this sort have been brought to bear especially when the allocation of funds is made to depend on some studied view concerning the feasibility of a postulated but not actual countermeasure. If some powerful enemy were able to put us in a similar situation, not by blockading the GIUK but perhaps by placing a wide ring around one of our submarine bases like Norfolk or Charleston, and if we did not control the surface of the ocean above this ring (a situation very difficult to imagine), then we would be faced with a similar countermeasure problem. There are conceivably other passages like the Malacca Straits, the Molucca Straits, or the Makassar Straits, which an enemy could mine, especially if we were interested in going through them but he was not. It is clear that the assessment of what is needed in the way of countermeasures must be made very carefully before we automatically embark on a countermeasure development or production which may be very costly and at the same time ineffective, or perhaps not likely ever to be needed.

Sweeping and clearing in Haiphong and Suez

The Vietnam minesweeping effort, called Operation End Sweep, is described by Rear Admiral Brian McCauley, who was in charge of it. His discussion shows to what extent the mining and minesweeping operations were related to the various negotiations both before and after the cease fire which was signed in Paris on 27 January 1973. Early in 1972, the Commander Mine Warfare Force was asked to assist in the planning of the minefields to be laid in North Vietnamese waters in May of that year. He says that "from the beginning the possibility of U.S. forces having to sweep the mines was a factor which influenced the types of mines used, their settings, and to a lesser degree their locations. As a result, when it came time to

Practicing at Subic Bay, Philippines, before deploying to Haiphong for Operation "End Sweep," 1973.

Close-up of the Mk 105 magnetic hydrofoil minesweeping sled.

A Marine helicopter is towing a Magnetic Orange Pipe (MOP) under outstanding scenic conditions in Operation "End Sweep," 1973.

With well deck awash in Haiphong Harbor roadstead, crewmen of USS *Ogden* wait for RH-53 helo to take MOPs under tow (Operation "End Sweep").

sweep, we knew everything about the mines and had purposely planted mines which could be swept easily and effectively by our mine countermeasures forces.... The vast majority of the mines were programmed to self-destruct and the remainder to go inert after a given time. Thus, even as the mines were dropped, the process of mine removal had been started." [30, P. 19]

The actual preparation for the operation began in July 1972 when it became apparent that minesweeping would be an important part of the "peace" negotiations. The basic sweeping device was the Mk 105 sled towed by a CH-53 helicopter and supporting an electric generator which streamed a standard magnetic tail astern. This proved to be rather tricky to launch and tow compared with a magnetized iron pipe filled with styrofoam for flotation (developed by the Panama City laboratory). These pipes, painted orange, were commonly known as MOPs (Magnetic Orange Pipe). To deal with acoustic influence mechanisms or the magnetic acoustic combination, noise makers were towed independently, or astern of either the Mk 105 sled or the MOP. To support the various helos, an LPH and two LPDs were used. There were 13 Navy HM-12 helos and 18 Marine CH-53 helos involved in sweeping. Other helos for control platforms and transportation ashore and between ships were based on the two LPHs according to space available. In addition to the airborne sweeping, there were ten ocean minesweepers (MSOs) used principally in deepwater approaches and as helicopter control ships. Further, says Admiral McCauley, there was a surface support force made up of two destroyers, two fleet tugs (later reduced to one), a submarine rescue ship, an LST for MSO support, and a "specially configured" LST to transit the Haiphong channel after sweeping had been completed, in order to demonstrate confidence in the thoroughness of the sweep. One can see immediately that in terms of logistics, support, and manpower this was not a simple or trivial operation. The order for this operation to deploy was received on 4 November 1972, and various units got as far as the Philippines when the Paris negotiations collapsed. However, the Protocol on minesweeping was signed in late January 1973 in Paris with the cease fire agreement. After various meetings and assurances that U.S. minesweepers would not be molested, the actual sweeping off Haiphong began on 6 February 1973 with four MSOs sweeping the areas in which the LPHs and LPD were to anchor. Actual airborne sweeping began on 27 February but stopped that night because of difficulties over POW exchange. It recommenced in the Haiphong area on 6 March and continued until 17 April on a daily basis.

On 17 April, because of difficulties in Laos and Cambodia, the sweeping force was again withdrawn. This period of inactivity lasted until 13 June when the Paris Joint Communiqué was signed. This communiqué said that the U.S. would resume sweeping within five days and would complete all sweeping within thirty days after that. McCauley says this was quite possible because by June all of the mines were well past their self-destruct date. "There was considerable statistical evidence to ensure that the vast majority of the mines would self-destruct and that any left would be inert and totally deactivated. Because of this, all sweeping after 18 June was exploratory sweeping. This is considerably less time consuming than full sweeping." [30, P. 22] In fact he says that prior to leaving Haiphong on 17 April the sweep in the Haiphong main channel had been completed. The demonstration ship, a modified LST (MSS-2) had made a number of runs through the channel, but not enough, he says, to declare it open. On 20 June 1973 the remainder of the runs were completed and the North Vietnamese were handed a signed memorandum stating that the United States had completed the sweep in the Haiphong main channel.

From this account it is certainly not clear how many mines were actuated by any sort of sweeping, how many self-destructed, and how many had become inert because of the passage of time. According to the design of the mines and the intent in laying them, it could be considered an entirely superfluous exercise to engage in pro forma sweeping nine months after the mines were laid, except that the Vietnamese did not know how useless it might be.

With regard to sweeping the inland waterways, these efforts were frustrated by the North Vietnamese who by now had learned enough to know that delay was the best method for sweeping the Mk 36 Destructors. When the Protocol dictated that the United States would assist the North Vietnamese in every way to accomplish inland waterway clearance, a training program was set up for about forty North Vietnamese to learn how to use the nonmagnetic remote-controlled boats supplied for towing acoustic and magnetic devices, how to maintain these equipments, and, yes, to learn the fundamentals of the Mk 36 Destructor which they were supposed to sweep. The North Vietnamese completed the course in about three weeks and said that what they wanted was more equipment, not more training.

It became apparent that the D.R.V. did not want any U.S. military personnel to help them inland. "As a result, the United States did no sweeping or supervising of sweeping in the inland waterways. Upon Task Force 78's return the second time in June, all of the mines had passed their self-destruct date. To our knowledge no sweeping was done in inland waterways. Negotiations continued to the end. The

North Vietnamese continually asked for additional equipment." [30, P. 22]

After comments on the mobility of helicopters and their increased speed of sweeping (up to 25 kts compared with 4 to 7 for MSOs), their limitations in deep-water sweeping, their current inability to deal with the mine hunting situation and the pressure mine, their logistic dependence on support ships or on shore facilities, Admiral McCauley says, "The most obvious conclusion one reaches in reviewing Operation End Sweep is the effectiveness, relative ease of laying, and the economy of the coastal mine campaign. It was an impressive sight on flying over Haiphong in the early days of End Sweep to see all twenty-six merchant ships at anchor behind the minefield. None had moved since May [1972] when the first mines were dropped. Few aircraft were lost during their emplacement. The effectiveness of this campaign demonstrates once again the vulnerability of a country which has little or no minesweeping capability to mining. . . . Thus the mining campaign provided a potent lever to U.S. negotiators both before and after the Peace Agreement." Finally he says, "Rarely will anyone in today's Navy argue against the effectiveness of mine warfare nor our vulnerability as a nation to its use by other powers. Yet the practical demise of the Mine Force in the U.S. Navy is already planned—a victim of other more sophisticated higher priority programs. We have relegated the Mine Force to a minuscule size. . . . There is no new surface minesweeper on the drawing boards and none is now contemplated when the wooden hulled MSOs finally expire." [30, P. 25] The admiral is concerned that the U.S. Navy may no longer be able to do anything about mines. However, the Phoenix of the Mine Force may still rise out of its scattered ashes by purchase of readymade and improved countermeasure ships from the British, or the French or the Germans, all of whom are busily at work on such systems (see the section on recent countermeasure efforts in Western Europe later in this chapter). Such purchases would be bad for our balance of payments, but would help exports from foreign countries, showing that there is a little good in everything.

Concerning the effectiveness of the mining operation, an interesting comparison is made by Admiral Hoffman in a recent article. [21] He compares the cost of the mining operation at Haiphong with the cost of the mine clearance operation. Although cost estimates to my mind are always suspect because it is difficult to know what costs have been prorated, the results are interesting. He says that both estimates cover just the expenses above those which would have been incurred by the normal operation of forces were they not involved in mine-laying or minesweeping efforts. "The cost of the mines and their delivery was $6.5 million. One A-7E aircraft valued at $3 million was lost to antiaircraft defenses. This brought the mining costs to about $10 million. The sweeping cost for operation, maintenance, and equipment was about $14.5 million. Three helicopters worth nearly $6 million crashed (i.e., were not lost to mines) to bring the incremental sweeping cost to about $20 million or double the minelaying figure. Significantly, the minelaying was an opposed operation conducted in enemy waters. The minesweeping, on the other hand, was conducted in a non-hostile environment and was to counter our own weapons." [21, P. 152]

With regard to the Suez sweeping which followed almost immediately after the Vietnam completion, it should be remembered that the helicopters were not alone at Suez. The 1967 and 1973 Arab-Israeli Wars had left substantial amounts of unexploded ordnance and mines in the canal's fairways, banks, and anchorages. The remaining explosives and mines had to be disposed of before the ten wrecks, sunk by the Egyptian army in 1967 to block the canal, could be removed. An interesting account of this salvage operation is given in reference 6 by Captain Boyd, USN, who was in charge of salvage.

The minesweeping operation started as a U.S. evolution beginning in April 1974 and being completed in early June. However, the disposal of ordnance continued right up to the reopening ceremonies on 5 June 1975. Land ordnance was cleared by the Egyptian army under the guidance of U.S. Army explosive ordnance disposal personnel. For the underwater side, a residual team of U.S. Navy EOD personnel remained until the reopening date. From April to December British and French mine-hunting forces were present and active. The British vessels and divers worked up to the reopening date. It was clear that many mines or unexploded bombs underwater were not responsive to the particular helicopter sweeping that had been done earlier and so had to be dealt with by mine-hunting and disposal techniques. These efforts serve as a peacetime example of international cooperation in using modern mine-clearance techniques to get rid of mines and ordnance laid by other powers—in this case, Israel and Egypt who did not have their own means for disposal.

Recent countermeasure efforts in Western Europe

The saying, "The best is the enemy of the good," attributed to Voltaire, applies particularly well to decision-making in the area of mine countermeasures. It is very difficult to decide on one system or another, because no single one is able to deal with all situations. A combination of systems is called for, and this of course introduces inevitable comparisons between value and cost and leads back to the original problem of choosing "the best." It appears that several countries, mostly in NATO, have in effect solved this problem by each pursuing its own version of a countermeasure system, each realizing that if a different system or apparatus turned out to be better, it could probably be purchased more or less on the open market from the other country. There appears to be a matter of common agreement, however; namely, that there must be more mine hunters vs. minesweepers, and that these new ships will be made out of fiberglass instead of metal or wood. These ships, unlike conventional minesweepers, will detect and classify mines by means of hull-mounted sonars without previously traversing the minefield. After detection of a mine or minefield the individual mine localization and destruction is accomplished by divers in rubber boats, or, in the new systems, by either surface or underwater remote-controlled unmanned vehicles. Information on these various systems has been obtained from Rolf Boehe [5] and Anthony Preston [37].

Recreation boats have been made for many years out of GRP (Glass Reinforced Plastics) or GFK (Glasfaserverstärkten Kunststoff) as the Germans would call it. The largest construction in GRP, however, was started in 1970 by Vosper-Thornycroft and completed in 1973 with the British minehunter HMS *Wilton*, 450 tons full load and 46.6 m in length. The advantages of simple maintenance and repair together with the lack of magnetic and electric potential fields make this type of construction ideal for mine countermeasures vessels. As early as 1973 the manufacturers were offering a 47-m minehunter made of GRP for sale, equipped with the Plessy 193M sonar and the Sperry CAT mine-locating gear. In parallel with this effort, the Swedish Navy has built an experimental sweeper, *Viksten*, 135 tons, of GRP and now has three of this class. Their plan is to build nine GRP minehunters based on their experience with *Viksten*. Recently, according to Boehe [5] the French Direction Techniques des Constructions Navales designed two GRP minehunting boats. A similar class of minehunters (chasseur des mines tripartite) forms the basis for a joint development program of France, Belgium, and the Netherlands, with the prototype to be delivered by 1979. Eventually the Federal German Navy will be interested in this program. Italy is also systematically preparing for the construction of GRP minehunting boats.

All of these minehunting boats have certain common features. Most or all will be equipped with sonar, hull mounted, with separated transmitters and receivers, operating at two ranges of about 600 m for detection and 200 m for classification. The minehunters will have special propulsion systems so that they can run at very low speed while keeping strictly on course. They must be highly maneuverable and quiet. For these purposes, in many cases dual rudders are equipped with electric or hydraulic propulsion, and sometimes an additional electric motor is used to operate the normal propellers when the main engine is uncoupled. Along with special propulsion it is necessary to have accurate navigation and positioning equipment on board. Apart from navigation radar, most minehunters use a transponder system installed in the ship, and in two reference buoys or landmarks. The handling of the data and the coordination of the location of the ship with the location of mines detected is done quickly by special computers for the purpose.

The French Navy's minehunter *Circé*, the first of a class of five built at Cherbourg between 1969 and 1973 as described by Preston [37], is an excellent example of a minehunting system embodying the general principles laid out above. The ship, 530 tons full load, 50.9-m overall length, is not made wholly of GRP but rather is built of laminated wood with composite wood/glass-resin superstructure. The planking has an external protective coat of glass-resin. The heart of the system is the Thomson-CSF DUBM 20A minehunting sonar. This elaborate installation, a French analogue of the British 193M, will detect and classify mines lying on (not in) the seabed up to 250 m away. The new part of the system is the introduction of the PAP-104 robot fish to locate and destroy the mines. The *Circé* carries two of these fish, called Poisson Autopropulsée, which can be launched for this purpose. A fish weighs approximately 700 kg, can operate at depths between 10 and 100 m, and runs at a speed of 5 kn. It is wire controlled, and transmits television pictures over 500 m so that the mine can be identified. When this is done, the PAP automatically attaches a 100 kg demolition charge to the mine and returns to the minehunter. The charge will be detonated by an electroacoustic signal from the ship.

I would say that if the demolition charge can be placed as accurately as this description implies, the size of 100 kg is a bit excessive. However, there is a comfortable margin in case of an excessive placement error. There are at least two advantages to this system apart from its inherent safety. The approach to the mine is made entirely underwater and hence is independent of surface waves which can be difficult for a small vehicle. Second, the approach of the PAP to the mine can be nulled on the sonar. The television (with its own lighting system) is used as a means of close up identification and fine-tuned location. If the water is too muddy to allow the television to work even at the close range of a meter or so, then the system is no worse off than a purely sonar-vectored system would be. If the PAP can be made to hover or backtrack, then it would have the advan-

A CH-53A Sea Stallion tows a Mark 105 magnetic sweep during trial off California coast, 1972.

Close-up of magnetic sled picking up speed.

Looking down on the sled. The amphibious transport dock USS *Trenton* LPD-14 stands by in support, 1972.

tages in this respect that the British CAT has. The CAT consists of a remote-controlled catamaran 4-m long and 2-m wide which tows a submerged platform on which a demolition charge is carried. Thus, the CAT can be vectored to the mine until the sonar location of the demolition charge coincides with the sonar location of the mine. The catamaran returns to its home base after leaving the charge in place. Either of these schemes has obvious advantages over the man diver approach. The French, Belgian, Netherlands tripartite development mentioned earlier will use the PAP system with an improved, less-bulky sonar, the DUBM-21A, and GRP minehunters. Boehe concludes his article by saying that minehunting is considered highly important by all Western European countries and their navies.

The German "Troika" system has been reserved for a separate discussion so as not to confuse it with the preceding purely minehunting systems. This is because the Troika is essentially a minesweeping system using many of the characteristics of the minehunting systems. The information on the Troika comes from an article by Klaus Bergmann in Wehrtechnik, November 1976, entitled "Minenabwehrsystem 'Troika'." [2] The essential idea of the Troika is to sweep mines by hardened robot ships which are able to supply magnetic and acoustic ship signatures and which are remotely controlled from a manned minehunting ship which stands out of the minefield and which operates its own mine-detecting sonar from a safe distance. The information so gained allows it to direct the robot sweepers, of which there are three (hence the name), along courses where they will do the most good. They can also be directed to avoid any moored mines shown up by the sonar.

The individual remote-controlled magnet skid sweeper is a self-propelled craft of 80 tons displacement and 27 m in length. Under its ship-like exterior lies a cylindrical, steel pressure hull with watertight compartments in which installations are shock proofed and made functionally secure. The craft carries magnetizing coils on the outside as well as noisemakers fore and aft. Each member of the Troika tries to make as much acoustic and magnetic signal as possible to simulate a target and induce magnetic/acoustic mines to detonate away from them. The Troika system is thus a remote-controlled Sperrbrecher operation, but is admittedly ineffective against pressure signature mines, especially in sheltered water. In concept, the system can also deal with moored mines. After the robots have dealt with the bottom influence mines, the control ship can run through the area with its kite sweeps deployed, avoiding the known moored-mine locations. One of the advantages of the system, apart from safety to the manned control ship, is that the manpower requirements are less, since the robots cover an area otherwise requiring two or three manned minesweepers.

The Germans have 18 Type 320 mine-warfare ships of 365 tons and length 44.7 m., which they are dividing so as to accomplish the best of all possible worlds. Ten of these units have been fitted out as minehunters (between 1975 and 1977) having Plessey Type 193 (British) minehunting sonar and PAP-104 (French) wire-guided minehunting fish. The remaining eight will be fitted with the Troika remote-controlled system between 1978 and 1980. The former units have an extended superstructure in place of the magnetic sweep cable reel and can no longer sweep. All of these ships are wooden hulled, and were built in the late 1950s.

6 Measures of Effectiveness

Operations research

It is interesting that the first operations research in this country started at NOL in 1940–41 with such people as Ellis Johnson, W. C. Michels, S. L. Quimby, Thornton Page, Francis Bitter, George Shortley, and many others. The question was how to use mines most effectively in warfare and particularly in that war. According to Johnson [23], the original purpose was to stimulate interest in operational questions among the technical staff. Later the craft of operations research grew enormously and could be generalized into asking how a series of operations could be conducted so that the results would be optimum. This led to questions as to the location and type of ship traffic, and the characteristics of mines that would be useful. Reports from Britain on how well the technical and operational people worked together in planning weapon design, performance analysis, and operational use were received with enthusiasm at the laboratory. Dr. Bitter proposed that an expanded Operational Research Group there gather information, evaluate proposed and actual operations, and specify operational requirements. On the basis of these requirements, materiel should then be designed and procured. Johnson goes on to say:

> However, the initiative of the Laboratory Operational Research Group was regarded unfavorably by the Bureau of Ordnance and the high command when the extent of its activities was brought to light in peculiar circumstances. An attempt had been made to ob-

tain information on the size and routes of Japanese shipping. This information was essential if the sensitivity and endurance of mines, as well as other characteristics, were to be intelligently determined. It was in full pursuit of this data that an enthusiastic civilian cut some well-regarded procedural corners and came a cropper. The coordination of operational information with technical development was considered to be too secret for study by such a low echelon and by civilians to boot. The Bureau of Ordnance therefore took action in July 1942 to disband the Laboratory Group. It was ordered to restrict its discussions to unclassified information only. It was also directed to make no further attempt to obtain or use operational information except with the specific consent of the Bureau. [23, P. 51]

All of this led to the transfer of some of the group and their proposed function to the Bureau and shortly thereafter, in November 1942, operational research in mine warfare was entirely under OP 30M. Thereafter, operational research gradually narrowed its scope to provide advice on how a given operational objective could be attained by more efficient use of the given materiel, rather than to visualize improvements in the materiel itself which would further these ends. It became a service to operators and not a guide for technical development. It became the province of mathematicians and statisticians rather than of physicists and engineers.

A typical example, in ASW, was to discover what method of search with given equipments—aircraft, ships, buoys, and sonar—would yield the maximum probability of finding or tracking a submarine. Inevitably in such analyses it would occur to someone that if a new capability did exist, such as a slight increase in speed or range or a new device of some sort, then the game would be very much slanted toward one side or the other. In the earliest days the operations research was not pure in the sense that it was useful not only to a commander trying to use his materiel, but also to the R&D side which was trying to decide what would be most worthwhile to develop. Later in the war the continuing question arose as to whether a given plan or calculation undertaken to improve effectiveness had in fact proven itself in practice. This raised questions which were difficult if not impossible to find timely answers to—even such apparently straightforward statistics as the number of ships sunk or damaged by mines, for example, what tonnage was involved, where these incidents occurred, what length of time was required for repairs, what cargoes were lost, and what importance was attached to them. The questions as to the effect of a given event form ever-widening rings—

like waves after an explosion—and the net effect becomes a statistical summation of many factors which are likely to be unknown.

The continuing need to make precise estimates of the effectiveness of mine warfare—the damage to the enemy per dollar spent, or some other measure—is symptomatic of the lack of emotional acceptance of mine warfare on the part of what Admiral Zumwalt has called the "unions" of the Navy. The extraordinary record of the mine campaign in the last five months of the war against Japan should be an ample statement: 5 percent of the Army Air Force bombing effort was diverted to aerial mine laying to blockade the major ports of Japan. This resulted in the loss of more Japanese ships than had been lost to air strikes throughout the whole war. According to some Japanese opinion, the starvation of the Empire's homeland and the loss of necessary materials and fuel would have led to capitulation without the bomb if the mining had been started a few months earlier. These are extraordinary statements, so much so that many people discount them as propaganda. Nevertheless, one would expect mines to be effective against an island almost wholly dependent on imports. The Japanese did not have the strength or resources left to mount a proper countermeasures campaign even if they had known what to do. They should have come out and shot down the mine-laying B-29s, but they had already been defeated in the air war. Today, air-space radar and surface-to-air missiles (SAMs) would probably prevent this kind of wanton mine laying. Nevertheless, after control of the air has been achieved it is clearly possible to lay the mines.

It is also clearly possible to continue the debate. No war is like its predecessor. A tactic which succeeded in one war could fail in another—often because of a change which has occurred in some area not previously recognized as critical; or it could succeed again but for entirely different and new reasons. An example of this kind of thing occurs in considering the convoy system which was credited with defeating the German submarine onslaught against merchant shipping in World War I. As Vice Admiral Hezlet says:

> The concentration of U-boats on independent shipping was not because of the danger or difficulty in attacking convoys, but because the U-boats could not find them and so had no alternative. The first effect of the convoy system was that the ocean suddenly seemed to the U-boats to be devoid of shipping. This was because, strange as it may seem, a convoy of ships was not much more likely to be sighted than a single vessel. A single ship will probably be seen by a U-boat lurking within ten miles of its track. A convoy of twenty ships is only two miles wide and so would be seen by a U-

boat lying within eleven miles of the centre of the track of the convoy. Five convoys of twenty ships each were not, therefore, very much more likely to be seen than five single ships and were obviously much harder to find than a hundred independents. The result was that the vast majority of ships when in convoy were never seen, and the greatest advantage of the system was the difficulty the U-boats had in finding the convoys at all. [19, P. 94]

Evidently these statements depend for their truth on the fact that the mobility of the submarines was poor compared with ships, certainly when they were submerged. They had to remain submerged because of the presence of escorts and antisubmarine patrols. Further, their detection range was limited to line of sight at that time. It was not so much a case of the submarine finding the convoy as it was the convoy stumbling by chance across the submarine.

Another factor, having to do with rate of firing, is discussed by Hezlet:

When all shipping sailed independently U-boats were presented with a long succession of targets at which to fire, and they had time to take deliberate aim and then reload before the next victim appeared. With convoy there would be only one chance to fire as the enemy swept by *en masse*. Whilst the selected ship was being attacked, the rest of the convoy would slip by unscathed and a second shot was seldom possible even if the escorts permitted it. With only two, or sometimes four, torpedo tubes, only one ship, or at most two, would generally be hit. Moreover the attack was complicated by the presence of escorts and the anticipation of the heavy counter attack which was likely to descend on the U-boat after firing. Convoy brought the escorts into the vicinity of the U-boats at a time when they had to reveal their position by firing. When the anti-submarine vessels patrolled at random they could never find the U-boats, but with a convoy system, the U-boats, when they did find the convoys, could not help bringing themselves into contact with the anti-submarine vessels. [19, P. 94]

It thus came to be doctrine that convoys were good. In World War II the system was again used, but here Hezlet says:

The British at once put an immense effort into defeating the U-boats, including the institution of convoy. A convoy strategy was again shown to be quite fundamental, and without it the U-boat campaign would certainly have won. Nevertheless, the U-boats had considerable success in finding and attacking convoys, and convoy could not have won on its own and would have been defeated had the other British anti-submarine measures not been successful. It was only because the British anti-submarine measures were able in the end to triumph round the convoys that the victory was won. [P. 188]

And so it appears that convoy was successful in World War II not as a means of hiding from U-boats as in World War I, but as a means of attracting them to expose themselves disadvantageously to attack. Their success lay in the success of the antisubmarine measures which had been developed and deployed. It is difficult to refrain from quoting Hezlet's final pronouncement on these matters:

> Not only were the British technically superior but, using operational research, were in the end tactically superior as well. They used all their new devices and ships and aircraft in the closest cooperation and with great skill. The German U-boats, after their invention of the wolf pack which made it possible to find and attack convoys in the open ocean, were unable to advance any further. They could not produce any solution to its fundamental weakness which was that when using it, U-boats ceased to be submarines and became surface torpedo boats. It was the British exploitation of these weaknesses which was the key to the Allies' success. [P. 189]

Hezlet also says:

> The U-boats were driven from the surface of the sea by radar in aircraft and so could not get into position to attack. The whole basis of the German wolf pack system was therefore countered and the U-boat fleet was defeated decisively and in a way they had never suffered during the First World War. [P. 185]

It would be easy to conclude that the German U-boat campaign in the Second World War was easily countered and merely confirmed the past history of the *guerre de course*, which showed that it could always be defeated by convoy. This would be a far too facile assumption. It failed only by a very narrow margin. Had the Germans built up their U-boat fleet quicker, or the British failed to adopt convoy, or develop Asdics and radar, or the Americans failed in their shipbuilding programme, the results could have been very different. The Battle of the Atlantic was won only because the Allies again put an immense military, civil and scientific effort into it. They had to do this for the very reason that the submarine, as a commerce raider, was a potentially decisive weapon of seapower and, if they had not, they would have lost the war. [P. 189–90]

Another way to say this is that the wolf pack was successful against convoys if the submarines could assemble at the convoy. They could do this only if they could travel fast, which required them to travel at the surface. Airborne radar made this hazardous for them at night. Therefore, airborne radar coupled with depth bombs may have been the most important factor in preserving Allied shipping against submarine attack. In today's world when almost all submarines can travel indefinitely underwater at high speed, the value of airborne radar is virtually nil in this context.

The lesson from this is that it is vital to continue analysis of the probable effects of continuing improvement of capabilities as applied to wartime situations and goals. It is essential to determine not only what was effective but why, so that any change in the underlying reasons will not lead to a sudden unnoticed reversal of the conclusions previously reached. It would appear very important that this kind of appraisal be consciously undertaken at a high level of experience and skill in the total area of mine warfare precisely for the reason that there is no "union," to use Admiral Zumwalt's word, to look after this subject, and because we have to believe that the subject is important. Surprises in mine warfare could be unpleasant.

Simulation

In these days of the new economics, computers, and math models the idea of simulating a complex situation by a set of equations or laws relating the variables, and computing what one variable must be—subject to these laws—when other variables are changed, is now fairly commonplace. The general idea, of course, is not particularly new, having started with Newton, whose particle dynamics and inverse-square law of gravitational force made it possible to calculate everything about planetary orbits. Partial differential equations due to Euler and Laplace made it possible to extend these methods to continuous media. The advent of electronic computers finally made it possible to accomplish incredibly long and tedious calculations in very short times so that, for example, the same calculation could be done over and over for different initial conditions and then a distribution of results could be obtained which of course was of a statistical nature. One could do this all over again for a different set of parameters which remained fixed throughout one whole set of calculations.

In January 1958 the first of what proved to be a long series of Mine Warfare Conferences took place at the Naval Ordnance Laboratory at White Oak. The subject of the first conference was "Theory and Simulation," which of course was a subject not very well defined in most people's minds. One of the main ideas was to review the status of minefield theory and to describe the first computer simulation by Weiss and Warner of a minefield. These men were trying to describe a minefield on paper using the properties of real mines, under real conditions, in the presence of postulated countermeasures, and subject to attempted transits by a variety of traffic. The question was whether one could compute a minefield threat from all this, and whether one could then vary conditions—like the number of mines, the rate of replenishment, the ship count, the sensitivity, the density, the traffic, and so on. This obviously got to be very complicated, and the mathematicians became more interested in the calculation itself than the result. This, of course, led to less interest in the subject on the part of nonmathematicians who, incidentally, made up most of the audience. Nevertheless, the idea of being able to *simulate* the presence of a minefield in a real situation and obtain a measure of its threat merely by a paper exercise, without actually having to do it and wait months or years for an estimate of the result, was an important idea—which in 1957 was perhaps a bit ahead of its time.

A renascence of this idea took place at Dahlgren in the late sixties, as I remember, when the computers there were able to perform increasing wonders of calculation. The difficulty with the results, which led to some criticism of the mine characteristics, was in part that the postulated minefields were made up without full knowledge of the mine materiel available, and so some of the conclusions coming from the calculations were not really valid. Similar difficulties are fortunately not common but also not unheard of. I can recall, at any rate, a story at the time of the Rolfe Committee investigation triggered by CoMinLant, which found that mine materiel was unsatisfactory. Admiral Sanders of the Mine Force said the mines did not work during certain exercises which he conducted in approximately 1955. He was correct in this statement. It turned out that some of the mines had been prepared for use without removing a wooden storage plug and replacing it with the water-soluble salt plug which in those mines was used to give delayed arming. The exercise had reaffirmed that wood does not dissolve very rapidly in salt water. The conclusion was that the instructions for assembling mines for use from their stored condition were too complicated—too many separate operations—and further that there were now better ways of achieving delayed arming which should be incorporated into the next R&D project. All of this probably indicates that there are ways to reach a valid conclusion even if the reasoning is somewhat dubious. Sometimes we have to reverse the adage and declare that two wrongs can make a right. (A further reference to this Mine Ex occurs in chapter 7.)

Suffice it to say that the use of computers for the simulation of particular minefields, taking great care that the inputs are reasonable and correct, is a powerful tool for the guidance of war planning, stockpile planning, and operational planning. The object, of course, is to have as much advance knowledge as possible in order to make the most effective use of the materiel at hand.

The effects of damage, delay, and dismay

It should be relatively easy to count up the ships or tonnage sunk by mines in a given field during its life and then to add up the value of these ships and their cargoes. This would give a quantitative measure of the loss endured by the enemy. The cost of the mines and their delivery would also be possible to add up, although because the delivery vehicles had other duties, some decision would have to be made about the share of their total cost which should be charged to the mines. If the value of the ships sunk was greater than the cost to sink them, then presumably the mining operation would be considered "cost effective." All such calculations and the conclusions from them are subject to almost interminable arguments as to whether the operation was decisive or not, whether its value was measured at all by the cost figures, or whether these figures are really irrelevant representing a past expenditure of effort, and whether the real significance of the operation should not lie in what it prevented being done, or in what it caused to be done that would otherwise have been unnecessary. It is not possible to settle arguments which are interminable. However, a few considerations on the possible effects of mining may well be in order.

The cost-accounting method of determining mine effectiveness has further difficulties with it—namely, that not all ships attacked by mines are sunk. Many, probably more than half, are damaged so that they have to be taken out of action and have to be either repaired or scrapped. One could say that a ship that is sunk is gone for the duration of the war. For that period its existence would have been of importance to the war. A ship which is out of action for a certain number of months might as well have been sunk for that period of time. For example, if a ship is damaged and out of action until the war is over, it might as well be counted with the sunken ships as far as the effect of mining on the number of ships is concerned. If, however, the ship can be back in action before the war is over, then it can be counted as only partially sunk as it were, or out of action for a number of months. These uncertainties lead to variations in the statistics wherein ships are listed as "sunk or damaged" without specifying whether the damage in some cases was severe or negligible.

But there is another measure of economic effectiveness which must be considered in the case of the delay of shipping due to the presence of mines. Many of these considerations are contained in W. C. Wineland's paper [50]. If a port, for example, is closed for one day while mine-sweeping or clearing operations are going on, or perhaps just because it seems prudent to the authorities not to risk ships until the extent of the mine threat is determined, then the delay in the movement of goods can be related to an equivalent number of ships. Suppose, for example, that on the average five ships go out of a port every day and five ships come in. Closing the port for a day means that the goods moved by ten ships are not moved on that day. If the war lasts for a thousand days, then the total amount of goods moved is 999×10 shiploads. One tenth of one percent of the goods that could have been moved have not been moved. The same result could have been achieved without the delay by one tenth of a percent fewer ships. If a thousand ships were involved in supplying the port, the day's delay is equivalent to the loss of one ship. Delay then may be related to an equivalent ship loss equal to the total number of ships involved in the long-term traffic multiplied by the fractional time lost by the delay.

In order to assess this in advance it would be necessary to know how long the war was going to last. Unfortunately, this information is lacking until the war is over. Perhaps the prime example of effective delay occurred with the mining of Haiphong Harbor, after which no shipping moved in or out until the war was over some months later. As a result of the mining, no ships were sunk or damaged. However, the permanent delay following the mining meant that these ships might as well have been sunk as far as usefulness in supporting their war effort was concerned.

Clearly, in the cost-accounting procedure it is necessary to count the cost of the mine countermeasure effort if there is one. In case there is an effective countermeasure which can be employed, then the duration of delay is reduced, but the effort put into the countermeasure work may be added up in terms of additional ships that could have been built if the countermeasure had not been built.

Now, all of this accounting may perhaps give an idea of the long-term effect of mining upon an enemy, but it is hard to see from this what effect it may have beyond spurring people to tighten their belts, work harder, waste less, and so forth. It would appear that there are many efforts in warfare that wear down the opponents but actually few situations which are clearly decisive in the sense of producing an immediate or short-term change in the enemy's will to conduct war or in his desire to capitulate. There are in fact very few such events in any war except at the end! The atom bombs at Hiroshima and Nagasaki clearly caused the Japanese surrender, although they were considering this step beforehand because of their disintegrating and hopeless condition. The approach of the Russian forces to Hitler's bunker finally triggered his suicide and the end of the Third Reich. Such decisive events are possible only as the final steps in a long process of attrition and loss, and so we are back to the realization that a victory in war tends to be the summation of many long-term actions, none of them decisive but some of them more effective than others in reaching toward the final moment of decision.

But occasionally there occurs in mine warfare a decisive moment which stands out above the continuing realization that certain losses are occurring which can nevertheless be endured, provided they do not go on too long. One such moment occurred in the mining of the Dardanelles by the Turks at the beginning of World War I. As told by Commander James A. Meacham, USN, [32], and as we have already noted, the French and British naval commanders having lost one battleship and three cruisers in one day at Kephoz decided that the key to Constantinople lay in a land attack over Gallipoli. This was a very far-reaching judgment which proved itself to be ineffective over the four-year duration of the war. Had they decided to continue their minesweeping efforts, admittedly dangerous under fire, they might possibly have succeeded if they had been able to bring in additional naval forces. No one can say whether the decision was right or wrong. The point is that the minefield caused an either/or decision to be taken which had immediate tactical effect as opposed to the usual long-term strategic effect due to attrition. In fact, there is an interesting possibility that could have solved their problem had they thought of it. The sweepers could not get to the mines because of the shore gunfire. The battleships could not deal with the gunfire because of the mines. If the commanders had decided to commit the sweepers and the battleships together, with the sweepers leading the ships instead of insisting on minesweeping as a separate and complete operation in itself, the chances are high that they could have transited the whole field without much damage. The concept of assessing the minefield threat was missing from the military thinking: it was all or nothing—either impassable or cleared out and hence totally free of danger.

Another and somewhat unimportant example of drawing a wrong conclusion from observed facts is described briefly by Hezlet [19] in connection with what he called the famous attacks in April of 1918 on the German submarine bases at Zeebrugge and Ostende. He says:

> It is true that after them [the attacks] few U-boats operated in the Channel and sinkings there almost ceased, but the entries and exits

from these bases show that they were blocked for only a few hours and that the relief in the Channel was entirely due to the increased effectiveness of the Dover Barrage. From this time on, the Flanders U-boats seldom tried to penetrate into the Channel and confined their operations to the North Sea. This increase in the efficiency of the Dover Barrage was due to the combination of patrols and the vast complex of new minefields. [P. 99]

It is very easy to fall into the fallacy of *post hoc ergo propter hoc*, and even doctors are prone to say, "I treated the patient; therefore he got well," rather than "I treated the patient and he got well."

Hezlet also makes the sage observation that a mine and patrol effort designed to block a passage must develop enough of a threat in the enemy's mind (by sinkings) to make him abandon its use. If the mine threat is less than that, then the effort put into it will be rewarded by whatever few enemy sinkings occur. A small increase in effort, however, might cause total abandonment of all attempts to transit, and the effectiveness of the effort would jump from 10 percent or 20 percent to 100 percent! Of course, in the Dover case the preferable alternative was to go around through the North Sea and this led, as we have seen, to the enormous effort to block off that whole area. It is also likely that analogous reasoning from one minefield to another is not necessarily valid. It does not follow that a North Sea mine barrage would be successful just because the Dover Straits barrage was. In the North Sea case the surface-control measures could not be as stringent as they were in the Dover Straits.

With regard to the influence of mine warfare on decision making in the conduct of a war, Meacham [32] makes several points and illustrates them with examples. He says that whether a given mine threat will or will not close off a passage depends on whether there is an alternate path. In the case of the Dover barrage there was the North Sea option open to the Germans. If there had been no such option and the choice had been either to accept the Dover losses or give up the submarine war and hence the whole war, it is highly likely that the Germans would have continued using the Dover Strait. An example of no option occurred in the mining campaign Operation Starvation against Japan in the Shimonoseki Strait. In this case there was in effect no alternate route, the movement of cargo was essential, the alternative being starvation, and therefore the ships tried to make the passage even though unswept or unsweepable mines were known to be there. Under this circumstance it was impossible to make the threat high enough to produce total blockade. This fact led the minefield planners to go for lower sensitivities so that the activation distances would be less and the damage when mines were activated would be more severe. In other words they went for sinking instead of scaring.

Another point made is that in all the operations cited—Dardanelles, Dover, Japan, and Korea—there is no evidence that any of the commanders thought of a threat in terms of a probability. The risk turned out to be either acceptable or unacceptable, and so the threat became in effect confused with the conclusion derived from it, which was not a probability matter—one decided either to go through or not to go through, either to sweep the channel for a longer time or not to, corresponding to an unacceptable risk (i.e., 100 percent) or an acceptable risk (i.e., 0 percent). The so-called psychological effects of mine operations are also of great influence in decision making, whereas in fact one's opinion of a situation does not alter that situation. The difficulty is that the real situation is not known and is merely sensed from various hints, observations, or intelligence. The violent destruction of a minesweeper with loss of all hands before your very eyes is a far more effective warning than merely reading a tabulation of losses in the morning bulletin. Meacham suggests that such an event at Wonsan led to several additional days of sweeping, thus delaying the landing of troops. The importance of that delay would have to depend on what advantage was taken of it by Chinese invaders in the north and would probably depend on other factors unknown to the local decision maker. For reasons of this sort we have to conclude with Robert Burns that "the best laid plans of mice and men *gang aft agley*." The rational course, however, would seem to be to estimate the probability of damage as best one can from what is known, to compute the cost of the resulting probable damage, and to compare that with the cost to you of events which the delay would cause or make likely. If there is no compelling reason to take risks, then it is foolish to do so. It is also foolish to avoid a lesser risk now if in so doing you ensure a larger disaster later.

7 Readiness and Delivery

Ready for what?

In linking readiness and delivery together we are implying that there is something that should be ready to be delivered when delivery is desired. The scope of "readiness" is considerably reduced when it is thus more or less limited to the condition of a stockpile—to whether the requested number of complete mines of a particular type can be assembled from the various components, and whether these can then pass whatever tests have been imposed in sufficient time to meet a schedule (imposed from the outside). In the early days of World War II as described by Ellis Johnson [23] the question of readiness was much broader, because mines not only could not be assembled in depots, they did not exist at all. Therefore, there was no readiness to carry out offensive mine warfare. In order for readiness of the simple sort to exist, it was necessary to visualize what sort of new mine could be developed and produced. Nothing could be done without effort and priority.

The control of this problem was ultimately in the hands of the Chief of Naval Operations, but because of the newness of this idea—the air-launched influence mine—in the framework of U.S. naval strategy, he was unaware of his needs. Clearly, in order for war material to be prepared for use there must be a stated need for this material and a plan for its possible use. There were no such plans, and in fact very little materiel in 1940–41 when the responsibility for operational planning was passed to the Fleets. The Commander-in-

Chief, Pacific Fleet, showed little interest in the development of strategic mining. Johnson says, "The Fleet's attitude was that depth charges and bombs, not mines, were really needed. It was stated that due to the great distances in the Pacific it would seldom be desirable for aircraft to use mines instead of bombs." [23, P. 45]

It was this hen-and-egg situation which had led to the beginnings of operations research on the part of Johnson, Quimby, and Michels at NOL in 1940, and did provide in that case an answer to the question whether developments are led by tactics or by technology. It would appear that the technology, if new, is initially unknown to the tacticians. In order for a requirement to be handed down, however, so that work toward recognized goals can proceed, it is necessary for the tacticians and operators to understand the issue and impose the requirement. Johnson says, "Early in 1942 Rear Admiral R. K. Turner, Assistant Chief of Staff (Plans), advocated the offensive use of mines by the Fleets. In response, Readiness pointed out that the mine program should be comprehensively reviewed and a specific program should be imposed on the Bureaus to facilitate procurement. This was the core of the problem. The materiel Bureaus could not develop mines without operational direction because of the tremendous competition in the field of ordnance with guns, torpedoes, depth charges, etc." [23, P. 48]

In March of 1942 with growing realization that the direction of mine warfare was fragmented among far too many agencies, a conference of interested parties—CNO, Fleets, Naval Districts, and Bureaus—was called. A consensus, probably influenced by the opinions of Admiral W. A. Lee (Readiness), was reached that a Mine Warfare Section should be set up within CNO. This did not actually happen, but the existing Mine Warfare Section of the Base Maintenance Division was assigned informally to be responsible for "coordination of all mine warfare activities." When a new chief of staff, Admiral Edwards, came in September 1942, he took steps to assign the responsibilities for various aspects of mine warfare as follows: to Plans Division, offensive employment of mines and review of Fleet plans; to Readiness Division, readiness of operating forces, particularly training, for offensive air and subsurface operations; to Operations Division, specific directives for offensive employment of specific types of mines for immediate tasks and areas; to the Mine Warfare Section, supply and production of materials, plans for defensive minefields, supervision of the Mine Warfare School, coordination of minesweeping matters, mine and bomb disposal, degaussing, and coordination of the activities of the Bureaus, Coast Guard, and Army in mining activities. Admiral Farber, the previous chief of staff, preferred to see the

Mine Warfare Section retain responsibility for the coordination of all mine warfare activities, with liaison officers from the CNO added. He commented that otherwise the situation would be "similar to that which obtained prior to the war, namely, divided responsibility for the employment of a type of weapon with great potentialities." Although Farber's memorandum was issued in opposition to Admiral Edwards's plans, it was never fully implemented—which Johnson says was fortunate since it would have brought confusion and duplication!

As these bureaucratic hassles over administration, cognizance, channels, procedures, authorities, and responsibilities continued, they fortunately did not always interfere with the "war effort." In fact, the enthusiasts and missionaries for mine warfare had done their work so well that they succeeded in annoying one of their own number, Captain McKeehan, who was also in charge of mine R&D in the Bureau of Ordnance. In response to charges that the Bureau had failed to develop and produce the necessary mines, McKeehan wrote that he was "tired of being told that we have not quickly enough passed from the stage when nobody wanted any aircraft-laid mines to the stage where abundant mines of various special types are ready to plant from all possible places." [23, P. 53] Readiness in this case seemed to be akin to the spoiled character in the play who said, "I want what I want when I want it." It is true, however, that the developers of a given product can pass very quickly from a position where much of their time is spent promoting their product to the happy stage when they spend their time explaining why it isn't ready sooner.

The obsolescent stockpile and training

The more standard and mundane problem of readiness occurred after the war was finished and there were a number of mines and spare parts left over. What to do with them? They were part and parcel of the stockpile, sent to one or more depots, Yorktown Naval Mine Depot, for example, where they were stored away in various states of preservation. It is true that this material can deteriorate with storage; some batteries have limited shelf life, and corrosion or just plain rust can keep parts from working. There is another sort of degradation which can make stockpiles less than attractive, and this is obsolescence. By this is meant the growing lack of effectiveness of the material when pitted against improved target types, or when compared with what could be accomplished by using newer ideas, materials, or potential developments. It is therefore very useful to have a way of sorting out the stockpile periodically and of using it up so that it does not remain as a permanent block to any new developments or procurements. An ideal way to do this in the absence of war is to conduct suitable exercises using the material for the purpose of training the operators and the planners and incidentally to check out the abilities of the mine depot organization to put working mines, such as they are, on the line. There is also another way to deplete stockpiles, and that is to sell them or give them to friends.

Frequently the conduct of training exercises can sound useful alarms as occurred, for example, in 1956 (as already noted) when Admiral Sanders, CoMinLant, complained to Admiral Withington, then Chief BuOrd, that he could not carry out his mission with the faulty mine materiel he obtained from Yorktown for his exercises. A study commission under Dr. Rolfe of Temple University was set to work to find out what the trouble was. Their findings led to some changes in the mine readiness responsibilities of the Depot, as well as a recognition that if you wanted more effort in checking, testing, and training, it would be necessary to plan for spending more money in these functions. There was a pregnant lesson, however, for the mine development organization, namely, that the mines should be designed to be more easily assembled—more like a store-bought television than a Heathkit. This was a long-range goal in that the existing stockpile had already been designed and produced and could not be altered. It consisted at that time mainly of World War II mines—the Mk 25, Mk 36, and some older mines. The Long Range Mine Development Program, started in 1949 with its goals of modular construction, interchangeability, and automated test sets had not yet produced any appreciable materiel for entry into the stockpile.

The Long Range Mine Program, in addition to these aims, had an all-important goal—namely, to be able to attack submarines. The influence fields of submarines are clearly smaller than corresponding fields of large surface ships, and the new mines would have to take this into account. This was done mostly by using combined fields, higher sensitivity, and more processing of signals. One could use rates of change, changes in the rates of change (first and second time derivatives of the signals), and different frequency bands to help in this effort. The mine exercise referred to suffered from semantic difficulties as well as material ones. One of the purposes of the exercise was to see how well the stockpile mines could function against submarine targets. Because the fields of these targets were small, a then current doctrine dictated making the mine settings more sensitive. This made them more vulnerable to the sweepers which were a part of the exercise. The result was that the sweepers got the mines, and the mines did not survive to get the submarines. In the presence of sweepers, the mines should have been made less sensitive so that the sweepers could not have accounted for all of them within the same sweeping schedule. There would then have been some mines left which still could have attacked the submarines when they came close enough—as some were bound to have done. It was apparent again that mine settings cannot be made solely with respect to the targets to be attacked but must also take account of the level and type of sweeping activity expected. The ensuing review of the whole situation at that time reinforced the idea that the old mines were not optimum for submarines and that the direction taken in the new long-range program was correct. One of the recommendations from the review was that research and development in new mine mechanisms should be increased.

At the time of the Rolfe report, the White Oak Laboratory pointed out to the Bureau that it would be cheaper to extend the responsibilities of the Laboratory to the stockpile itself than it would be to start up a new quasi-technical organization at Yorktown and provide it with duplicate test and simulation equipment, more people, and so on. We pointed out that the necessary technical background, development experience, and awareness of potential difficulties in mine equipment were available for free at White Oak; that it was a matter of knowledge, experience, and advice that was a large part of the missing ingredient. In other words we were suggesting that the mission or responsibility for the development organization should be extended to cover the life of the product in Fleet use. We felt that it would cost less in the long run, that it might have a beneficial feedback effect on future Laboratory developments, and that we would rather look after our products than see them possibly neglected elsewhere and not achieve their proper usefulness. Furthermore, the public relations aspects of such failures were not easy to live with.

This extension of Laboratory responsibilities toward service to the Fleet later became a great talking point in some areas and was abbreviated as a "cradle-to-grave" responsibility. On this matter of nomenclature I always preferred "womb to tomb," in spite of its possibly crass connotations, mainly because I felt that the Laboratory should also have a responsibility or a role in the conception as well as the birth. In the particular case of the mine readiness problem, it was the Bureau decision in 1956 (or '57) that the expansion of Yorktown would be preferable to a formal extension of Laboratory responsibilities. In this way the R&D effort would not be diluted (provided funds were not taken away), and responsibilities would be clearly assigned. Many people at the Laboratory thought this was a mistake, but in a sense a happy one because it relieved the Laboratory of a rather unpleasant responsibility. It must also be remembered that there is a natural tendency for Bureaus to divide responsibilities. If everything were assigned to one field activity, the Bureau would have less to manage.

For several years after this it was felt that readiness was improved, but at the same time there were fewer mine exercises. In this period, the late fifties, there was considerable urgency attached to the development of so-called practice mines for use in exercises. These were dummy-loaded real mines fitted with releasable floats which would come up if the mine were actuated either by a target or by a sweep. These floats were also provided with pyrotechnic devices—flares of various colors and durations—with requirements that they should be visible at night or in fog or in certain sea states. This work led to the development of a sizable section of researchers at the Laboratory looking for better pyrotechnic materials that would burn longer, be brighter, and have the right color. There were nearly as many projects in this section in response to various requirements as there were in the whole mine program. Much of the expanded pyrotechnic program years later was moved to the Crane Depot in Indianapolis on the ground that it was not in the NOL mission.

Vehicles and ordnance

Another consideration of growing importance was the logistic one vitally related to readiness. How, for example, would one get the mines to the airfields, and in what condition? Would there be advanced bases set up for offensive aerial mining, or would the final assembly be done in a depot here? These questions were very difficult to answer, of course, because there were so many possibilities and also because there was no real pressure to arrive at an answer such as would exist in wartime. Nevertheless, it became clear that the stockpile mines were not necessarily optimum for the aircraft that might have to carry them. If the mines were to be carried internally, as in the proposed Martin P6M, it turned out that the box-shaped tails which provided stability during the free-fall part of their trajectory interfered with the dense packing of the mines in the aircraft. These planes could carry the weight of more mines than could be loaded in. Consequently, the box tails were taken off in a retrofit design and an aerodynamically equivalent fin tail was fitted to the mines. The fact that the P6M never materialized as a plane is of course irrelevant. The increase in speed of the newer planes over the World War II planes, e.g., over B-29s, led to considerable retrofitting to develop "high-speed flight gear." Projects included parachutes that would stand higher stress, fairings to reduce drag in external carriage, lugs and reinforced cases to match the mines to the release racks (for external carriage) and other devices (so-called arming wires) to match the mine to the aircraft.

The appearance of higher-speed aircraft had at least two other effects with considerable R&D implications. One was the existence of a higher stagnation temperature to which the mine and its explosive load was subjected; and the other was the appearance of higher water-entry velocity and hence of higher impact shock which had to be dealt with by shock-hardening techniques. The high-temperature problem, particularly for bombs, led to the development at the Laboratory of a new explosive (DATB = Diaminotrinitrobenzene) which was not limited to the melting point of TNT, because it contained no TNT, and was stable well above 80°C. The water-entry shock problem was dealt with by a number of schemes, including nose-shape alteration, and the use of shock-absorbing materials and mixtures which also found good use in other ordnance projects such as Subroc and the Destructors. It always seemed somewhat anomalous that a given new (and hence high-performance) aircraft would be designed, built, and tested without too much attention being paid to the weapons that it was supposed to deliver. Frequently when the ordnance load was attached in the form of bombs, mines, or rockets, the high-performance features in terms of speed or maneuverability disappeared.

In fact, it was considerations of this sort which motivated the ill-fated joining of BuOrd and BuAir to form BuWeps in 1959. When (in 1966) the bilateral Navy became unilateral with new Systems Commands under CNM reporting to CNO, rather than directly to SecNav, it was necessary to dismantle BuWeps. In this process, in which BuWeps was pulled apart to form the Air Systems Command and the Ordnance Systems Command, the air-weapons part of BuWeps stuck with the new Air Systems Command, leaving ship- and submarine-launched weapons for the new Ordnance Systems Command. It was only a matter of time and the next reorganization before the OrdSysCom was absorbed into the ShipSysCom, the organization which provided the vehicles (ships and submarines) to carry the ordnance. The new combined organization is called the Sea Systems Command and is thus analogous to the Air Systems Command. And so now the Navy has no weapons organization at all, or it has two—depending on the point of view. Under this arrangement it is possible that future mines could be designed by different organizations, depending on whether they were to be dropped from an airplane or laid from a submarine.

A Mine Conference at White Oak in 1966 on "Mine Delivery Problems" noted that mines must be designed, tested, manufactured, and assembled to live through the delivery environment, whatever it may be, whether delivery is by aircraft, submarine, ship, or even by torpedo. Yet all this is in a sense peripheral to the problems of getting mines from the stockpile to the target area. These relate to plans for mine laying and the availability of vehicles for such missions. It would seem that the laying of mines is such an occasional sort of job that it is not realistic to say that it is the main job of particular vehicles. Vehicles are going to be used for something else when not laying mines. The problem is to see that they are not always doing something else and hence are never available for mine laying. I said on that occasion, "I have heard it said that since there are no plans to deliver Captor mines (which incidentally don't exist yet), there is no need to have them (either plans or Captors). This is a transparently specious argument but doubtless has always confronted a new weapon which might mean a change in the old way of doing things. If we cannot have special-purpose mine-delivery vehicles, then it is necessary to provide mine-laying capability to as wide a variety of vehicles as possible and hope that they will be so used by the naval commander when mine laying is called for. This is why it comes back to the appreciation of what mines can do, what plans have been made, and what materiel is available. The best mines in the world (and I think we have them) are useless if they sit in the mine depots when they could instead be measuring out the hours against enemy traffic." Other comments: "Mine warfare to be effective must be considered as a system—and it always comes to that no matter how hard we try to carve it up into pieces."

An interesting submarine-delivery suggestion was made which, it turned out, was not new. Some of the first German submarines and indeed the first Russian submarine in World War I were visualized and designed as minelayers. The British E class in 1915 had mine-laying tubes in their saddle tanks. "Because the Laboratory is concerned not only about the technical potentialities of mine warfare, but also the ultimate utility of the product to the Navy, we have perhaps overstepped the strict lines of mine development and asked, for instance, why a submarine could not be configured in such a way that it could carry large numbers of mines externally, strapped around its waist like cartridges in a belt, and lay them as an aircraft does bombs from its bomb racks. What law of nature says that submarine mine laying must be conducted through tubes which are exactly twenty-one inches in diameter?" Diagrams of the Russian Krab minelayer which carried sixty mines and passed them out the stern, German UC boats which carried mines which could fall through nearly vertical shafts, and the E class British minelayers may be seen in Anthony Preston's book [38], from which figure 3 is adapted. Duncan [12] describes the fitting of special moored mines, the Mk 11, to the USS *Argonaut*, the only U.S. submarine minelayer before World War II. The submarine, ordered to Hawaii, laid no mines because it was turned into a submarine cargo carrier early in the war, and subsequently was sunk by the Japanese. Its mine-laying machinery with rotating cages was considerably more complex than the earlier minelayers. Whether it was better is not clear.

The British view of the utility of the submarine as a minelayer has, I think, consistently been more affirmative than the U.S. opinion on this. Captain Cowie, in a message to the 1958 Conference on the Naval Minefield, wrote—perhaps somewhat paradoxically—first, that the minefield will remain one of the most effective antisubmarine weapons ever devised; and second that the submarine, in areas where she can operate at all, will, as a mine planter, remain the most effective means of exploiting the value of the minefield. In 1977, in commenting on U.S. mine-laying vehicles, *Jane's Fighting Ships* points out that "Although the covert nature of submarine operations makes them preferable in certain mine-laying situations, modern U.S. attack submarines have only four torpedo tubes and a limited number of reload spaces, severely restricting their capacity for tube-launched mines. The large B-52 Stratofortress bombers of the Strategic Air Command can also plant sea mines. The use of B-52s in the mine-laying role presupposes the availability of the aircraft for this purpose, the proximity of suitable air bases, the suitable location of mines, and no interference from hostile aircraft in the target area." All of which shows that there may be alternate solutions to the general problem of laying mines wherever and whenever one chooses, and there probably is no single solution that may be counted on generally.

Full circle?

The foreword to the Mine Advisory Committee (National Academy of Science) report on Project Nimrod graciously states that the decision to undertake this study was due "in large part, to the steadily deteriorating conditions within the mine stockpile and the overall mine program, as pointed out by Dr. G. K. Hartmann, Technical Director of the Naval Ordnance Laboratory, at the Minefield Conference in 1964 and to his recommendation that a broad-based long-range study be conducted to assist the Navy in deciding the role that mines should play in naval warfare during the forseeable future." [33, P. viii] The need for such a study, as seen from the readiness side of the house, was reemphasized in 1966 by Rear Admiral W. H. Groverman (ASW and Ocean Surveillance, OP-32) when he reported a CNO study on U.S. Mining to the Committee and scored the "excessive cost in manpower, maintenance, logistics, and storage required to keep the present mine stockpile in various stages of readiness, and the critical implications this could have on the operational utilization of the stockpile because of an excessive time-to-ready which characterized most service mines." What I actually said to the Conference in 1964 was somewhat reminiscent of the situation which Ellis Johnson deplored in 1940 (although his text did not become available until 1973). In effect, the developments dictated by the Long Range Mine Program had been completed, and the main question was to decide somehow what more could be done and to decide whether a requirement for it should exist. It was therefore essential to involve operational people in the study.

Some of my remarks from the 1964 Conference on "The Past and Future of Mine Warfare" summarize the status at that time, particularly with regard to the attempt to stimulate planning:

> Today I would like to say something discouraging about mine warfare—namely, that I am not sure that it exists! It is always considered as a part of something else. It is very difficult to find anybody who is making it his business to plan for the use of mines on a national or Navy-wide scale. Hence the title of our conference today is appropriate. It is of fundamental importance to understand the objectives of mining in a very broad sense before starting to lay out specific development projects, or before having detailed discussions about the best way, for example, to eliminate the effect of wave swells on pressure mechanisms. We have assumed all along that the basic objectives were understood and that there was a known and planned use for mine material. In fact, there always was a formally stated requirement for everything which has been developed.
>
> I am afraid, however, that our meetings today and tomorrow are not going to answer the question, "Who wants mines, of what sorts, and why, and how are they to be used and by whom, and when and under what circumstances of warfare?" That, by the way, is all one question and it needs to be answered. Our conference this year will provide another volume to add to the background material. It is important, of course, to consider what new vehicle types may be added in the future to the present list of targets for mines. It is important, also, to look at R&D objectives and other more specific questions. But the mine program in the Navy, whatever it is, ap-

pears to be falling apart because it is not being considered in its own right as an integral part of warfare on a global basis and as part of this country's limited war arsenal.

I would like to review very briefly the history of mine warfare. I will neglect everything up to the day in November 1939 when the first German magnetic mine was used. By 1942, degaussing—a British idea—had been developed into an effective countermeasure by the Admiralty and the U.S. Navy. Other influence mines appeared and in the next four years all sorts of tricks with corresponding countermeasures. In this country degaussing was developed to a very high degree of complexity; and in parallel with this, the Naval Ordnance Laboratory (NOL) developed the U.S. influence mines which were used against the Japanese Empire in the last stages of the war. These mines and the earlier ones were credited with sinking over $2\frac{1}{4}$ million tons of shipping and on any cost-effectiveness basis were extremely effective. They were not as dramatic as Hiroshima, but they probably had as much to do with the defeat of Japan. Who can say? This was the opinion later of many Japanese.

The war ended. Mines had been proven to be very effective against shipping. They could be laid from aircraft, from ships, and from submarines. They had galvanic, magnetic, acoustic, and pressure mechanisms in almost all combinations. What more, if anything, was to be done?

For one thing, the existing mines were rather tricky to assemble —they looked like experimental laboratory equipment and called for experts more than sailors. So one thing to do was to put the mechanisms and other components—clocks, sterilizers, etc.,—into neat modular packages and provide unique and foolproof cabling harnesses for assembly. Manuals needed to go with this, and the knowledge had to be developed of the firing widths and resistance to both environment and countermeasures. But the most important new point was that the old World War II mine mechanisms were not designed to respond to the slow submarine as a target. Thus was born the Long Range Mine Program in 1947.

The R&D program at NOL successfully attacked and solved the problem of mining the small slow submarine, and simultaneously improved the mine logistics, operability, maintainability, and all the rest. The Long Range Mine Program is over. We have as a result the Mines MK 52, 55, 56, 57. It is interesting to note that during the course of this work (which was never funded at an optimum level and which hence took about ten years to accomplish), the concept of the submarine threat for which this program was initiated shifted from a small, slow, conventionally powered submarine creeping out of its home port on the surface, to the large, fast nuclear-powered submarine deeply submerged in the high seas. The idea of mining has been extended to this newer target.

The Nobska study of 1956, which considered the impact of the atom on the submarine, also generated the idea of the String Mine as a means of dealing with the deep submarine. This, in turn, gave rise to the random buoy field for surveillance—a concept explored in the Atlantis study (1958)—and generated other ideas for large sonobuoys like Lolita and Nutmeg. NOL has studied them all, and a few more—the nuclear mine, and the Captor mine. We recommended the Captor as the best feasible cost-effective deep-sea mine. But now we are finding that there is still doubt about whether we should even want to mine the deep sea, or possibly have barriers, or possibly seal selected ports by a ring of deep-sea mines.

When we saw the end of the Long Range Mine Program and also the reoriented effort on new mine mechanisms suggested by the Rolfe Committee in 1956, the Laboratory (in 1960) asked itself, what now in mine R&D? We felt that extending the possible mining depth to say 400 fathoms would be useful—would open up possibilities for mining the GIUK line, the approaches to Murmansk and Archangel, the deep channels off Petropavlovsk, etc. We felt that the Arctic Ocean might be important. We felt that a small-craft mine for use in shallow water would be useful in limited war situations. We felt that mooring problems in deep water should be investigated for mine-like surveillance devices or for oceanographic data stations. Accuracy of delivery was important, leading to elimination of parachute flight gear and the development of shock resistant bomb-like mines. We felt that the role of the submarine as a minelayer should be examined. Does this call for a mobile mine or does it call for a Captor mine which could be laid with considerably less risk? Is there sufficient reason to have both or either? These are questions that the Laboratory cannot answer alone. And I feel they are questions that probably cannot be answered without Laboratory help. But the R&D program cannot be planned or executed without answers.

Today, as I look from White Oak out toward the west bank of the Potomac, I see, to modify a current phrase, a few isolated spectral line items, but not a broad spectrum in mine warfare. I am referring to the current roster of studies—namely:

a. Cyclops II, which considers the force levels inherent in a convoy passing back and forth across the ocean. There are various minefields and/or barriers postulated at either end of the trip.

b. A Center for Naval Analyses (CNA) amphibious warfare study—with mines in amphibious situations as a sub-study.

c. An Assistant-Secretary-of-the-Navy-R&D-requested study on R&D goals in ASW with ASW mines as a small annex in this.

Each of these studies considers only a part of the mine program as a subordinate part of a bigger and more general category. It is either a part of R&D, or a part of ASW, or a part of amphibious warfare, and it is still in its historical original role, a part of antishipping warfare, an instrument for blockade, a method of providing defense for ourselves in selected places and at selected times, as well as a weapon of offense, deliverable by any and all elements of the Fleet. It is a potential means of obtaining surveillance data, and intelligence data on target characteristics. It can be used in biological and chemical warfare. It can be used against helicopters as an antisweep weapon. It can be used as an expendable convoy protector, as a sabotage weapon, and to discourage swimmers. It can be used by submarines to inhibit aircraft search. It can be used against mobile land targets as an alternative to shore bombardment. It is all things to all men; it intersects nearly every aspect of warfare from broad national policy on down. And yet, it is fragmented and neglected.

In Fiscal Year 1963, the RDT&E Mine Funds were reduced by $2 million, i.e., cut from $8 to $6 million. In Fiscal Year 1965 we see another $2 million cut from $6 to $4 million in spite of the fact that $9 million was requested by the Bureau of Naval Weapons. If it is indeed true that the further development of mine types and capabilities is not desired or required, then this decision should be taken rationally, in full view of all the pertinent factors—but it should not be done as a matter of default. In these rational days, the allocation of national and Navy resources is made on the basis of cost/effectiveness. I believe that many aspects of mining will stand alone on this basis. What is needed is some attention to the whole question. What is the future Navy involvement in all aspects of mine warfare?

In order to answer all these questions, I submit that a Navy-wide study is required on a mine program basis, initiated by the Chief of Naval Operations and with adequate representation and schedule to permit meaningful work to be done. The study should consider national objectives, strategic and tactical matters, targets, force levels, operational planning, organization and funding, training, logistic support, and R&D—both continuing and new system starts. In order to do this effectively, the whole mine community must participate—namely, CNO (OP-09), CNA (OEG, NavWag), OP-03, OP-06, OP-07), CoMinLant, CoMinPac, BuWeps, BuShips, ONR, Mine Advisory Committee, National Academy of Sciences, USW Committee, DDR&E, NMDL, and, of course, NOL, White Oak. And I may have overlooked somebody. We must have planners, directors, users, researchers, and developers.

Nothing good would come from such a conglomeration, you will be saying. On the contrary, nothing good will come without it. There is no mine czar or program director who has the responsibility for mines and a large mine budget in his hands and who will fight for it. This may be an organizational weakness, but it may also be a strength because it tends to prevent an ingrown and parochial viewpoint and gives the opportunity for many minds with diverse backgrounds to contribute. But if we do not take advantage of this opportunity, what indeed do we have? Perhaps NOL has been too long in the mine business and has become ingrown on the subject. Perhaps NOL and NMDL should exchange mine missions. I understand they have recently invented a new mine mechanism which we overlooked, and I am convinced that mine hunting is the only way to deal with mines and that sweeping is futile. Or perhaps the time has come to put an end to mine warfare in a future of deterrence and disarmament and beat our mines into swords or plowshares.

Does the unexpected always happen?

While all this planning and studying was going on, the unexpected processes of technical serendipity were also going on quietly in a remote corner of the Laboratory. The thin-film magnetometer was coming into being and as a result it was now possible to build a magnetic mine-detection mechanism into the size of an eight-ounce beer can. It was owing to the efforts of a very few people, principally Gene Beach and Charles Rowzee at the White Oak Laboratory, that the so-called Destructor Adaption Kit (Mk 75) was developed. This kit in effect turned the ordinary low-drag aircraft bombs of the Mk 80 series into magnetic mines merely by inserting the mine-mechanism package instead of the fuze-mechanism package. The key component was the Firing Mechanism Mk 42. The resulting mine was called a Destructor for use in Vietnam, presumably because it was all right to destroy but not all right to mine. Although the mining of Haiphong Harbor toward the end was widely advertised, it was actually done with 11,000 Destructors and only about 100 Mk 52 mines. The use of Destructors before that had exceeded a quarter of a million.

This was more than all the offensive mines used previously in all our wars. A total of about a third of a million Destructors had been manufactured. These Destructors could also be dropped on land as well as in canals and rivers. Their principal targets were truck traffic and river boats. In performing interdiction they were later assessed as effective. They became decisive when they were given larger ships as targets. In the early days of their operational use their very ease of assembly militated against their most effective use as mines, because it was easy for their users to view them as bombs with delayed fuzes rather than as mines. If the tactic was to hunt out the target and attack it with Destructors, then the benefit of mining would be lost. The enemy would know where the mines were and would try not to stumble across them. Later on the Destructors were laid at random times at night to defeat mine spotting. Precisely what tactics were used and what effectiveness was achieved in specific instances were examined after the war by the Defense Intelligence Agency which confirmed the effectiveness of the Destructors. However, the remarkable thing, I think, is that the goal of easy, immediate mine assembly with complete compatibility with delivery aircraft was achieved overnight and with very little comparative effort, although the efforts on the part of a few, as usual, were huge.

In the meantime, a debate continued about the desirability of building a new mine (called Quickstrike) for riverine uses, which would have a larger explosive weight fraction than the old general-purpose bombs, and which might therefore not withstand ground impact, but which would be more suited to water-use options. There were, of course, cognizance arguments as to the propriety of the Navy carrying on ground strikes. The aerial land-mining mission seemed to be a new military option which clearly belonged to the Air Force. After the initial breakthrough leading to the simple "destructor," it became clear that a new sort of weapon for land warfare was feasible —namely, one which would truly have the characteristics of a mine operating underground instead of underwater. This weapon would have to be shockproof so that it could have a free fall, and an underground trajectory so that it would not go too deep nor make a big cavern where it hit. It would have influence mechanisms, actuation counters, sterilization devices, and so on. Its use would be governed by mine warfare considerations, not bomb warfare considerations. It would not be necessary to find the traffic to attack it, only the routes the traffic would take. Land mines would be dropped in locations where defenses were at a minimum. The potential of this concept—

aerial land mining—was not developed at the time of the Vietnam War. I have no doubt that in time weapons of this sort will be developed by the Air Force or the Navy or perhaps jointly, and added to our various other capabilities. Such weapons should be quite useful, for example, to NATO.

8 Strategy and Tactics (or Vice Versa)

Purpose

The fifteenth Minefield Conference held at White Oak in 1972 was entitled "Strategy and Tactics." These topics probably at one time had clearly different meanings, but nowadays the advent of nuclear weapons has muddled these meanings. Possible changes in mine capabilities have caused further confusion. It has been suggested, for example, that mines could be more useful if they had a sort of tactical versatility—i.e., if they could be laid at one time and turned on at another, not in a prescheduled fashion (called delayed arming) but on demand by some sort of communication to the minefield. And while we are about it, the field could also be turned off on demand. This would allow all sorts of "tactical" events to occur. John Keegan in his interesting book *The Face of Battle* [24] comments on the usefulness of military history in understanding events and the determinants of men's behavior in battle.

As a first step, let the student get hold of the distinction between strategy and tactics, say some academicians. But, says Keegan, this distinction is as elusive as it is artificial. This is comforting news to an amateur who had reached the same conclusion. In thinking about the use of mines, some have said that dropping a mine from an aircraft is not a very exciting military mission—nothing goes boom; there are no immediate results, and in fact what results there are appear much later in the form of statistics (many of which can be argued!). The delivery crew do not have to make an overt attack against a target that can fight back. This might suggest that a contact with the enemy demands the exercise of tactics, whereas an action carried out without such contact is relegated to strategy. In 1972 I said, "I am not sure that any given event or series of events can be clearly classed as belonging to one or the other. It seems clear that 'strategy,' however, is related to a long-term plan, related to attrition, depletion, and is what you do to win a war; whereas 'tactics' refers to short-term events, the achievement of specific objectives, or what you do to win a battle. The use of the term 'strategic' applied to nuclear war tends to confuse the issue slightly, because the war in this instance will, we presume, be over quickly. Perhaps a nuclear war would be both strategic and tactical at the same time." On this topic George R. Lindsey says:

> Today the term "strategic" has been appropriated for the concept and operation of nuclear deterrence. The ultimate sanction of deterrence, the all-out nuclear attack on population and industry, is certainly "strategic." The weapons to deliver the attack, the intercontinental ballistic missiles, long-range bomber aircraft and ballistic missile-firing submarines (and their missiles) are "strategic offensive weapons," while the systems opposing them such as ballistic-missile defences and continental air defences are "strategic defensive systems." Those anti-submarine systems which can be used to detect, track and destroy ballistic-missile-firing submarines are also classified as strategic defensive systems, at least when employed for that purpose. However, the very same anti-submarine systems can be used to protect merchant convoys, troop ships, amphibious assault groups, or carrier strike forces against anti-shipping attack by submarines. These operations are now classed as "tactical anti-submarine warfare" (ASW).
> The major difference between tactical and strategic ASW lies in the time, place, and circumstances of the encounter rather than in the equipment and methods employed. [26, P. 32]

From this it would appear that the designation is almost subjective depending on the intentions of the attacker, or in some cases on the unexpected outcome of an attack. Returning to the area of mine warfare, even though now and in the foreseeable future mines will not use nuclear warheads, principally because their wide destructive underwater range can be achieved by giving mines a limited mobility, it seems unreasonable to exclude mines from the area of strategy, whatever that means. The nomenclature tends to become rather confused and is possibly best avoided altogether. Deterrence of nuclear war by threat of nuclear retaliation is strategic in nature. If it should fail, then the strategy has failed with it. In the meanwhile we have "tactical nuclear weapons" for air defense at sea, and for battlefield use. Therefore, the equation "all nuclear = strategic" is not correct. Neither is the equation "all strategic = nuclear" correct, as the example of mines shows.

The development of longer-range detection and classification mechanisms and the attendant possibilities for communicating with mines or for mines to communicate with us, have led to some intermingling of the strategic and tactical labels. The possibility of minefields that can be reliably turned on or off by remote command, the use of mine technology to report the detection or destruction of a ship or submarine, providing mines with motors so that their destruction radius will continue to match their detection and classification radius, the attendant possibility of using fewer mines and hence of laying a field quickly for a specific or tactical purpose—these are examples of the extension of mine warfare into the tactical arena.

Having gone this far, it may be useful to remember the very first proposal to use mines in Europe, made possible by the early development of insulated wire associated with the telegraph. In 1812 Baron Schilling placed a gunpowder charge on one side of the Neva River with the firing switch located on the other side. The idea was that when the enemy came up to cross the river, the defenders could command the mine to fire from a location removed from the immediate action. This would be classed as a tactical operation, planned in advance, and made possible by a remote command and control system. Whether a command and control operation can be done at great distances or in deep water depends on its technical feasibility. If it becomes technically feasible with suitable reliability (and safety), then someone will doubtless find a clever use for it, perhaps in a situation that was not anticipated.

Matching mines to targets; the calculation of threat

In planning to put an actual minefield in a specific location, it is essential to decide what the targets are and what sort of result is to be achieved. It has been pointed out that if the targets are warships rather than merchant vessels, a smaller minefield threat is likely to hold up operations. Commanders are loath to risk their important warships, because they are relatively scarce and irreplaceable. It is also possible to alter the type of damage which will be incurred. If, for example, it is desired to cause the sinking of large vessels rather than to merely damage them, then the mines will have to be set coarsely—i.e., set to require a larger signal to initiate them. This will mean that for the same number of mines there will be fewer explosions, and some small ships will be able to get through unscathed. The total harm done to the enemy, however, will be greater than if the same number of mines were set with high sensitivity, provided, of course, there are some large ships in the traffic.

This kind of fine tuning, in adjusting the kinds of damage, is to be carried on after the basic threat of the minefield against its principal target has been determined. Such a determination will enable the planner to know how many mines at their basic settings he will need to meet his tactical (or strategic) objectives. The threat calculation is very simple. Assume there is a channel of breadth B and undetermined length through which a ship must pass. Assume there is one mine at random in this channel with an influence range of R feet. The charge weight in the mine is large enough so that the explosion at R feet will sink the ship. If the ship comes within this range, the mine will go off and the ship will be sunk. If the ship is outside this range, and the mine has a perfect discrimination, it will not be actuated and the ship will proceed scot free. The width 2R is then the same as the firing width w associated with unit probability of firing discussed in chapter 3. If the ship's track through the channel is also at random, then the chance of actuating the mine is $\frac{w}{B}$. There will have to be a correction for the beam of the ship, and of course the ship does not pass randomly through the channel, because it will avoid the extreme edges even if everything is of uniform depth (which it is not). Ignoring such corrections, the chance of avoiding this mine is $1 - \frac{w}{B}$. The chance of avoiding n randomly placed mines is $(1 - \frac{w}{B})^n$, and the chance of actuating at least one mine is $1 - (1 - \frac{w}{B})^n$. This is then the chance of being sunk and is also the threat of being sunk. If the influence range is taken as somewhat larger, associated with a smaller degree of damage than sinking, then in general we can write

$$T_x = 1 - \left(1 - \frac{2R_x}{B}\right)^n$$

where T_x is the threat associated with a degree of damage x (sinking, critical damage, moderate damage, light damage, shock damage) and R_x is the horizontal range at which such damage occurs and is related to the influence range that has to be chosen. The sensitivity settings to achieve this influence range depend on the charge weight for the mine, on the charge depth, and on the ship size. The width of the channel is B and the number of mines in the channel is n. After the settings have been determined, the threat can be increased only by increasing n.

If one wishes to calculate the threat for submerged submarine passage, the same method of calculation holds except that the submarine sweeps out a vertical area of the passage rather than a horizontal line as occurs with surface ships.

In *Mine Warfare, History and Technology* [16], an estimate was made of the threat of the North Sea Barrage in 1918. The threat of the American section of the Barrage, which was 134 nautical miles across and contained 56,611 mines, was estimated for submerged submarine passage. This threat for a one-way passage turned out to be .45, assuming that all the mines planted were actually working. This was known to be an overestimate. Some 4 to 8 percent of the mines prematured soon after laying. If any of the automatic features failed to work, then a floating mine, a sunken mine, a dead mine, or a premature firing could occur. All of these things were observed. Nevertheless, the field sank six submarines in the last months of the war, says Hezlet [19], and damaged at least six more which had to return to port. Submarines that made it to the Atlantic were delayed by picking their way through the weak areas of the field.

It is important to note that the threat calculation could be entirely irrelevant if there were a safe alternate path known to the Germans, and there was. Surface passage on the Norway side and on the west near the Orkney Islands was entirely possible without risk from mines, provided the mines stayed moored. The British, who were in charge of this 70-mile width of the Barrage, felt that the passage should not be entirely shut off to surface traffic. Mines were also kept out of the Orkneys proper so that passage between islands for British units was safe. Consequently, there should be no connection between threat estimates for a portion of the Barrage and the actual record of casualties. According to Hezlet there were 121 U-boats at sea up to the end of the war. If we assume the tour at sea for them was three months, then there should be forty coming back every month and forty going out to replace them, or eighty transits per month. If the effective operating life of the minefield was the last two months of the war, then 160 transits would be expected whereas about twelve casualties occurred. This gives an actual threat of 8 percent compared with 45 percent estimated for a part of the field. There is no reason why these numbers should match as we have said—and indeed it is hard to see why there should have been any casualties at all if the alternatives were known to all the German submarine commanders.

Observation and replenishment

It should be clear from all that has preceded that in fact no two mine operations are exactly alike. Nevertheless, we like to arrive at generalizations no matter how narrowly they may be applied. In any minefield, whether defensive or offensive and whether in narrow waters or in broad areas, the mines must be present and working in the numbers necessary to maintain the desired threat. If an offensive minefield, for example, is laid by air (which means, incidentally, that the laying forces have at least local control of the air), then it will be necessary to have an idea of the attrition suffered by the minefield itself under the combined influences of countermeasures and traffic. It would be very useful to know how the traffic itself has been cut down and how much sweeping activity there has been. This information would be useful in assessing the immediate effectiveness of the operation (without waiting for a postwar post mortem). It would also help in deciding how many mines needed to be added to the field and how often to maintain the desired effect. In other words there is a definite need for minefield surveillance, especially in areas which are far from home or nominally under enemy control.

If the measure of effectiveness is a decrease in surface traffic, it is possible to monitor this traffic by means of satellite surveillance, now available to the minefield planner—an example of the influence of a technology which arose completely outside the perceived needs of mine warfare. If, however, the purpose of the minefield, say a barrier of Captors, is to stop submarine traffic, then some undersea surveillance method must be used. Such systems presently abound— consisting of bottom-mounted listening arrays that are wired to signal-processing devices and print out information by the yard. Such systems are not portable and require considerable lead time for installation. However, it would be possible to have smaller and more localized systems that could be portable, need not be bottom-mounted, and that could yield information on demand as certain large sonobuoy systems have been visualized to do.

In designing such a system it would appear that the mine technology of deep mooring, acoustic detection, and information processing would be valuable. The coordination of equipments and information is difficult if there are security-tight compartments separating mine development, mine operations, surveillance, and intelligence. The use and effectiveness of the Destructors in Vietnam was not generally shared with the development agency during the height of their use. Perhaps the matter of replenishment can be handled as a matter of policy—i.e., every so often on a slightly random basis a few sorties of mine-laying planes will go forth to replenish perhaps 1 or 2 percent of the mines in the field. In fact, in the starvation operation against Japan in 1945 it was found that frequent small replenishments were more effective in reducing traffic than a few large ones. This was because the Japanese tried to do some sweeping after every mine-laying mission. It is even possible that one could lay a few easily sweepable mines each time, so that after they had been detonated and nothing else happened the enemy would be lured into sending a large or valuable target through the field. In this campaign it was possible to keep careful track of the mined areas by extensive aerial photography, which furnished information as to the extent of the sweeping effort and whether the areas had been opened to ship traffic.

Law of the sea and other matters

As worldwide demand for food and mineral resources increases with growing world population, men are being driven to the sea in search of some of these things. The doctrine that the sea should be open and that it belongs to everyone or no one is slowly giving way to a more complicated one. Considerations such as the proximity to the coastline, the existence of prior or extensive undersea investment, the possibility of attack or defense, the ownership of shore establishments, and the existence of treaties all play a part in developing the new body of agreement or opinion which is called the law of the sea. That there have been now dozens of sea conferences among half a hundred nations for months at a time without having really solved all the problems attests to their difficulty. Should nations with no present coastline be forever excluded from sharing in the world's supply of undersea oil? How far from a coastline should the fish be considered the property of the adjacent country? If one asserts that the deep ocean floors mark the part of the earth's surface which should be shared by all mankind, then it can be argued that the continental slopes extending from the shore line down to the ocean floor are actually and geologically part of the continent to which they are attached. The exploitation of mineral resources within them should be done from the continental base and under the control of the continental power unless, of course, it is unable to do so and signs a lease to someone who can. It is fairly clear, for example, that the United States would not want a foreign power (other than Mexico) to begin drilling oil wells off the Louisiana coast even 200 miles out, whereas we might take a less rigorous view of a foreign power catching fish 200 miles off the coast of Ecuador.

If and when agreements are reached on some of these questions, there will be a protective or regulatory function assigned—presumably to the Navy—to see to it that our rightful investments are rightfully protected. It is thus conceivable that naval mines and mine technology in the sea environment may have a role to play in this effort. In other words, we will have to examine the effect of undersea mining on mining and vice versa! The point was enunciated in the Truman Proclamation of 1945 as follows: "Self-protection may compel the coastal nation to keep close watch over activities off its shores which are, of the nature, necessary for the utilization of the mineral resources lying reasonably beyond the shelf." Self-protection becomes shelf protection (we will have to be selfish to protect our shellfish). It seems that the Geneva Convention of 1958 reached a definition of the term *coastal shelf*, which is taken to include all the sea bottom between the shore line and the 200-meter depth contour, and this applies to islands as well. In the case of two or more nations having coastal shelves that overlap under this definition, the median principle would hold—i.e., the boundary between adjacent shelves would be the locus of points equidistant from the coasts of the nearest nations. The median-line principle is useful for settling ambiguities which arise in the North Sea, or perhaps the Arctic or in the Sea of Japan, and is also a useful rule to apply to our own Atlantic, Pacific, or Gulf coasts. We have very recently seen, for example, that there is a junction of fishing rights between us and Cuba which has led to the opening of discussions between the two countries.

In 1969 the Stratton Report of the Commission on Marine Science, Engineering and Resources entitled "Our Nation and the Sea" suggested some modifications in the definition of the coastal shelf—namely, that sovereignty of the adjacent nation over the seabed extend out to the 200-meter isobath or fifty nautical miles, whichever gives the larger area subject to the median-line principle in case of overlaps with adjacent countries. This, for example, would extend our Pacific coastal shelf area from 25,000 square miles to the order of 100,000 square miles. The Stratton Report suggested that the slope between this coastal shelf zone and the sea floor or the 2,500-meter isobath which is roughly where the sea floor begins, be designated as a zone within which the adjacent nation would control the registration of claims on resources. Beyond this zone a proposed international commission would control and record claims on a first-come-first-served basis. In a rough sort of way, one can say that the typical coastal shelf is about fifty miles wide, goes out to a depth of 200 meters, and has a slope of about one quarter of a degree. The typical continental slope is also about fifty miles wide, goes from a depth of 200 meters to 2,500 meters, and has a slope of about three degrees.

However the law of the sea turns out, it is clear that there have already been several billions of dollars invested in offshore oil wells and pipelines alone. These undersea installations are vulnerable to sabotage or to underwater demolition of a covert nature. A first requirement for these properties is adequate surveillance and protection. If the intruder is quiet or hard to detect, many short-range detectors are necessary instead of a few long-range ones. As the signal becomes smaller and the ambient background, both natural and man-made, becomes relatively larger, the range of the intended long-range (passive) detectors becomes less. It is, I believe, a losing battle to try, by increased array size and more subtle signal processing, to maintain long-range detection of targets which can become quieter and quieter. One is forced by both economics and the desire for a decent probability of detection to the use of locally dispersed systems. The mine technology makes possible a ready-made, portable, deployable, low-cost surveillance system that will cover any specified spot or area in any water depth, up to 2,500 meters, certainly, with any specified probability of detection. It will do this with local signal processing, thus eliminating the massive problem of transmitting signals to a central shore-based processor.

The idea of remote sensors on the battlefield was tried in Vietnam in the so-called McNamara line at a cost of more than one billion dollars and with the impetus of top priority. The first sensors used, oddly enough, were taken from underwater use in sonobuoys and used to "bug" the battlefield. There were difficulties in the concept in that initially an aircraft summoned to a place by a signal from a detector usually arrived long after the person or troops detected had departed. The same ideas could, however, be developed for bugging the continental shelf or slope in areas of large investment, and this would be better than having no information at all. Mine material designed to work under water would be readily available. Information gained by these means could be relayed by bottom-laid cable or by radio link from buoys back to mainland command posts where appropriate action could be taken. In short, one could hook up the mine mechanisms to transmitters rather than to detonators.

There is another very important potential interaction between sea mines and the general use of the oceans, and this lies in the preservation of our sea-based deterrence. The first sea-based deterrent system, Polaris, had missiles with a 1,500-mile range. It was therefore criticized on the ground that the submarines had to come close to a continent to reach all interior targets, and were therefore more vulnerable than if they could release missiles from a wide area of ocean. The later system, Trident, now uses much longer range missiles, and incidentally much larger submarines to carry them. This means that our deterrent force is more invulnerable in that the submarines are harder to find in the broad ocean. It is also possible for them to lurk off our shores where they can be protected by more active measures. It should be possible to use both safeguards—i.e., the invulnerability due to the fact that Tridents are hidden, moving, and not targetable, and the invulnerability due to the protection afforded by having a large sanitized zone from which enemy submarines are excluded and within which our own deterrent force remains both untargetable and unassailable. The key to this situation is deep mine protection along the outer edge of the continental slope to exclude enemy submarines in conjunction with active ASW measures beyond this belt. There would, of course, be no ASW activity within the zone where our own submarines would be safely cruising. A stable nuclear deterrent of this sort, representing a combination of technologies and of interests would, I believe, meet all the requirements for deterrence and be a very cheap alternative on a national basis to what we are now doing.

9 A Catalogue of Specialties

Harbor defense

One of the first examples of harbor defense occurred in 212 B.C. at Syracuse when according to one story, Archimedes focused sunlight onto the Roman ships when they were at bowshot distance and set them on fire. A more likely version is that large catapults designed by the sage had hindered the Roman attack by hurling quarter-ton stones into their ships. Thus was born the coast artillery which flourished with the development of guns and cannon in the Middle Ages and right up to our Civil War. Fort Monroe at Old Point Comfort, Virginia, is but one of a series of coastal defense forts built before 1860 to French design, and was meant to cut short and out-gun any attacking naval force which might seek to land. Indeed, during the Civil War itself the North held the fort, and yet was never able to make an attack on Richmond or points south using the access of the James River.

We have seen that Samuel Colt spent a great deal of effort and enthusiasm to show that harbor defense could be accomplished using control mining rather than coastal artillery. He tried to demonstrate that control mining was technically feasible, and he tried to argue that it would be cheaper and just as effective. He failed in these attempts for various reasons, his excessive secrecy for one, and the natural skepticism of any group asked to divest itself of a large well-established system in favor of a smaller unproven one. In fact, it took a great many years for the U.S. Army to embrace the use of control mines as an adjunct of harbor defense.

Eventually there appeared at Fort Monroe a mine-development activity whose business it was to improve the Army's harbor defense mines! Up until World War II these mines were moored contact mines which could be armed from shore but which required an impact with the target ship to set them off. These mines were troublesome to friendly traffic, because there was a tendency for the mooring cables to get fouled up in the propellers of the ships which occasionally ran across the mines. The Army Development Laboratory at Fort Monroe had the project of replacing the moored mines with bottom mines equipped with magnetic detectors. This group worked closely with the Naval Ordnance Laboratory's staff and developed a controlled mine with a magnetic detector which could either signal to the control station that a ship was present, or could fire the mine. This development occurred during World War II. The mines were not used extensively as far as I know. Nevertheless, the policy in the army was to use control mines for harbors rather than coastal artillery.

At some time after the war, the cognizance for harbor defense mines was passed from the Army to the Navy, and this started a new development in control mines which was completed in 1956. This new system developed by NOL at White Oak was called the Control Mine Mk 2 and consisted of bottom mines, improved magnetic detectors, cable networks, and control centers. The system successfully went through a rigorous technical evaluation and then was put into production to the extent that harbor defense systems were available for New York, San Francisco, and several other ports. These systems were stockpiled—i.e., stored, never used—and now are obsolete and even may be junked. The latest and best control-mine system went the way of Archimedes' catapult. The increased range available to weaponry, and to the detection of surface ships, which came about with the development of aircraft, led to its oblivion. In this respect the control mine and the huge gun mounted on the huge battleship suffered the same fate and for the same reason.

This is not to say that defensive control mines in the hands of many of the world's countries could not be useful. Many such countries are not blessed with modern aircraft and their weapons. For them the control mine could be an important item for defense. In fact, NATO interest and concern about harbor defense is also, understandably, much more lively than ours.

Riverine and land mining

The first examples of mining occurred in rivers—the Delaware at Philadelphia, the Neva at St. Petersburg, the Mississippi at Vicksburg, the James, the Yazoo, and the Red. Inland waterways are important means of communication, even though in the United States, for example, canals were largely replaced by railroads (which in turn have been partially replaced by trucks). In World War II, however, the rivers and waterways of Europe formed a worthy mine target. Throughout the summer and fall of 1944 American Liberator bombers and British Wellingtons laid mines in the Danube, 1,200 by October, sinking over 200 ships and barges and damaging many times that number. Information from a Mine Warfare Training Center course on Aerial Mine Warfare [13] points out that this is the highest casualty rate per mine ever achieved, but does not mention what mine was actually used.

The combatants in the Vietnam War also depended heavily on inland waterways, which were pathways to ease communications if one wished to travel along them, but were also obstacles if one wished to go across them. The idea was generated at White Oak around 1965 that a riverine mine planted from the air in canals or rivers at such crossing points could serve to block both the waterway itself and the highway or ferry going across it. Such a mine should be free-fall in order to increase the accuracy of delivery and decrease the chance of discovery. The Destructors, which were low-drag bombs fitted with compact magnetic detectors, were the natural result of the conjunction of the tactical need and the technical possibility. Of course, in actual use a great many Destructors were dropped on land at river crossings.

The idea of a land mine configured for the purpose so that it would bury itself in the ground without damaging itself was an immediate possibility. This sort of mine could then be used along highways and could be definitely set for the proper sensitivity to respond to trucks. A few new disciplines were necessary—namely, terradynamics, i.e., the dynamics of earth entry and penetration, and the determination of the signatures of vehicles and their vulnerability to nearby buried explosions. This was not an entirely new field for the laboratory, because back in World War II there had been a small effort to build a railroad mine. This was an explosive charge to be buried along a railroad track and equipped with an influence device which was in effect a trembler switch. The idea was that as the train approached, the oscillations of the switch would increase until the necessary level was reached just as the locomotive was nearest the charge. This project, never at high priority, was not pursued with the vigor necessary to overcome its many inherent shortcomings.

In the late 1960s the Laboratory again was carrying out feasibility studies and doing some experiments to determine a suitable landmine shape, in order to generate some official interest in what could be a very important new capability in warfare. For example, instead of having an elaborate sensor surveillance system (McNamara line) to summon aircraft to the scene of action, it could be far more effective to use the aircraft to lay land mines and let the traffic find them. In this way the delay between detection and destruction would be a certain few milliseconds instead of an uncertain half hour. Apart from the doubt about the ultimate practicality of this idea and the feeling that a new development of this sort would not be available in quantity before this already overly long war was concluded, there was the conclusive cognizant difficulty that a Navy outfit was trying to do something that was an Army responsibility to be delivered by the Air Force. Furthermore, because the idea was new, there were no plans for the conduct of the development and worse, no budget even to pay for making plans.

Mobile mine

The mobile mine is certainly a specialty among mines, partly because it is designed exclusively for submarine delivery and partly because the number of them is very small. The use of submarines for mine laying is also a rather special thing, because it can be well argued that a submarine would spend its time better if it carried torpedoes and used them to sink ships at the rate of two or three torpedoes per ship rather than to carry mines and get only one ship for every ten or twelve mines. In World War II when operations researchers discovered that our submarines were returning to port after months at sea still carrying unexpended torpedoes, it was possible to argue that a submarine should carry a few mines with its torpedoes and leave them behind in suitable places rather than to come home with unexpended weapons.

The mobile mine, however, has a more special purpose than that. It is addressed specifically to the mining of a channel or port which is not accessible to our aircraft because of the control of the air by the enemy at that location. If mining is to be done in this case, it must be done covertly by submarine from as safe a distance as possible. The mobile mine is in effect a mine which is launched from a submarine torpedo tube and is carried by a torpedo-like propulsion system which need not be high-speed, but which should be quiet, stable, and programmable. The mobile mine, for example, should be able to go along a channel and then make a programmed turn to arrive inside a harbor. It should not make a disturbance when it reaches the end of its path. Another advantage for offset mining of this sort is that a submarine should be able to go back at a later time to replenish the minefield without running the unacceptable danger of encountering its own mines. There are several problems in this type of mine delivery. For example, if there are strong currents in a mine-laying location, then the mine can miss its intended destination by appreciable distances—as large as perhaps 20 percent of the total range. A simple inertial guidance system designed to operate only a quarter hour or so would take care of undesirable drift problems, but the added complexity and cost has to be measured against the real needs for dealing with drift.

The desirability of having adequate mobile mines in order to block harbors which otherwise would be secure has led to some concern about the small and inadequate stockpile of obsolete Mk 27 mobile mines released for production by White Oak in 1950. In this particular case, it is interesting that one obsolete weapon can be used as a good replacement for another obsolete weapon. The electric-drive Mk 37 submarine-launched torpedo is useless as a torpedo against modern submarines because of its slow speed and inadequate homing system. These features, however, are no drawback for conversion to mobile mine carriers. The slow speed—much faster than present mobile mines—and quiet propulsion are a distinct advantage. The torpedoes would have to be junked if they could not be transformed into mobile mines. Their initial use as submarine-compatible ordnance makes such conversion relatively easy. This fairly recent Laboratory proposal was readily seen to meet a number of useful objectives, and led to a useful project.

Deep-water mines

Back in the early 1950s the Hughes Aircraft Company submitted a proposal for a mine which was ahead of its time. The proposal was called the Shark. The idea was that a buoyant body would be anchored near the bottom in deep water. When a ship came along, its influence would release the body from the anchor. It would then rise rapidly and explode under the ship. The idea required the acoustic field of the ship to trigger the mine, because this was the only influence with long enough range. The White Oak Laboratory rejected the proposal as infeasible, because it would be too easy to sweep all such mines by making appropriate loud ship-like sounds from a safe distance.

The problem of providing antisubmarine mines which could be used in deep water, say between Iceland and Scotland, or which could be used to make sanitized sea lanes to keep submarines out, has received a lot of attention. There are basically only three possibilities —the String Mine, the Atom Bomb Mine, and the Torpedo Mine currently exemplified by Captor, now designated the Mine Mk 60.

These concepts for deep-sea mining were studied by the Mine Advisory Committee at Bowdoin College in 1958.

The String Mine evolved from Project Nobska and later became a random-buoy field for surveillance in Project Atlantis (done at M.I.T. in 1959). If it is desired to detect submarines, why not put cheap contact switches at intervals along a vertical anchored line held up by a float which could also be submerged? If these detectors were a submarine diameter apart then it would be impossible for a submarine heading for the line to escape encounter with one of them. If there were a large number of such vertical lines located at random, then the average distance a submarine could travel without detection is analogous to the mean free path of a molecule in a gas, depending on the total number of particles and their cross sections. This analogy delighted the physicists involved in the study and lent a certain respectability to the concept. If one provided each detector with a small but nevertheless seriously damaging explosive charge, the presence of the submarine would be signaled, its location could be determined by sound-ranging techniques, and the submarine itself would probably have to surface if indeed it had escaped destruction. Assuming all of these matters could be satisfactorily agreed on, then also assuming that logistics, reliability, and costs were right, one could imagine a viable system for protection against submarines.

The second scheme, which was considered and rejected at Bowdoin, was to explore the use of a single very large charge instead of a large number of small charges. This scheme places a greater burden on the longer-range detection and classification device which would be necessary, and also increases the danger arising from premature explosion and all the undesirable effects of such an event.

The third idea was to use a detection and classification system to release a self-propelled body, a small torpedo in effect, which would travel a relatively short distance straight to the submerged target and would also be equipped with homing. One advantage of this idea is that the torpedo spends very little time hunting around to acquire the target. The torpedo is started and released only when the submarine is at a suitable distance. If a submarine is detected but does not come within this distance, where it will be a certain victim of the torpedo, the mine decision mechanism will not release the torpedo and will instead shut down and wait for another submarine. The thing which made the Captor idea possible, whereas the Shark idea had not been, was the development of accurate bearing information using passive acoustic correlation and low-power digital techniques. This technique, incidentally, had been developed at White Oak into a passive ranging sonar system for submarines called PUFFS (Passive Underwater Firecontrol Feasibility System) and later called BQG 2/4 Sonar.

PUFFS developed from an idea that started back in the early 50s at White Oak, when Herman Ellingson was trying to identify seismic signals from underground explosions when these signals were largely masked by the random seismic background. Signals from an explosion arriving at two different locations were cross-correlated, i.e., multiplied together with their product being maximized by shifting the arrival time of one with respect to the other. The accuracy of the resulting bearing determination was dependent on the resolution of the time-shift interval. With accurate bearings from two pairs of detectors, it was possible to determine two bearings and hence to calculate the range to the source. Although Captor did not attempt to use passive range determinations, it did start by passively establishing an upward-looking cone, extending from the mine up to the surface. If anything making submarine-like noise crossed the cone, then the mine went into a further routine, sending out high-frequency active pings to see if anything was there, how big it was, how deep it was, and even how fast it was moving. If all criteria were met, then the mine would decide that a submerged submarine was within torpedo-attack distance.

There were clearly many misgivings about the sweepability of this mine and many debates among proponents and opponents. The most obvious countermeasure is to drag a deep sweep wire through the ocean and cut off the mooring cable. If the mine is in deep water or is near the bottom, this is not easy to do—it is a very slow thing to drag a deep cable, and not easy to provide it with underwater towing motors. There are more sophisticated countermeasures which can be visualized, such as towing an echo repeater, which will provide the correct lengthening of the echo, and a suitable amplitude and Doppler shift. All of these schemes are slow, costly, depend on detailed knowledge of the mechanisms, and take time to build and provide in quantity. Furthermore, they do not all work with 100 percent effectiveness. But to my mind the most awkward part of conducting countermeasures against Captor is that the enemy will generally be a long way from his home base, and will have to operate on the surface. All such countermeasure operations are easily detectable and are vulnerable to air or surface attack. The main counter to the countermeasure is simply to sink the tug or the sonar echo-repeater ship. The idea of carrying out such countermeasures from the would-be transiting submarine, in order to do them covertly, is very risky indeed. The submarine may easily become the target of its own sweeping operations. It is also possible that the submarine would have to carry so much sweeping gear that it would have to displace a large number of its own weapons.

With modern acoustic schemes it is now possible to have buoyant mines moored at the bottom which release themselves, travel toward their target, and may not need to be equipped with motors. Such devices would be, in effect, a realization of the old Hughes Shark and would form a good example of the impact of technical development on military capability. It would have been difficult, I dare say, to try to justify Ellingson's work on passive correlation of seismic signals by saying that it was really part of the mine program, and yet it has had significant use there and in antisubmarine warfare, as well as in the detection and location of underground nuclear explosions. In this connection, part of the business of justification is similar to the beauty which exists in the eye of the beholder and depends on his powers of visualization.

There is another idea for providing an interim mining capability in deep water. This is the oscillating mine, intended to work against submarines for a limited time. The mine would be unmoored and of approximately neutral buoyancy. It would have a power supply in it which would alternately increase and decrease the buoyancy so that the mine would slowly rise and then sink, repeating this cycle. Such an oscillating mine would cover all submarine depths periodically. It would have a relatively limited lifetime because of the power demands, and it would also follow whatever ocean or local currents existed in the area. Whether such a mine would be at all practical or would fit into the Hague Convention agreements on drifting mines is perhaps not clear. But such uncertainties have not prevented the development of oscillating mines—the Leon of Swedish design and the Sandford mine by the Vickers Company. In January 1918 the British destroyer *Ferret* laid a field of oscillating mines in the Heligoland Bight, the only time such mines were used [10].

Under-ice mines.

The Arctic Ocean is not a useful region for commerce. Even the incentive to use the Northwest Passage for tankers from Prudhoe Bay was insufficient to prevail over the overland Alaska pipeline. The possibility of using mines in arctic waters and, further, the possibility that these mines would have to function under ice which would form after they had been laid is therefore remote. Nevertheless, there are now many vessels of the class of true submarines which could make a practice of lurking in the Arctic Ocean where there is no other traffic at all. I am referring to nuclear deterrent (strategic) submarines which must spend their days at sea—hidden, out of harm's way. This applies certainly to U.S. and Soviet deterrent submarines and possibly to the few other vessels of this sort which the British and the French have. The Arctic may indeed be an ideal habitat for Soviet deterrent submarines because of the proximity to their northern ports, assuming that problems of communication present no difficulty.

A second problem to be solved by the submarines, and surely not beyond modern design, is to have access to the surface either through gaps in the ice (polanyas) or through upward icebreaking techniques. Either method of access would allow missiles to be launched with little or no delay. In view of this possible exploitation of the Arctic it is perhaps natural to consider how mines might be fashioned to operate under the ice. Problems of ice penetration by specially designed air-launched mines have been examined. The study of electromagnetic background as a function of latitude including the Far North has been going on for decades. The observation of acoustic background and sound propagation under ice poses some interesting problems. The breakup of ice because of temperature changes and ocean currents has been recorded by NOL in Nova Scotia at Minas Basin, where the listening hydrophones in the Bay of Fundy were connected by cables to a shore station.

All such information may be of use if under-ice mining should ever become a necessity. There is, however, a basic incompatibility between strategic deterrence and mining: the former is a state of affairs persisting over a long period of time, even forever, during which no actions by or against strategic forces are allowed. Mines, on the other hand, while generally requiring a long time to be effective, are used in wartime against traffic and nonstrategic forces. Placing mines in the Arctic against possible strategic submarines would be a long-term action, with the effect of carrying out a preemptive strike against missile-launching silos, although it could be done covertly rather than within view of all. If deterrent submarines do not return from their cruises at sea, neither side will be anxious to publicize this accident. Each side has to be quite sure that the loss was in fact an accident, however, and was not caused by hostile action. The conclusion is that there has to be at least a tacit agreement on both sides not to molest strategic submarines at the risk of seriously disturbing the mutual balance of fear. I suppose it would be possible at some future SALT talks to agree not to use the Arctic at all for strategic submarines and to agree, as a pledge of good faith, to the deployment of mines there to enforce the agreement. Such a possibility may someday be considered as a reason for having an under-ice mine. In the meanwhile, the development of these mines may be purely as a limited-war weapon against the use of occasionally icebound passages. A more straightforward position would be to agree not to mine the Arctic so that deterrent submarines could be kept there, safely out of the way.

Miscellaneous

Under this heading are mentioned a number of mine-like devices which have been used in the past and which may again be of importance. The idea of covertly placing an explosive charge under an anchored ship in some harbor may seem unlikely in a far-ranging total war. Nevertheless, such circumstances do recur and form a part of the reason for swimmer-trained forces and minisubmarines. So-called limpet mines which can be silently attached to ships by hand and which explode later form a part of their arsenal. Drifting charges launched by swimmers were a continuous worry for U.S. ships in Saigon and other ports in the Vietnam war. In the absence of torpedo nets and booms for protection, attempts were made to develop anti-swimmer charges which could be lobbed off periodically, to discourage any possible undetected enemy swimmers.

Obstructors are another class of explosive devices which are associated with mine operations. The obstructor is a shaped charge attached to a moored mine cable in such a way that, if a mechanical wire sweep encounters the cable, the obstructor will sever the sweep cable before the sweep cable cutter can sever the mine cable. This is an anticountermine device. I am not sure whether the sweep people have attempted a counter-obstructor device which would then be in the counter anticountermine department. The sprocket wheel sweep evader invented in 1911 by Assistant Paymaster C. Bucknell, Royal Navy, designed to let the sweep wire pass through the mine mooring rope, is another ingenious anticountermine device [10].

Mines have been designed specifically to effect the destruction of hydroelectric power-generating equipment. Although the British in World War II built especially large bombs which would be air-launched against dams and which were designed, on the basis of model experiments, to be big enough to destroy the dam, the lesser goal of wrecking the power equipment was also considered worthwhile. The U.S. Mk 19 mine was a free-floating mine that could be air-launched above a dam near the water inlet of the penstock. It would then fall into the turbine and explode there. No use was actually made of this limited concept. The actual design, using old motor-car relays, is of course obsolete.

The Mk 49 mine was also rather unusual in its day in that it was designed to be shock-resistant and free-fall. The absence of a parachute meant that it could be more accurately placed. However, the high velocity at impact of a bomb-mine with conventional nose shape when laid from high altitude would frequently result in deep burial, especially in very soft or muddy bottoms. The Mk 49 mine therefore had a lopsided nose, one side being cut off at a slant so that on water entry the mine was given a powerful lateral force which caused it to turn rapidly and come down sideways to rest on the bottom. This mine also did not see extensive production or use. Mines, to be laid accurately at high speed and low altitude, must be equipped with an antiricochet nose.

More recently, some NATO-inspired considerations led to the idea of laying mines by rockets. A possible application was to have a number of bottom mines at the ready, each mounted on its own rocket motor and launcher, so that the turn of a switch would project them to their final resting place in a channel which it might be desirable to close at very short notice. Examples of such places might be the Kattegat, the Bosporus, or possibly the Red Sea. The feasibility of the emplacement of mines by rockets was established. Whether such an operation was sensible was considered outside the scope of the study. For example, it was not established that placement by small boats would take longer once the mines were ready for delivery. The problem of readying the mines and of keeping them at the ready seemed to outweigh the mere matter of physical delivery over the short distances involved.

Finally, under "miscellaneous" should be mentioned the so-called indigenous mine which can be designed in such a way that it can be assembled by local allies from locally available materials. These materials might be oil drums, recovered explosives, fertilizer explosives, miscellaneous cables, improvised anchors, flashlight batteries, triggering switches, and the like. Whether this sort of application of mine technology is beneficial in the long run can be argued; nevertheless, it has been known to occur from time to time.

The word *mine* seems to be capable of very wide interpretation, reminding us of the dictum of Humpty Dumpty to Alice. He said, "When I use a word it means just what I choose it to mean—neither more nor less." Thus, our catalogue of mine specialties would be incomplete if we did not recall the Mine Mk 29, which turns out to be a towed explosive hose for the protection of ships against torpedoes, and the Mine Mk 24, which was the first acoustic homing torpedo, also affectionately called Fido. After World War II the torpedo quickly lost its mine sobriquet, and the explosive hose, transferred to the Bureau of Ships, became Project General, and after many years of decline leading to demise, was resurrected for a while under the title Phoenix (not to be confused with a much later air-to-air missile of the same name, nor an earlier bird of antiquity).

10 R&D Policy

The best policy is honest

R&D policy, and indeed any policy, seems to be a state of mind existing in the head of the policy maker. It is hence easy for policies to change either by a change of heads or a change of mind. In the one case, there may be a change of administration; in the other, a change of opinion—a reassessment, a balancing of options, and so forth. With regard to states of mind, a General Services Administration spokesman, when asked one spring why the airconditioning had not been turned on, replied, "It is our policy that the current hot spell will be over in a few days." What he meant, undoubtedly, was that their policy was not to turn on the airconditioning for a rise in temperature which they believed to be short-lived. The result was the same in either case. But sometimes confusion due to illogic can result in real differences. One would like to believe, however, that in a subject as realistic as engineering development there are some views which are based objectively on experience and which are not entirely as flexible as a mere matter of opinion would seem to be. There are, in fact, more wrong ways to do something than right ways. It is better to be safe than sorry, if you have any choice. Intense hope and optimism that something will work is no substitute for actual tests. Tests have to be fair and statistically valid. There is always a conflict between the urgency that gives birth to a development and the time required to carry out the work.

In this case, management has a grave responsibility not to disregard technical discipline under the shield of the broader knowledge of the project and its urgencies—undoubtedly available only to management. Some managers are so eager to meet their own schedules that they will cut necessary corners to show an apparent good performance. Some managers inherently believe that the technologists have to be disciplined, because otherwise they will spin out the problems to show how expert they are. All of these situations exist, and many more, because the conduct of R&D in producing military hardware is endlessly complex, varied, and fascinating. When people get hooked on this heady brew, it is very hard to get them out of their addiction. Everybody becomes an expert, and obviously not all the experts agree.

In general, R&D can be said to be a subject which very few understand but in which there is no shortage of experts. Some people believe, however, that those who have actually done it know more about it than those who have merely watched or have read about it. Perhaps there is an analog to this in the relation between the artist and the critic. For our purposes here it is not necessary to consider the whole wide-ranging subject of R&D policy. Nor should we become entangled in such questions as to whether research should or should not be pure if it is funded by mission-oriented activities. If it is relevant to the mission, then it will lose some of its long-range purity depending on how relevant it may be. The convenient custom of lumping the words Research and Development together into a pair of initials, R&D, equivalent to one word, implies a process in which scientific and technical activities are pursued through a spectrum ranging from initial technical ideas to the final testing of emerging products—all dedicated to solving some problem or providing a way to do something which could not be done before. There is no need to clutter up the nomenclature with further descriptions of subactivities within the general framework of R&D. Unfortunately, over the years this has happened, driven by budgetary, bureaucratic, and even semantic considerations. Thus we have many kinds of research—basic, fundamental, foundational, applied, relevant, even useless—as the British are said to term what we call basic. There are many kinds of development—exploratory, advanced, engineering, operational. In addition there are the many kinds of tests: environmental tests, simulated tests, tests for technical evaluation, operational evaluation, and fleet evaluation.

All of these interfaces have to be defined, responsibilities and missions have to be assigned, funds have to be programmed, schedules have to be developed; all of this has to be monitored, reported, adjusted, directed, and managed. Priorities and dates are required to settle conflicts among projects in their demands for services. In principle, the smallest project has all the elements of the largest project. Only an uncommon common sense can keep the management of projects from overburdening and submerging all but the biggest projects, and even here project personnel can spend much of their time coordinating with others what they do not have time to do, and explaining why they are late with their work. All of these complexities are contained and covered by the simple designation, R&D. The unifying concept here is that the R&D is undertaken in pursuit of some goal. The complexity and magnitude of the R&D effort depend on what that goal is. The kinds of skills, technologies, and facilities depend on what that goal is. However, there are common threads which run through all R&D endeavors, and I suppose these could form the basis for various statements of policy. It is not our purpose to attempt this, however, except with respect to the experience of the mine program—a group of R&D projects related by their actual or potential relevance to mine warfare.

In the last decade the Navy R&D laboratories have been reduced in number by combining some into so-called Centers and actually closing others, shifting their remaining work to the surviving Centers. It was felt necessary by some to differentiate among these Centers on a mission basis. It would have been unacceptable simply to say that each Center existed to perform R&D as needed. Such a statement would lead to continuing competition among the Centers for various developments or parts of developments. Competition was widely regarded as good in moderation in order to avoid complacency on the part of laboratories, but if it got out of hand the laboratories would become too duplicative and probably too independent.

The problem was solved by assigning missions to Centers on a platform basis. If a weapon was to be carried by a surface ship or launched from a surface ship, then it would be assigned to the Surface Weapons Center. Similarly, air weapons and submarine weapons were assigned to the NWC and NUSC respectively. There were a few Centers left over under the scheme, and also a few areas of endeavor which did not fit nicely into the platform system of classification. One of these areas clearly was mines, which are air-launched, sub-launched, and some of them can even be ship-launched. Nevertheless, in the new scheme the mine development laboratory (White Oak) turned out to be in the Surface Weapons Center, concentrating on ship-launched weapons. This awkwardness could easily be remedied by making an exception to the rule. There were other exceptions made in this system in order to avoid duplication. For example, most weapons, wherever they are launched, carry explosives. It is probably not necessary or economical for each weapons center to have explosive development expertise. The same thing presumably goes for other common elements of naval development such as fuzing or materials.

We have said that R&D is motivated by the goal of solving a problem or answering a question. But it is important that the right question be asked. For example, if we ask what can be done about the submarine threat, we will get more new possibilities out of a generalized R&D community than if we ask, instead, what can be done to improve this or that antisubmarine device and then assign the problem to the proper mission-oriented laboratory. To use an example somewhat closer to home, if a company asks what can be done to improve the removal of beards, it is possible that the electric shaver will be developed, whereas if it asks how shaving can be improved it is likely to be bogged down in new shaving-cream developments and the metallurgy of sharper blades. Achieving an important new capability is certainly more important to the user than questions as to who should do it or who is supposed to do it. When urgency, danger, or national emergency appear on the horizon, nobody questions whether a good idea should have arisen in one place or another. It has always seemed to me that the assignment of responsibilities to laboratories is a valid way of getting attention directed to key issues, but it should not be a way of excluding consideration of these issues by others. That would be a form of thought control, does not use the best ideas or a synthesis of the total technical establishment, and in any case is abandoned in time of emergency. Because the questions to be answered are so vital to the answers, it is clear that the questions should be asked not only by users and managers, but also by the technical community that is supposed to provide at least part of the answers.

The platform mission assignments in the technical community, while providing guidelines to the assignment of work and responsibilities for bureaucrats and managers, do solidify the cognizance of thought along the lines of vested interest, which have been called "unions" by Admiral Zumwalt in his book *On Watch* [53]. The existence of unions defending this or that system for waging the war at sea may be useful in peacetime debate over force structure, promotions, or budget allocations. However, such rules about mission do not facilitate transitions from one problem solution to another. Suppose, for example, that it is necessary to reduce the time of travel of a torpedo in order to increase the effectiveness of the weapon. More horsepower will do this but at the cost of increased noise which limits the acquisition range of the torpedo. This situation could lead, and indeed has led, to an almost endless sequence of studies on noise, noise control, noise reduction, and so forth. Another solution might be to avoid the underwater trajectory and instead to fly it through the air silently and gracefully. However, such a solution will not be favored by the torpedo people if it means a sudden switch of responsibility, support, or money to another activity, which is, *a priori*, in charge of items traveling through air. It is not productive to play a game in which the players become ineligible if they show signs of winning. Clearly no one with any sense would play to win in such a game. No matter how a system is organized, there will always be defects in it. These can be overcome by the proper exercise of enlightened authority. Hence, the outcome is generally as good as, but not better than, the people who run the system. Nevertheless, a system which appears reasonable to the knowledgeable people in it is a better system for it. The best policy is honest.

Requirements and invention

The important role of a stated requirement has already been discussed in previous sections. It is obvious that unless an item is desired by those who would use it and, incidentally, who control the funds needed to buy it, the item can neither be produced nor bought. This was fully illustrated by Ellis Johnson's discussions [23] of World War II mine development. There is, however, another facet to the requirements business, and that is the formalization of desires on the part of the military that may be quite fanciful to achieve. Some even have said that the setting of a goal which is far distant will stimulate efforts to reach it. Others feel that a goal, after all, is by definition something to be striven for but not necessarily reached.

Many of these views are fallacious when applied to R&D and can lead to misdirected effort and the waste of funds. It is, for example, foolish to set a goal which would be in violation of the more prestigious laws of nature. A levitation machine may be beyond the state of any art. A metal detector "requirement" which ignores the inverse 6th power law holding for the detection of induced dipoles is due for some rough sledding. The area of requirements extends beyond mere technical invention or design, but also stipulates a level of reliability that a given design should reach when it has been mass-produced and used by the Fleet. It is very important not to overstate a reliability "requirement," because the cost of an acceptable failure rate may be less than the cost of establishing a lower failure rate. This cost may be measured in terms of delay and also in terms of difficulty in determining by tests what the reliability actually is. If the reliability is set at 99 percent, for example, instead of at 95 percent, then the number of tests to establish this will be of the order of one hundred rather than twenty. The cost of the evaluation program can rise fivefold just by a slight change in desired reliability. In days past there have been tendencies to say that nothing is too good for our armed forces, or that the goal of perfection will be an incentive against sloppy workmanship. In fact, such requirements can make it far more difficult for the armed forces to have any effectiveness at all while waiting for this superlative product to appear. There is, therefore, a proper balance to be struck which can be done if the decision makers understand the elements of the problem.

Requirements related to safety, on the other hand, have to be set much higher. Safety should be designed into the weapon so that it depends on a series of environmentally connected events rather than on the attention of a human operator. In other words, weapons should be designed so that they cannot become armed until conditions are met which are physically connected with the intention to use the weapon. As a simple example of this, a depth bomb should not have its explosive initiation train lined up so that it can fire unless a hydrostat in the bomb has been put under the hydrostatic pressure associated with a minimum depth of operation. This environmental safety device would then, for example, prevent the bomb from exploding if it were accidentally dropped on a ship instead of into the water. Use of many different safety devices—all of which have to be activated—can make overall safety very high even when the individual devices are not highly reliable and hence may be inexpensive. The failure of such devices can yield a corresponding failure rate or dud rate for the weapon as a whole which may be acceptable provided the safety rate is high.

The design of a weapon must result in a high degree of safety as computed from a so-called relative accident probability (RAP) analysis, invented and practiced by John Armstrong at White Oak. The design must also result in the desired reliability as determined by test and computation on both the weapon as a whole and its components. The overall reliability includes the dud rate determined from the failures of safety devices to accomplish the necessary arming when environmentally commanded to do so, or from other failures. The setting of criteria for all these things into the requirements is a rather technical matter which may produce gross perturbations if done cavalierly. When the first free-fall antisubmarine nuclear depth bomb (Lulu) was developed by the White Oak Laboratory in the early fifties, we were very happy to draft the requirements in terms of what we thought could be done and to suggest a price package at the same time. It was relatively easy for the CNO to decide whether he wanted to buy this or not. I should point out that in those days there was comparatively little intervening bureaucracy between the Laboratory and the CNO, at least in the new area of nuclear weapons.

We have seen that it is important for the requirements to be in consonance with technology and at least not to contravene the laws of physics. But we must realize that frequently requirements are set that call for the development of a new technology or the application of an old one in a new way. This has given rise to the question whether a new requirement can or should drive new technologies or whether, in fact, the technologies are already there to be applied in a new manner toward meeting a new requirement. This is a rather tricky question, which I suppose is the reason that it is always good for a debate. It is by no means purely academic, because it is at the base of arguments about the funding and motivation for "research." I suspect that in solving a problem the first step is to realize what the question is, and the second is to hunt through the technologies to find something that can be used in a new way or for a new purpose. This is then called having a new idea. In order for this to work, it is necessary (a) that the technologies exist, and (b) that they be known to the person trying to solve the problem.

Perhaps Archimedes gives us the first example. If it is true that he burned up the Roman ships by focusing sunlight on them, it is certainly true that he knew about the properties of paraboloids or ellipsoids in advance of the Roman invasion, and that in developing the knowledge of these properties he was probably not thinking about a military application. Such thoughts would probably have prevented him from thinking about conic sections. Thus, in looking at these and other historical matters, it is clear that if a technology exists, it can be applied toward meeting a new tactical need, if someone is lucky enough to think of it. The mere desire for a given capability, on the other hand, does not lead necessarily to the development of the technology which would be used to provide it. The objective of a tactic is obvious, but where a piece of technology may be applied is not clear in advance. An unexpected or peripheral piece of technology may become the solution in an unforeseen manner. Pasteur said it best when he said, "Chance favors the prepared mind."

We still have to decide how best to prepare the mind, and this is part of the argument about relevance in research. The concept of ordnance sciences arose out of discussions with Vice Admiral Ed Hooper, then captain in charge of Research and Development in the old BuOrd, as an answer to guide the development of freedom in research. An ordnance science was defined as a field of investigation within which things likely to be useful in ordance were to be found.

It was easy to agree that the study of explosive sensitivity and detonation, for example, would be considered an ordnance science, whereas the study of the cross-fertilization of wild flowers was not. Even here, however, one has to be careful. What a surprise it was to find all the planes on a carrier at Pensacola immobilized because of a living organism that grows in high-octane gasoline and clogs up carburetors!

The interplay of tactics and technology is a feedback process in which progress may be made more rapidly by reducing the feedback time. Ways to *increase* this time include the reduction of R&D funding, reduction in technical meetings, reduction in travel and telephones, reduction in recruiting, lengthening procurement time, and emphasizing petty economies. Progress can be greatly reduced by these means without actually saving very much money. During the crisis of war, however, such measures are generally not practiced. There is another way to inhibit creativity which should be added to the foregoing list, and that is to insist on the observance of preconceived missions, and on orthodoxy of thought as well as of action.

The development of the concept of the ordnance sciences back in the fifties between the R&D branch of the Bureau of Ordnance under Admiral Ed Hooper and the major Ordnance Laboratories led to some interesting results. It became pretty clear, for example, that the study and investigation of magnetism was a legitimate activity for a laboratory interested in magnetic mines, magnetic targets, and magnetic background. The magnetic materials work, starting with Parabonol, which was used in magnetic detectors in moored and hence moving mines, produced Alfenol, which found use in magnetic tape-recorder heads, and Nitinol, which is being used in railroad brakes to prevent wrecks due to hotbox seizures. These "spinoffs" from ordnance science work are not considered to be a black mark on planning, but rather a serendipitous benefit illustrating the basic thesis that not all benefits can be planned or anticipated in advance. The fact that ordnance sciences have produced results useful outside the field of ordnance in some cases is certainly not an argument against the conduct of such work in the first place. The temptation is great, if some innocent spinoff of a laboratory's work leads to a breakthrough in another laboratory's field of responsibility, to go ahead with it without attempting an immediate transplant. Work in acoustic mine mechanisms leads naturally to acoustic signal processing, to passive sonar, to torpedo homing devices and to anechoic coatings for ships and submarines. But most of these topics are outside the newly stated mission of the place where they arose. My point here is that the concept of the ordnance sciences most relevant to mine development leads to a list of accomplishments which are also relevant to many other weapons, however they may be carried or launched.

There is a unity of science that insists on existing, no matter how devotedly one tries to separate it into categories. The basic question is whether we want solutions to problems or whether we are satisfied with preconceived categories of effort which presumably are more pleasant to manage no matter how counterproductive they are. An example of an ordnance science activity which became inadvertently useful to mining was in the study of the solid state and the deposition of thin films of semiconductors on substrates. This work, undertaken initially as a means of increasing computer memory capacity, led to improved optical and infrared photoconductors, and in another direction to thin film magnetometers which made an order-of-magnitude decrease possible in both the size and the power requirements of mine mechanisms. And so in these mysterious ways is progress made possible in spite of the best laid plans.

In summary, requirements are necessary tools; they should be forged with care and in conjunction with the people who are going to try to meet them. The conduct of R&D should not be overmanaged. In particular, the practice of making too many categories and distinctions where there were none before, and of making general rules to cover unique cases, gives useless work to useless people. The growth of knowledge and the application of ideas should be encouraged as a means of problem solving, not inhibited by insisting on a box for everything and everything in a box. It is impossible to organize the unknown. Nothing new can occur if it is immediately discarded as not fitting our known structure. Ideas exist in the minds of men, not on charts filled with mission statements. A wise policy, therefore, in R&D matters is to conduct the business in such a way that the best and the brightest will stay with it. They will do this if they see opportunity for creativity, if they believe that competence is given more than lip service, and if they have some sort of respect for the management. The real payoff will not be in charts made or dates met or good conduct, but in productivity.

The sophisticated buyer

The concept of the sophisticated buyer was first enunciated in the Bell Report on Contracting in the Kennedy administration. The idea was that the government, in pursuit of its constitutional functions—such as providing for the common defense—was necessarily involved in the purchase of many items, of course using taxpayers' money (present or future). It could easily be cheated by clever salesmen or shoddy goods unless it had its own knowledge of these matters, knew what it needed, knew what these things ought to cost, and knew whether they worked as advertised. Here is where the in-house professional service, including government laboratories, came in. These institutions were to perform an advisory watchdog service based on their unbiased yet informed opinion.

The Navy quickly developed a list of additional reasons why in-house laboratories were necessary. Skilled technical establishments were necessary to be able to respond without delay to emergency situations. They were also necessary to provide long-term coupling with modern technology. They could be used for the economical and skilled conduct of specific developments and to make specifications for contractual development. They could provide technical service to the operating Navy and give support, troubleshooting, and technical opinions on new problems. They were a useful insurance policy to have around and could do useful work as well between emergencies. All this they could do in addition to providing the basis for the evaluation and acceptance of manufactured items.

But there was, in addition to all these functions, still another which really antedated all these but which in later mission statements tended to get left out. That function was actually to develop the hardware from concept through final acceptance, especially when there was no skill or experience in such work in the civilian economy. Such developments were carried out from the beginning in the mine programs, with the laboratory designing a mine, using one contractor for one part, another for another, and sometimes using components already developed or available from earlier developments. This was thought to be cheaper than would have been the case if a prime contractor had been given a job to deliver as best he could a certain number of complete mines. Another prime contractor on another mine could come up with other parts and other designs that should have been common to several mines. The interchangeability of parts was helped by the laboratory approach, because groups of mines could be considered as a system rather than as a set of separate items. The so-called prime contractor, on the other hand, would deliver whole mines which would be stored in depots, ready (or almost ready) for use.

The laboratory approach resulted in components which arrived at the depots and were then assembled into mines. The laboratory operated in this fashion from the beginning, because there were no mine-wise system contractors in existence, and the laboratory in effect provided all the missing functions. The lack of prime mine contractors in the fifties may have been due to the esoteric nature of mine design, the fact that almost everything about it had no industrially useful counterpart in civilian life, or may have been simply due to the fact that the mine development and procurement budget was small and vacillating compared with business in the ship, tank, aircraft, missile sector. This situation came in for its share of criticism later when it was noticed that the readying of a batch of mines for use in an exercise took quite a long time and needed a lot of men skilled in the arts of mine assembly and testing. Why couldn't mines be simply ready and waiting on the shelf? The ready round, analogous to a loaded projectile or bomb, became a desirable goal for R&D and many felt that having a mine delivered by a prime contractor would solve all the Navy's problems with mine readiness.

It was very well known to the laboratory, of course, that the assembly of the Mk 25 and 36 mines used against Japan was a very labor-intensive and skill-demanding business. There were wires, ammeters, connectors, wiring diagrams, innumerable measurements, test settings, and so on, after the fashion of an engineering school's physics lab experiment. The Long Range Mine Program which emerged out of much laboratory and Navy planning was a specific attempt to avoid these problems. Technically the program was oriented toward the quiet submarine as a target operating in deeper water. The design was modular to allow for easy substitution of one mechanism for another. It required standard cable harnesses that could be quickly plugged in—in only one way. It called for standard test sets and simple procedures so that an assembled mine could be quickly checked out to determine if everything was working properly. In this way the problems of cost, compatible design, interchangeability, and readiness in the stockpile were addressed in, if you will, a systems engineering manner. Nevertheless, it was the result of this effort which also fell under the criticism of poor readiness. What had formerly taken an hour to do and now took ten minutes would have been much better if it took only one minute—or better still if it did not have to be done at all.

Later on in the sixties, the prime-contractor concept became possible because fairly large developments (for mines) like Captor, for instance, were put out for bid as a system, and contractors who had formerly worked on parts of the system or who had partial expertise would join together and select one of themselves to be the prime, and the others to be subcontractors. These combines of contractors would then compete against each other for the award. Actually, in the case of Captor, where the laboratory had worked for prior years on critical parts of the system—deep automating anchors, sound systems, torpedo ejection, case stability, data processing, and so on—the laboratory was designated the project manager for the system with contract authority.

This was an unheard-of innovation. The assignment of authority and funds to a laboratory and not to a headquarters project office was justified as an experiment in management. Although the project stayed within its budget and overcame many technical problems while keeping well to its original schedule and maintaining cordial relations of mutual respect with the contractors, the delegation of project management authority was rescinded about halfway through by the Chief of Naval Material on the ground that the project was not managed the way other projects were. The Laboratory is continuing, however, in its role of providing sophistication to the buyer and also of providing insight, experience, design, development, and evaluation skill at each step along the way. In the mine program as a whole this type of guidance seems to be more naturally provided by an in-house technical capability than is the case in many other sorts of development.

The Long-Range Mine Program which was planned, accepted, and set in motion in 1948 forms an example of this sort of in-house service. The objectives have already been mentioned. The program consisted of four antisubmarine influence mines—two bottom mines and two moored mines. The air-launched bottom mines, Mk 52 (total weight 1,000 pounds) and Mk 55 (2,000 pounds), were released to production by the lab in January 1954 and November 1955 respectively after certain vicissitudes which will be discussed later. They did not become operational until 1961. The moored mines, Mk 56 (air-launched) and Mk 57 (submarine-launched), were both released in 1960 and became operational in 1966 and 1964 respectively. Some have claimed that these mines took an excessively long time to be completed, i.e., something like twelve years. This illustrates the effect of low priority and vacillating funding. I used to keep what I called a pathological funding graph of money vs. time for these two moored-mine projects. Every few years the funding and priority for these mines would be interchanged—I suppose depending on whether the mine officer were an airman or a submariner. When the design disclosures of both were released to production in 1960, there were no funds immediately available to undertake production, and these mines did not become operational until several years later.

Evaluation—How much and who says?

Emphasis on evaluation—a formal process of testing to determine whether an item will function under the various conditions to which it may be subjected—probably arises out of the special status of ordnance. Webster's 2nd Unabridged Edition (1934) says that ordnance means military supplies "which include all artillery with mounts, carriages, equipments, and ammunition; small arms; horse equipments and harness and in early times included armor, arrows, etc. . . . In the U.S. Navy it includes all material for fighting including guns, gun mounts, torpedoes, submarine mines, armor, range finders, etc." A second meaning includes "arms or engines for throwing missiles of any kind, as catapults, bows, arbalests, etc." These archaic definitions with their pregnant "etc." served very well to describe the generality of work associated with weaponry and allowed extension to all matters associated therewith such as fire control, fuzing, propulsion, guidance, homing—the lot.

It is curious that the Naval Ordnance Laboratory in spite of its name never developed or worked on any naval artillery or gun. Nevertheless, in conjunction with the remissioning of laboratory sins it was felt necessary to drop the archaic term *ordnance* and to substitute the term *weapon*. Incidentally, a later Webster says that a weapon is "an instrument of offensive or defensive combat; something to fight with." Under this definition an argument is a weapon and certainly an airplane is a weapon. If, however, we consider the class of objects normally called weapons, which contain high explosives, which are expected to detonate against an enemy and not against a friend, and which therefore must complete their cycle of operation in an unmanned state, then we have the requirements for a more rigorous form of evaluation. A manned vehicle such as a hydrofoil or an aircraft is not expected to blow up, is not expected to be entirely self-sufficient in its operation, and will not have to explore the extremes of environment that a weapon may be subjected to.

As has been discussed elsewhere, the explosive danger of weapons places safety requirements on their design which must be proven by proper evaluation. This is qualitatively different from the statistical arguments and tests which must be carried out to establish reliability at whatever level has been agreed on, preferably in advance. Clearly it is easier under the pressure of schedules, money, or other urgency, for an authority, be it manager or laboratory, to relax reliability requirements than to relax safety requirements. There were, however, differences of opinion and of policy between the Army and the Navy, or the Atomic Energy Commission and the Navy, as to the calculation of the likelihood of an accident, the means for limiting that likelihood as a matter of design, or the likelihood that is considered to be acceptable. If an accident will kill a thousand people, it should be made less likely to occur than one which would kill only one or two.

Apart from such difficult questions, there is still another technical matter relating to the evaluation of a development. This is to decide whether a set of design drawings is adequate to disclose to a manufacturer the information necessary to build the item so that it will function as designed. The Long-Range Mine Program provided a series of examples of these problems. In the 1950s the Laboratory and in fact the Bureau of Ordnance worked on the theory that one should be able to buy "hardware" including weaponry, components, or systems, to the greatest advantage to the government by letting industry compete for the business. The lowest price, together with a few other contract-selecting criteria like competence and reputation, would determine which manufacturers would receive the work on a fixed-price basis. It therefore became very important to have a complete and accurate specification against which companies could bid. If the item was a new mine, for example, that had been developed and evaluated, then it was necessary to use the set of drawings and other information (suggested processes and so on) which actually had been used to produce the evaluation hardware as modified by lessons learned in the evaluation itself. The presumption was, then, that a company previously unconnected with the development could, using these drawings, nevertheless manufacture and produce hardware that would pass all the required performance tests.

As developments became more complex, and as the concept of prime contractor arose, the idea of competition at the beginning of production (where, incidentally, expenditures took an order-of-magnitude jump) gave way to the idea of selecting a contractor earlier in the development cycle after a competition based on his potential to perform. In this way the importance of the Design Disclosure which could be interpreted by any competent production contractor gave way to the idea that it would take too long and involve too many mistakes to transfer the necessary development experience to a new production contractor. Nevertheless, even when we were locked in to a production contractor, it was incumbent on the Laboratory to extract from this prime contractor the statement that "yes, he understood the drawings," which, of course, were meticulously kept current with development experience, and that "yes, he could perform the actual manufacturing which they called for." In other words, he stated that the drawings represented a producible design. If trouble developed in the manufacture or performance later, he could not blame the drawings or claim that they called for unreasonable tolerances or impossible processes.

The need for evaluation to determine performance was recognized very early at the Laboratory. The next step was to insist that the evaluation be carried out by a group that was independent of the developers. Although this policy was designed to provide the very best unbiased and impartial evaluation, it nevertheless carried some risks with it. As evaluators became more fussy to find something wrong, their tests became in the eyes of the developers more unrealistic. Often the item was tested for something it was not designed to withstand. Evaluators would ignore all previous tests done by developers as possibly biased. Developers would do more and more preliminary tests to ensure against their products being sent back as failures. And finally this situation would lead to an impasse within the Laboratory where the evaluators would refuse to release an item which the developers said was ready to go. This situation in fact came to pass in 1954 and 1955 with the Mk 52 mine, the first in the Long-Range Mine series. It was obvious that a refinement on the policy of independence in evaluation was needed.

The new policy was that the judgment to be made from the evaluation was not delegated solely to the evaluation segment of the Laboratory. Their judgment was to be independent, but they were not constrained to ignore other test evidence, for example, obtained by the design engineers in the course of the development. The new policy required a concerted attempt to plan in advance, at an early stage of the project, what sort of evaluation program was needed, how many samples were needed to meet stated reliability requirements, and how much money and time this might require. Once these things have been laid out, it is clearly inappropriate for evaluators to make unilateral changes in the evaluation program. Their independence should not reach that far. It is true, however, that the development and the test have to proceed in an iterative manner. Neither can spring full blown out of the head of Jupiter (or Mars either).

One cannot test produced items until they are produced, and one should not produce items until their prototypes are tested. With this contradiction, the only way to proceed is step by step with small initial lots, correcting their design for the next lot on the basis of tests until one can make a large enough production to provide statistically valid evaluation data. The purpose of the final tests is more to establish performance than to alter or improve design. All of this process is fraught with peril—there are possible huge delays, huge expenses, and huge risks in failing to carry out the process correctly. There are also huge opportunities for logical confusion on a grand scale. Under these circumstances, the original policy (that evaluation must be done independently of the developers) is perhaps too simple. If one reaches an impasse where the evaluators insist that the item fails too often and therefore does not meet the operational requirement, and the developers say that the test being applied is unrealistic or too rugged or wrongly interpreted, then there must be a loophole through which the light of sweet reason may creep. As usual in such situations, the matter is passed to the next level of authority where a decision is made. To insist that evaluation is independent does not mean that it is omnipotent nor omniscient. To make it too independent, incidentally, is to make it more expensive and also open the important possibility that by the time the evaluators have discovered a failure, they will also have a fix for it and will eventually turn into redesigners.

All of these issues were raised and churned about in connection with the release of the Mk 52 mine. The Laboratory management finally reached a decision on the conflicting views and formally stated to the Bureau that the evaluation was complete, that the mine met the requirements, and that the design disclosure package was released to production. After all this confusion, however, the Bureau was somewhat skeptical about the development and did not rush into production. Whether this was due to lack of funds, failure to program funds, or simple caution, I do not know.

Producibility and competition

At about the same time the release struggle was going on in the Mk 52 mine development, a change of command occurred in the Bureau mine desk. It was taken over by Captain Eli Reich, later to become admiral and later still Comptroller of the Department of Defense, who had achieved great distinction as skipper of the submarine *Sealion II* which sank the Japanese battleship *Kongo* and its escorting destroyer *Urakaze* in the Formosa Strait on 21 November 1944. He instituted a scrutiny of the Mine Mk 52 drawings on the theory that the Laboratory had perhaps created an improvement in mine performance but that after all the scientists and engineers there did not know much about production and production engineering. They could well have created something that no one could build, and that might not be a fair specification for American industry to bid on.

The question to be looked into, then, was whether the design was producible. The criterion of producibility became the watchword of the day. The Laboratory had an easy answer to all this. Of course the design was producible, because it had already been produced in two or more lots for design and evaluation purposes. Selected contractors had carried out these limited or "pilot" productions using these same drawings to which they had contributed their production engineering experience. Besides, the Laboratory, though not itself directly involved in production, was still quite knowledgeable in this field because of its many contacts and long experience with industry. In fact on occasion, when difficulties arose in production, it was quite frequently the Laboratory which helped with the solution—by better control of processes, or by suggesting modifications to processes, or by improving quality control on materials, and so on.

The Bureau, however, felt it prudent to turn over the drawings to the production facility then existing in the Gun Factory in the Washington Navy Yard. This facility had undertaken some ordnance production other than guns and had a good reputation for competence, although they had not recently built any mines. They were assigned the job of assessing the drawings for producibility. Did the drawings represent the kind of disclosure, with regard to accuracy and detail, on which the average contractor could bid comfortably? After a few months' work the Gun Factory asked for additional manpower and assistance, and so at this stage an outside contractor (A. E. Smith), specializing in producibility review, was called in.

A change in drawing format that was decreed at this time by the Bureau resulted in hundreds or thousands of deviations between the older drawings and the newer standards, and it became difficult to tell a real error in the drawings from a noncompliance with the new format. In addition to this, the new drawing contractor undertook to make new drawings for items like clocks which formerly only clockmakers would bid on. The idea was that with properly specified drawings any competent contractor could make a clock. Some other components were redrawn and in some cases redesigned. All of this took many months and resulted in the Bureau's sending all the altered drawings back to the Laboratory with the request that the changes be reviewed to see whether there would be any impact on the proper functioning of the design. The so-called producibility experts felt that the mine could be built, but they were not so sure that, as so built, it would work. Some of the changes, principally those in format, were approved, but for some of the others the Laboratory said that the only way to determine their effect was to run an evaluation program on them, of course, and this meant to manufacture a test lot.

We were back in the middle of a design and evaluation program again. The number of mines needed had to be sufficient to match the reliability requirement, although the Bureau felt that the number could be less than this—and so a compromise was reached which really satisfied neither side. But it turned out to be fairly academic because the resulting cost was "substantial," as financial types say, and it had not been included in any prior budget. Further, to carry out this new program would have added several more years to the availability time of any new ground mines. In the end the matter was resolved by accepting the original design release as adequate. The experiment in producibility review had consumed on the order of five millions in dollars and five years in time.

It was a matter of some interest that not long after this the West German Navy, through a Mutual Assistance Pact or Bilateral Agreement with the United States, acquired the Mk 52 drawings and went into production with them in Germany, without any modifications, in order to restock their mine stores. They thought that these would be the best and most quickly available mines. It is understood that they had no cause for disappointment and had no difficulty in production.

The remaining mines in the program, the Mk 55, 56, & 57, were eventually put into production using the design disclosures released by the Laboratory at the respective completions of their developments. These were the last mines or large developments to be done without a formally recognized prime contractor. Later developments of greater cost and complexity such as Subroc, Captor and the Mk 48 torpedo were done with a prime contractor who was selected at the beginning of engineering development. The discipline of the Design Disclosure was continued in order to have a complete and accurate set of drawings which specified the system in question. Although initial production was assigned to the prime contractor who then needed essentially no learning time, it was always possible, in theory at least, for later production—perhaps on a less urgent basis—to use the drawings in a general competition with fixed price bids to get the cost down. Without this option the government becomes the prisoner of industrial monopoly.

11 Limited War and Terror

Relegated to a position of strategic and tactical inferiority in the U.S. Navy throughout much of the post–World War II period, as was the case before the war, mine warfare nonetheless emerged during the 1980s as an effective tool of the disadvantaged states, "terrorist" groups, *and* great powers intent on political intimidation. Naval mines were used by the Argentines in the Falkland Islands War of 1982; in early 1984 mines laid in Nicaraguan waters, reportedly by the U.S. Central Intelligence Agency, damaged several ships; later that same year mines were planted in the Gulf of Suez and the Red Sea, presumably at the direction of Libya's Muammar Qaddafi; finally, during 1987 and throughout 1988 "obsolete" mines created significant difficulties for merchant and naval traffic in the Persian Gulf. In the last conflict, particularly, the very real threat of mines was brought home to the United States in April 1988 by the near-tragedy of the frigate USS *Samuel B. Roberts* (FFG-58). Only the most heroic of damage-control efforts by her crew saved the *Roberts* from breaking in half and sinking after she struck a World War I–era mine that had been laid in the main shipping channel by Iranian Revolutionary Guards.

In his excellent treatise *On War,* Clausewitz noted "that war is a mere continuation of policy by other means." [64] One might also say that the use of naval mines is war carried on by other means. In effect, one does not directly attack the enemy with the use of naval mines; rather, one attacks the medium and routes that he must use. It is not known when

This chapter was prepared for the updated edition by Dr. Gregory K. Hartmann and Dr. Scott C. Truver, Director of National Security Studies, Information Spectrum, Inc.

the victim will appear or who he will be, but it is known that there will be a victim sooner or later. Indeed, it is this rather ambiguous character of mine warfare that has lately captured the imagination of terrorists seeking to inflict the greatest psychological and physical damage on their targets without necessarily incurring significant military or political costs in doing so.

Mines are extremely versatile in the purposes they can achieve. Although most mines have no motors or propulsion, they do have brains and patience. Time is on their side; they are truly weapons that wait. If they cannot overtake a predator, they may catch him in a trap. Mines can be used in shallow water, in rivers, in harbors, in the Arctic as well as the tropics, on the continental shelf, and in the deep oceans. They can make blockades, interdictions, barriers, and bastions. They can be delivered by aircraft, submarines, surface combatants, merchant vessels, barges and ferryboats, and even rockets and human swimmers. Their targets are anything afloat on or below the sea; even low-flying aircraft and helicopters are susceptible to naval mines. In the minds of the potential minefield penetrators, the threat of a minefield, once known or suspected, is frequently greater than actual. Furthermore, once planted a mine normally has no national loyalty or, in the absence of sterilization devices, no predefined shelf life. The record—perhaps apocryphal—is held by a U.S. ground mine planted in the Inland Sea of Japan in 1945. Twenty-five years later it finally detected a target that satisfied its sensitivity settings and blew a hole in a Japanese freighter large enough to drive a Toyota through. [66]

What's past is prologue?

The historical development of naval mines was coincident with and parallel to the development of the science of electricity. Bushnell's tactical idea in 1777, to destroy ships remotely by underwater explosions, had no practical result until the nineteenth century saw the development of insulated wire, batteries, electric detonators, and high explosives. Apart from Samuel Colt's experiments in the 1830s and the sporadic effectiveness of moored contact mines in the U.S. Civil War, there was very little activity in mine development until the twentieth century, when in 1904 the Japanese and the Russians taught each other that mines were indeed something to be reckoned with. The extensive wartime experience with mines is in reality quite brief, ranging from 1904 to 1990, a period of only eighty-six years that includes the Russo-Japanese War, World War I (Dardanelles, Dover Straits, North Sea Barrage), World War II (both Pacific and European theatres, with limited German use off the coast of the United States), the Korean War (Wonsan), the Vietnam War (inland/coastal waterways and Haiphong), the Falklands War, and the Iran-Iraq War. The use of mines in wartime operations and low-level conflicts has increased significantly. The advances that have made mines increasingly useful are the development of influence devices for target sensing and detonation, the use of aircraft as minelayers, the miniaturization of mine mechanisms, the development of long-lived batteries, and the extended coverage to quieter targets and deeper waters.

To these twentieth-century *wartime* examples should be added the *peacetime* clandestine/terrorist use of mines during the early 1980s in the Nicaraguan port of Corinto (allegedly by the CIA, during the Reagan administration's fulminations against the communist government of Daniel Ortega) and in the Red Sea/Gulf of Suez (advanced Soviet "export" mines laid from a civilian merchantman by Qaddafi's henchmen).

The Value of Mining

In his foreword to Duncan's *America's Use of Sea Mines* [12], R. D. Bennett, wartime director of the Naval Ordnance Laboratory (NOL), said, "The effectiveness of the submarine mine has not decreased with the coming of the space age. So long as cargo ships cross the sea, this unspectacular weapon will remain a major factor in control of the approaches to harbors, and the shallow straits between seas." Indeed, the inspiration behind the invention of the mine rests on the simple observation perhaps first made during the reign of Queen Elizabeth I (1558–1603) that a ship is more vulnerable to explosive attack below than above her waterline. Although the post–World War II preeminence of the United States' and other developed countries' merchant fleets has declined, to be replaced by third-world merchant marines and flags of convenience, surface ships will continue to be the principal means of transport of most of the world's commerce, far outstripping the capabilities of aircraft. And while some analysts and strategists have argued since the late 1980s that the nuclear-powered submarine has overtaken the aircraft carrier as the capital ship in the world's navies, surface combatants will undoubtedly remain the only true means by which command of the seas in future wars will be effected. Furthermore, in the postwar era extensive mine developments have been primarily—especially in the United States—directed at ways of countering advanced submarines. So the continued existence of targets for mines seems assured, in the form of cargo ships, aircraft carriers, and surface combatants, while submarines, no matter how independent of the surface they evolve to be, will not become less immune to attack by naval mines.

A further question exists, however, and that is whether mines have now benefited as much as possible from the post–World War II technological revolution, whether they have reached a plateau in development beyond which advances are not likely to go. Following the Navy's consternation at Wonsan, Korea, in October 1950, when a handful of Soviet mine specialists aided by unskilled North Korean labor using sampans laid some 3,000 World War I and World War II mines, planting a defensive minefield that kept 250 ships and 50,000 Marines at bay for seven days, the Office of Naval Research and Catholic University established a Mine Advisory Committee to help the Navy maintain an effective mine warfare capability. Transferred from Catholic University to the National Academy of Sciences in 1955, the Mine Advisory Committee operated continuously until it was replaced by the Naval Studies Board in 1974. During that twenty-four-year period, the Committee conducted more than *seventy* mine and mine countermeasures studies, principally in the countermeasures area.

From 1955 to the mid-1980s, however, only *four* mine-specific studies were undertaken: (1) the Deep-Sea Mine Study of 1957, leading to the Captor (Mk 60) mine development program of 1961; (2) Project Pebble of 1965, in which the development of the Mk 36 Destructor mine was promoted; (3) Project Nimrod of 1968–69, which thoroughly reviewed mine technology, operational utilization, and the psychology of mine warfare, and which initiated several research and development programs such as the remote control of mines, enhanced mine delivery for cargo aircraft, and solid-state electronics for mine mechanisms; and (4) the 1979–81 mine warfare study that reemphasized the requirements in all these areas and called for the development of a universal mine mechanism the size of two beer cans. Moreover, the Navy's *Quo Vadis* and the Chief of Naval Operations' *Navy 21 Study* of the late 1980s only "touched on" mine warfare requirements and technologies of the final decades of the century. Some well-placed observers noted that the Navy's focal points in these most recent analytical efforts were "not really" on mines, mining, and mine countermeasures, but on "more important" missions for the U.S. Navy.

If there is going to be a continuing development in mine warfare in the U.S. Navy, the Navy will have to maintain a high measure of commitment to its mine and mine countermeasures R&D programs, a commitment, however uncertain, that has often been frustrated by insufficient funding and higher naval warfare priorities placed elsewhere. As one well-placed analyst of the Navy's post–World War II mine warfare programs has observed, mines have not kept pace with technology or evolving operational requirements, and the R&D program of the 1980s ("such as it is") did not anticipate the changes in naval warfare that actually occurred. [66]

Nevertheless, mine warfare *has* become an unavoidable concern of the Navy—whether it likes it or not, and whether it is using mines against an enemy or defending itself from mine attacks. This was apparent in the statement of Rear Admiral Byron E. Tobin, Commander Mine Warfare Command, during the 1987–88 Persian Gulf crisis: "Mine warfare is an absolutely critical element of Navy combat strategy. It has played a significant role in every major hostility. Some people had forgotten about that for a while until the activity in the Persian Gulf jogged their memories a little bit." [61, P. 6] Perhaps improved planning and programs will constitute the future breakthroughs in mine effectiveness, as has been urged by a small coterie of individuals inside and outside of the Navy and in the Congress, but that is doubtful without an enduring commitment to allocate the funds that are needed to carry out these efforts. Admiral Tobin's sentiments notwithstanding, when one reviews the record from the mid-1970s on, Eliza Doolittle's lament in *My Fair Lady* is particularly apt: "Words! Words! Words! First from him, then from you! Is that all you blighters can do?!"

The most likely denouement of the Navy mine warfare contredanse—mine warfare continuing to be something of a "backwater mission" in the U.S. Navy—is incongruous, especially in light of the naval warfare requirements of a maritime country like the United States *and* the desirable functions of naval mines, functions that no other weapon offers in the same degree. Based on the original observations of Duncan [12], which have been updated to reflect recent experiences, these functions are summarized in the following discussion.

Mines lie in wait for the enemy without accepting a return threat. Insofar as the delivery of bombs and torpedoes is concerned, the attacker is always subject to the possibility of counterattack. This statement may be less true in the current era of long-range, standoff weapons, but when an adequate counter to long-range attack is available, then the balance of risk will again be restored. The only real concern for the minelayer is that he be able, for the short time necessary, to exercise local sea control. This concern is alleviated somewhat by the development of long-range mobile mines (e.g., the Mk 67 SLMM, or Submarine-Launched Mobile Mine), which can be launched from submarines and then travel under their own propulsion and guidance to the area to be mined.

Mines may win battles passively—that is, they may influence an enemy to retire without attacking. During World War II the Japanese abandoned several areas in the Pacific without a fight rather than mount major efforts to clear the minefields. Obviously, the risks and costs of the mine clearing outweighed any benefit thought to be gained.

Even in "undeclared" wars, such as that in Vietnam, an openly declared and laid minefield can cause a reversal in global public opinion about the responsibilities of the participants. By publicizing the existence of the minefield, the onus of accepting damage lies with the side attempting to breach the mines. As opposed to an offensive air strike or naval bombardment, which can result in many "innocent victims" and an outpouring of world opprobrium against the attackers, a well-publicized minefield does not have to result in any victims at all to be successful. And those ships that do fall victim to mines bear the ultimate responsibility for their folly.

Mines may keep ships in constrained areas where they may be attacked by other means. The most outstanding example of this principle is the World War II U.S. mining of Palau Atoll when there were thirty-two Japanese ships inside the lagoon in late March 1944. [58, P. 112] Unable to pass through the mined exits, twenty-three ships were sunk by bomber or torpedo aircraft, like sitting ducks! The Japanese later lost to the same mines three more ships sent in to salvage the wrecks from the earlier attacks. At this point, they abandoned the port entirely. Later, the major Japanese naval base at Truk was closed by mines.

Mines may cause ships to take longer alternative routes, thus reducing the number of effective ships available to the enemy. An example is the final closure of Dover Strait in 1917, which required the German submarine force to transit the North Sea to reach the Atlantic, thereby materially reducing time on station.

Mines are a continuous menace to enemy morale. Whether mines are actually present or not, the mere threat or supposition of mines having been laid in important areas becomes a form of psychological warfare against the enemy. As two analysts have described this facet of mine warfare, "The psychological impact of a minefield is often given a solemn nod but rarely emphasized because of its slippery nature. The specific causes of any decision are obscure, and the decision to avoid a minefield or to risk it will be influenced by many factors. The decision maker's perception of the minefield is one of these factors." [60, P. 58]

This was most recently underscored by the seemingly indiscriminate —but actually very precise—use of mines in Persian Gulf sea lanes by Iranian Revolutionary Guards in Iran's desperate war against both Iraq and anybody who was foolish enough to continue trading with Baghdad and its allies in the region. Although marine traffic continued despite the real threat of mines, clearly the mines entered into the decision-making calculus of ship owners, operators, and insurers: "The real effect of a minefield derives from the more subtle influence of an exaggerated fear. Minefields work more on the mind than on ships." [60, P. 60]

Mines can attack targets that human controllers cannot see or hear. Prophetically, in the early 1960s Duncan stated, "Great effort is being expended to learn how to locate a submarine some distance away. As submarines are improved with increases in underwater speed and with the ability to stay submerged for longer periods, the value of the mine is relatively increased." [12, P. 165] The Navy's Long-Range Mining Study of 1947 identified the Soviet submarine as the prime mine target in future wars and promoted the development of the 50-series mines. The idea of a mine that would employ an encapsulated homing torpedo was first proposed in 1942 and was strongly advocated in Ellis Johnson's long-suppressed book, *Mines Against Japan*. The analysis performed by the Mine Advisory Committee in 1958 clearly supported the superiority of the encapsulated torpedo compared with alternative-design open-ocean mines, and Captor (Mk 60) became a funded program in 1961, finally reaching the fleet in 1980. Still, some mine warfare technologists warn that Captor will become less effective against the newer submarines, which are becoming more quiet, sophisticated, and capable each year.

In light of then–Chief of Naval Operations Admiral Carlisle A. H. Trost's 1987 statement that antisubmarine warfare (ASW) was "his highest war-fighting concern," and acknowledging dramatic improvements in Soviet nuclear submarines' operational characteristics, especially the era of near-equal submarine quieting ushered in with the Soviet Akula-class SSN in 1985, *more,* not less, emphasis should be placed on U.S. ASW mines in the future. Stated boldly, mine warfare is terribly important to the United States' ability to conduct effective ASW! The significant improvements in submarine quieting, speed, depth of operation, and endurance make the ASW mine even more valuable than in the past, because it operates at close range and in the submarine's own environment, where disguise is impractical if not impossible. Yet almost as if these considerations were being deliberately ignored, the Navy's mid-1980s R&D and acquisition programs for enhancements to the Captor and SLMM (Mk 67) mines, and several programs for advanced medium-depth mines—one of the last in an aborted joint effort with the Royal Navy—have repeatedly been canceled or delayed by a lack of funding commitment. The ideas and proposed weapons are there, but getting them into the hands of the operators has proved immensely difficult to achieve.

Mines not only sink and damage ships as other weapons can, but their effectiveness is also measurable in terms of the delay created in enemy operations. This effect can be measured by the concept of equivalent ships lost. If a ship is unable to perform its function for a certain period of time because of damage or threat of damage, it might as well be counted as lost for that time. If a ship is held up for half the duration of a war, then the number of equivalent ships "lost" is one-half, as far as waging that war is concerned.

Moreover, operational plans can be stymied for critical periods ranging from days to months, allowing the enemy to strengthen or regroup his forces. During World War II German operations aimed at taking Leningrad were stymied by the Soviet mining of the Baltic Sea. The Soviets laid some 14,000 mines off their Baltic coasts and in the Gulf of Finland, effectively stopping the resupply of German forces by sea. In the case of the U.S. amphibious operation at Wonsan in 1950, the 400-square-mile minefield kept the landing forces frustrated for a week, cutting circles in the sea while the minesweepers, in what came to be called Operation Yo-Yo, accomplished their dangerous task. In all, seven minesweepers sank from hitting mines. According to one account, the U.N. forces had just barely averted a similar fate at Inchon a month earlier, as the Soviet-tutored Korean forces were within a day or two of being capable of mining Inchon Harbor when MacArthur's amphibious landing took place. Indeed, during the first two years of the Korean War, all Navy ship losses and 70 percent of all Navy casualties could be attributed to mines. [78, Pp. 44–47; also 70, Pp. 123–35]

Mines can force the enemy to expend much effort and materiel on countermeasures that would otherwise not be necessary and, ceteris paribus, are otherwise not productive. If a sufficiently great mine threat could cause all of the enemy's effort to go into countermeasures against it, there would be no effort remaining to wage the war, whether the mine threat were real or not, and even whether it had been used or not. While clearly an exaggerated case, the compelling requirement for countermeasures does impose a difficult problem for the enemy: how much is enough? The mine countermeasures effort expended by the United States, Great Britain, Japan, and Germany during World War II totaled some 3,230 ships and 146,000 men—for example, the German defense against Allied mining involved 46,000 men and officers, 1,276 sweepers, 1,700 boats, and 400 planes, whereas the British defense against Axis mining required 53,000 men and officers and 698 sweepers. [13] Meanwhile, the proliferation of U.S. Navy degaussing efforts during World War II almost depleted U.S. copper stockpiles. After the war the Navy realized that it was impossible to degauss a large ship (6,000 tons or larger) to such a degree that it could escape a sensitive mine; it thus became policy not to make the attempt. Duncan estimated that the Allies spent billions of dollars degaussing their ships, "much more, probably, than the cost of all of Germany's magnetic mines." [12, P. 165] In eras of squeaky-tight fiscal resources, this must continue to be a concern for the world's navies.

Mines are individually very economical compared with many other weapons. More important, the targets of mines are so valuable that when one is successfully attacked, its loss is usually many times the cost of all the mines laid to attack it, including the cost of the mine laying. Furthermore, the use of mines has been very cost-effective in terms of inflicting damage on the enemy. The Allies, principally the United States, planted 25,047 mines in the Pacific during World War II, accounting for 1,075 ship casualties. (All told, approximately 101,000 mines were laid by the United States and Great Britain during the conflict, resulting in some 2,660 ship casualties.) In the Pacific theatre, 3,000 mines were laid by surface ships and about 600 by submarines; the rest were planted by aircraft. The United States lost fifty-five aircraft and no ships during these mine-laying operations (some submarines on mine-laying missions, however, failed to return to their bases). Air-delivered mines accounted for nine casualties per million dollars of mine-laying expenditure;

submarine-laid mines, five casualties per million dollars of expenditure. Submarine-delivered torpedoes, on the other hand, accounted for just 1.2 casualties per million dollars of expenditure, a function of the fact that more often than not the U.S. torpedoes, at least through 1943, failed either to run straight or to detonate if they did hit the target. Table 5 provides another perspective on the comparative "economic efficiency" of air-delivered mines versus submarine-delivered torpedoes during the Pacific campaign in World War II.

The idea of an unsweepable mine is desirable but not necessary to achieve. It is a mistake to label a mine useless as soon as the enemy learns to sweep it. A sweepable mine can cause the enemy far more trouble to sweep than it costs the miner to plant or to develop a new unsweepable mine. Furthermore, mines can be made temporarily unsweepable by equipping them with delayed arming devices, or clock-controlled/preselected dead/live periods, using ship counters that require several actuations before actual firing, or designing the mines to go after the minesweepers, whether surface ships or helicopters. To meet these challenges, sweepers will have to work overtime and the enemy will have to accept the delay caused by the necessity to sweep. All this can be achieved by using mines that are, in principle, sweepable.

An excellent example of this principle came in 1987–88 when Iranian Revolutionary Guards began a deliberate mine offensive in the Persian

TABLE 5.
ECONOMIC COST PER ENEMY TON CASUALTY: WWII, PACIFIC*

	Inner Zone Mining	Submarine Campaign
Duration of Campaign in Months	4½	44½
Number of Craft Employed	40 (B-29)	100 (SS)
Number of Crew per Craft	11	85
Total Craft Lost	15	52
Total Crew Lost	103	4,000
Cost of One Craft in Dollars	500,000	5,000,000
Ship Casualties to Enemy in Tons	1,250,000	4,780,000
Enemy Casualty Rate in Tons per Month	280,000	110,000
U.S. Ship Investment per Enemy Ton Casualty, Dollars	16	100
Tons of Enemy Casualty per Crew Member Required	3,500	560
Tons of Enemy Casualty per Crew Member Lost	12,000	1,200
Cost of U.S. Loss per Enemy Ton Casualty, Dollars	6	55

* From Reference 23.

Gulf. The use of obsolescent, but still effective, Soviet-design World War I–era contact mines created significant confusion, damaged several ships including a U.S. Perry-class frigate, and in the end sparked a loosely coordinated mine countermeasures effort among the littoral states, the United States, the United Kingdom, Belgium, Italy, France, and the Soviet Union. Although by all accounts obsolete and imminently sweepable, these mines constituted a real and significant threat to maritime traffic in the region. They could not be ignored.

If in the future all mines do become unsweepable, then there will be no need for antisweep devices such as ship counters. If ever a reality, unsweepable mines will be used with the idea that they can be dealt with if need be by advanced mine-*hunting* techniques or by remote command and control. The element of bluff in mine warfare will thus be reduced. Remembering the old policy of not using a mine unless a countermeasure were available for it, a requirement of the Hague Convention of 1907, Cowie remarked, "Had the British used the Oyster pressure mine before the Germans did, it is of interest to speculate whether the enemy would have refrained from using his version on the grounds that the British would have never done so unless they had developed a satisfactory countermeasure. Some such suggestion was in fact made at the time." [10, P. 186]

Defensive mining offers tremendous war-fighting "leverage." Although it is difficult to tell from existing sources exactly what numbers of mines have been used defensively rather than offensively, an overview of how many mines were used in various conflicts and their effectiveness in sinking or damaging ships is enlightening, especially as many pundits question the mine's usefulness in naval warfare of the twenty-first century. Mines used defensively around ports or at marshaling areas have in most cases a record of no enemy sinkings. For example, in World War II, of the some 300,000 mines laid in all theatres by the United States and the British, only about 100,000 were laid as offensive minefields; Cowie [10] estimated that 185,000 defensive mines were laid by the British, while Duncan's research [12] identified some 18,500 defensive mines laid by the United States.

What is extraordinary is the fact that in all the U.S. defensive minefields, there is no record that any enemy craft were sunk or damaged, nor is there any record of any enemy craft even passing through the fields. As far as the British are concerned, Cowie noted that "over 185,000 British mines were laid for protective purposes—the effectiveness of these minefields must be judged chiefly by the degree of protection afforded to shipping and in many cases [was] the sole reward for many months of slogging and unspectacular effort of the minelayers." [10, P. 133] The defensive minefield designed to keep attackers out is successful even if no ships are sunk. In this sense it reminds one of the "Australian McGubbin," which was explained as a device for keeping elephants out of Devonshire. "But," said a puzzled inquirer, "there are no elephants in Devonshire." "Aha!" exclaimed the McGubbin salesman, "you see, it works!"

On the other hand, several American ships—merchant and some naval ships—were sunk or damaged when they blundered into U.S. minefields off Key West, off Hatteras, in the Chesapeake Bay, and at other locations. Although the ultimate value of these particular defensive fields, so far away from the fighting in Europe and the German bases of operation, was uncertain, no one at the time was prepared to do without them. And perhaps with good reason, as it was certainly true that German submarines operated off U.S. shores and did in fact lay some 350 offensive mines off ports from Halifax to Trinidad and Panama, as shown in table 1. These mines resulted in the closing of many of the fifteen ports involved for several days each—a total of forty days' closure for all ports—and in sinking or damaging eleven ships. To this mine-effectiveness number could then be added American ships sunk or damaged in U.S. fields that had been planted because of the supposed continued presence of the German submarines.

In future naval conflicts, however, defensive mining could easily hamstring offensive naval operations. The undisputed Soviet capability of mining the approaches to key naval areas, in the Norwegian Sea and Northwest Pacific, for example, directly counters the forward, offensive operations envisioned in the Navy's Maritime Strategy. As expressed by Admiral Carlisle A. H. Trost, the Chief of Naval Operations, at the July 1989 Commander Mine Warfare Command change of command, "Soviet mining capability poses a threat to every aspect of the maritime component of our national military strategy. Resupply of our European and Pacific allies could be impeded or stopped at either end of the pipeline. The Soviet inventory includes sophisticated deep-water mines that can even threaten operations of our submarines, and open ocean operations of carrier and surface ship battle groups." The Soviets clearly embrace defensive minefields, and thoughtful observers would do well not to belittle their ultimate effectiveness.

Yet given the available data, perhaps it is still more appropriate to examine only offensive minefields if a meaningful measure of mine effectiveness is desired. Thus, it is necessary to find out the number of mines used and where they were laid, to separate sinkings from seriously damaged or slightly damaged ships, and to determine the types and tonnage of individual ships—combatants, auxiliaries, merchant vessels—so affected in one way or another by the mines. In most cases, these data are difficult to find.

Other important variables are the method of mine laying and the types of weapons used, which are important for the effectiveness of the field. In World War II a total of 576 Mk 12 and 82 Mk 10 mines were laid by submarines in thirty-six different fields in the Pacific. Of these, 421 mines in twenty-one fields sank twenty-seven ships (approximately 63,000 tons) and damaged twenty-seven more (approximately 57,000 tons). One ship was sunk or damaged for every eight mines laid, and the average ship size was 2,200 tons. [12] In Operation Starvation in 1945, on the other hand, the 12,128 mines planted in Japan's home waters by LeMay's 21st Bomber Command resulted in 294 ships sunk, 137 damaged beyond repair (making 431 ships—an estimated 900,000 tons of shipping—lost to the Japanese), and an additional 239 ships damaged. Therefore, eighteen mines were laid for every ship attacked; twenty-eight mines were laid for every ship lost, at an average cost of about $3,000 per mine. More important, however, this effort brought Japanese shipping traffic to a standstill and, in the opinion of some analysts, if it had been carried out earlier than March–August 1945, it would have been sufficient to end the war with Japan, thus avoiding the "necessity" of using atomic weapons against homeland targets.

Comparing the two methods—aircraft and submarine delivery—for the Pacific minefields, it took about twice as many mines per ship lost when laid by air as when laid by submarine. This might mean that submarine delivery of mines was more accurate in placing the weapons in recognized shipping lanes, or that it takes more mines to respond to smaller ships because their influences do not reach as far, or that there was more sweeping activity—or a "damn the torpedoes" mentality that resulted in many ships "being minesweepers, once"—later in the war in the case of air-laid mines, or some of all three effects. The fact remains that even with well-substantiated data there are rather wide variations in the number of ships lost per mine laid from one operation to another.

Another example of this fact is the Danube River mining by American Liberators and British Wellingtons during the summer and fall of 1944. According to the "Fundamentals of Mine Warfare Planning" [13], 1,200 mines were laid in the Danube by air during this period, sinking more than 200 ships and barges while "damaging many times that number, the highest casualty rate per mine ever achieved"—one ship sunk for every six mines laid. If twice as many more ships are counted as seriously damaged as were actually sunk, a not unreasonable assumption, then one ship was attacked for every two mines planted. The constrained geography of the minefield area—solely the Danube River—and the high density of shipping traffic even at the end of the war could account for the "highest casualty rate per mine ever achieved," if indeed it was that high.

In the interest of simplicity the data can be aggregated to arrive at average numbers that will balance the highly effective mine operations against those that were dismally unproductive (at least in terms of ships actually attacked). The total shipping tonnage lost per ton of mine laid can be used to see if there is a common thread from one campaign to another. In the Pacific during World War II, for example, the total number of Allied offensive mines laid was about 25,000, resulting in approximately 2 million tons of shipping lost (sunk or seriously damaged). This gives 80 tons per mine laid, or 107 tons of shipping sunk or damaged per ton of mine laid. Excluding the data from Operation Starvation, 750,000 tons of shipping were lost for 9,700 tons of mine used—a kill ratio of 77 tons of shipping per ton of mine delivered. If, however, "damaged ships" data are also included with data for ships sunk and "seriously damaged," and some figures for the number of ship-days of operations lost to repairs are therefore at least implicitly taken into account, the actual total effect of the mining efforts is greatly enhanced.

Comparisons of mine events among various campaigns is made more difficult by the fact that between World War I and World War II there was a fundamental change in the damage potential of mines. For example, the standard moored contact mine of World War I contained 300 pounds of TNT, while mine developments in World War II included ground influence mines containing, on the average, 800 pounds of high explosive, in many cases aluminized and hence the weight equivalent of about 1,000 pounds of TNT. The damage width for bottom mines varies as the square root of the charge weight. The damage width for a 300-pound charge on the bottom in shallow water is roughly equivalent to the beam of a ship passing over it, and for a moored contact mine the damage width is also the beam of the ship. The comparative damage width of the 1,000-pound charge on the bottom, therefore, will be the ship's beam times the square root of the ratio of the charges (1,000/300), or 1.82. Therefore, it should have taken fewer mines in World War II to damage a ship than in the earlier conflict, in about the ratio of 1 to 1.8, all other things held equal.

But all other things, in reality, are hardly ever equal, and it is difficult to compare meaningfully the mine data from the two World Wars. Nevertheless, table 6 provides data for both conflicts gleaned from several sources, and table 7 shows the experience in World War II. Analysis of the available data indicates that even though the average ship target in the European and Pacific theatres during World War II was for all purposes essentially the same (1,900 tons vs. 2,000 tons), it took twice as many mines to attack a ship in the European theatre as in the Pacific (forty-seven mines vs. twenty-three mines). One might conclude from this that the American mines used in the Pacific were better than the British mines employed in Europe, or that German countermeasures were better than Japanese, or both.

Comparing the two World Wars, the data seem to indicate that the effectiveness of the mining effort was much better in World War II than in World War I—an average of thirty-seven mines used per ship attack compared with sixty-seven mines per ship, a ratio of 1 to 1.81, not too far different from the ratio expected on the basis of charge weight. Of course,

TABLE 6.
MINES AND DAMAGE IN TWO WORLD WARS

Mines Laid		Ships Lost or Damaged		Mines/Ship
WWI				
British mines offensive & defensive (not in barrage)	114,000	Central powers* 129 warships ? merchant ships		?
Allied North Sea Barrage	70,000	German subs	~12	
Central powers* (of these 45,000 were German)	51,900	Allies* 586 merchant ships 87 warships Omitting: 2 minelayers 45 minesweepers 105 patrol boats	673	67
WW II				
British offensive mines in Europe	75,000	German (in Europe) 250 combatant 800 merchant ships sunk	1,590	47
Allied in Pacific	25,000	Japanese	1,075	23
Total offensive U.S. & U.K.	100,000	Axis & Japanese	2,665	37
Defensive U.S.	23,000	Axis	0	
Defensive U.K.	185,000	Axis. Zero to "several U-boats sunk"		
German* (offensive & defensive)	223,000	British ships sunk 280 combatants; 296 merchant**	576	

* According to von Ledebur, Reference 25. He does not distinguish between offensive and defensive mines.
**Reference 33.

this must be regarded as something of a happy coincidence—but on such sorts of evidence are many operational and managerial decisions based.

Table 8 provides a summary of the numbers of mines used by various countries in the twentieth century, and table 9 lists "noteworthy" mining events since the American Revolution. Figure 19 shows the locations of British and American minefields in Europe, 1914–18; figure 20 illustrates British minefields in Europe, 1939–45. These indicate the mineable waters of those periods and show the increased spread of mining operations in World War II, at least from England's perspective. The best map of the mine war in the Pacific is reprinted on the endpapers of this book. [49] (The numbers of mines used in previous wars, such as the Crimean War, the U.S. Civil War, the Paraguayan War, the Austrian-Prussian War of 1866, the Russo-Turkish War of 1877, and the Spanish-American War, are not known, but in all probability could not have

TABLE 7.
WORLD WAR II IN TWO THEATRES

Offensive Mines Laid	Sunk or Damaged beyond Repair	Damaged but Repairable	Total	Mines per Ship Sunk or Damaged	Own Vehicles Lost in Laying
	German ships				
British in ETO					
55,000 by air	864	483	1,347	41	?
17,500 by surface ship	124	50	174	100	13 minelayers 2DD
3,000 by sub	59	8	67	45	4 subs
75,500 Total (Data from Ref. 13)	1,047	541	1,588	47	
	Japanese ships				
Allied, mainly U.S., in Pacific					
9,254 by air	219*	120*	339	27	40 aircraft
12,135 by air (Operation Starvation)	431	239	670	18	15 aircraft (mostly not by enemy action)
~ 3,000* by DD	8	4	12	250	No losses
658 by sub	27	27	54	12	No losses
25,047 Total	685	390	1,075	23	

*Estimated. Other data compiled from References 12, 23, 33, 49.

exceeded 10,000 for the total.) These are examples that had a definite purpose, location, and result, and probably understate the actual totals. In most cases involving the use of mines, however, the specifics are unknown. How did the USS *Coolidge,* a troop ship in the Pacific theatre, encounter disaster? And what happened to the HMS *Britannic* in the Aegean?

A new twist: Terror and "low-intensity conflict"

In the post–World War II period mines have been *extensively* used in only one "limited war" campaign—during the Vietnam War. The May 1972 mining of Haiphong Harbor and its approaches saw the only use of Mk 52 influence ground mines, although "thousands" of Destructors (DST, Mk 36/40/41) were used in the initial and successive mine strikes at Haiphong and in massive mining efforts aimed at closing inland waterways. Reportedly, the entire inventory of Mk 52s in the Pacific Command was depleted in the initial strike of this one operation, on the order of only

TABLE 8.
MINES USED IN THE TWENTIETH CENTURY
(Offensive and Defensive)

Country	Russo-Japanese War	WWI*	WWII*	Korean War	Vietnam War	Post-Vietnam Crises
Germany		45,000	223,000**			
Austria-Hungary		6,000				
Turkey		900				
Japan	~1,000		51,000			
Finland			5,500			
Bulgaria			1,000			
Romania			6,500			
United Kingdom		129,000	263,376			
U.S.A.		57,600	44,000		~333,000	~1–3
U.S.S.R./Russia	~1,000	52,000	40,000			
Italy		12,000	54,457			
France		5,000	?			
Denmark		1,200				
Norway		400	5,000			
Sweden			4,346			
Greece			?			
Poland			110			
N. Korea				3,000		
Iran						~100–150
Libya						~20
Argentina						~10
Unknown		600				
Total	~2,000	309,700	~700,000	3,000	~333,000	~130–200

* Data from References 25, 57, 67, 68, 73, 74, 76, 78, 79.
** Reference 13 gives 126,000 mines and 32,000 obstructors.

130 weapons—so much for the adequacy of U.S. mine stockpiles and readiness! Destructors were also extensively planted by the U.S. Navy and Air Force to "mine" land routes suspected of being used by the North to resupply communist units operating in South Vietnam. The magnetic influence mechanisms were strengthened to withstand the shock of delivery on land, and the sensitivity of the target detection device was such that small vehicles passing nearby were sufficient to detonate the mine. (Another indication of the sensitivity of these early bomb-conversion kits is that the Haiphong Destructor field was actually swept by a solar magnetic storm in August 1972.) Something on the order of 333,000 mines—principally Destructors—were dropped during the Vietnam War; some 11,000 Destructors were used against Haiphong alone.

Mines and other ordnance were used to a limited extent by the Egyptians and the Israelis in, and to block the approaches to, the Suez Canal during the 1967 and 1973 Wars. No data are available in the public domain regarding what types and how many mines were actually laid, however. The U.S. minesweeping efforts of 1974 and 1975, dubbed Nimbus Star and Nimbus Moon, saw the second operational use of helicopters to clear suspected mined areas. The first, the End Sweep operation to clear

TABLE 9.
NOTEWORTHY MINING EFFORTS

Date	War/Incident	Location	Mines Used (and by whom)	Results
1777	American Revolution	Delaware River at Philadephia	"Several kegs" (U.S.)	One rowboat sunk by accident.
1861–1864	American Civil War	Mobile Bay Charleston Red River Mississippi River James River, etc.	Moored Control or Jacobi Chemical contact Singer with percussion caps (Confederates)	27 Union ships sunk (more damaged) by Confederate forces.
1904	Russo-Japanese	Port Arthur	Moored Inertial contact	16 warships sunk.
1915	WWI	Dardanelles Kephoz	350 moored contact 20 mines south of Kephoz (Turks)	3 warships sunk. 1 beached. Campaign abandoned.
1917	WWI	Dover Straits	5,000 moored Hertz horn (U.K.)	Closed Strait after "many losses."
1918	WWI	North Sea Barrage	56,033 Mk 6 Antenna mines ~14,000 H-2 Hertz horn (U.S., U.K.)	6 submarines sunk and ~6 damaged.
1940–1943	WWII	Orkney-Iceland	110,000 Antenna Mk XX (U.K.)	Defensive. "Several" U-boats sunk.
1944	WWII	Palau	78 Mk 10; Mk 25 mines (U.S.)	Closed atoll. 32 Japanese ships trapped & sunk. 3 more sunk later by mines.
1945	WWII	Shimonoseki Strait & Japanese Islands	12,128 Mk 25 & 36 (U.S.)	670 ships sunk or damaged.
1972	Vietnam	Haiphong Harbor and approaches	11,000 Destructors ~130 Mines Mk 52 (U.S.)	One plane lost in delivery. No further ship movement until war ended and the Destructors sterilized or self-destructed.
1982	Falklands	Port Stanley	~20 "WWI"-design (Argentina)	RN swept after conflict ended.
1984	Nicaragua	Corinto	1–3 "mines" (Unknown)	2 ships damaged. Claimed laid by CIA or *Contras*.
1984	"Mines of August"	Red Sea/Gulf of Suez	~20 "99501" mines (Libya)	16–19 ships damaged. 1 mine recovered.
1987–1988	Iran-Iraq	Persian Gulf	100–150 M-08/39 Hertz horn	Numerous mines laid by Iranian Revolutionary Guards. At least 15 ships damaged, several sunk.

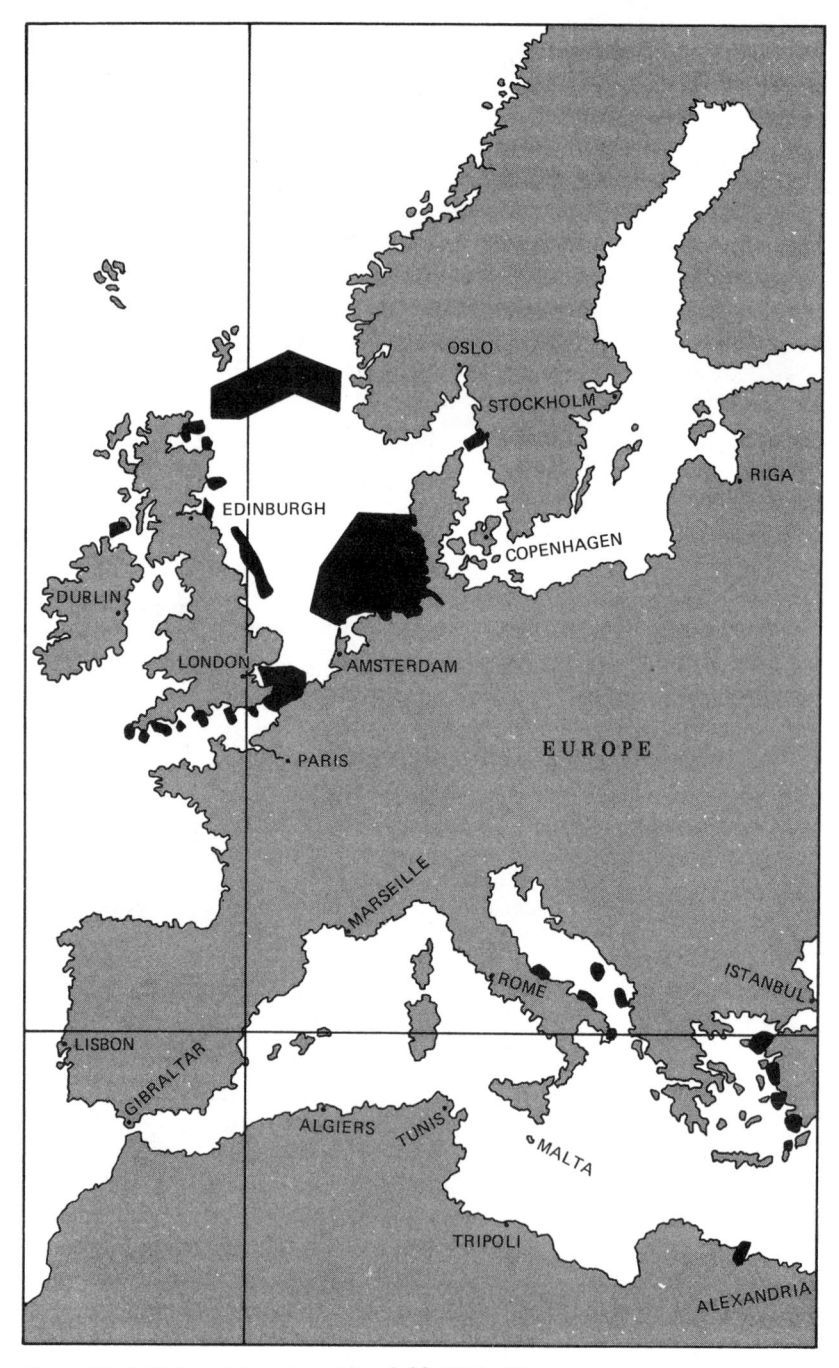

Figure 19. British and American Minefields 1914–18.

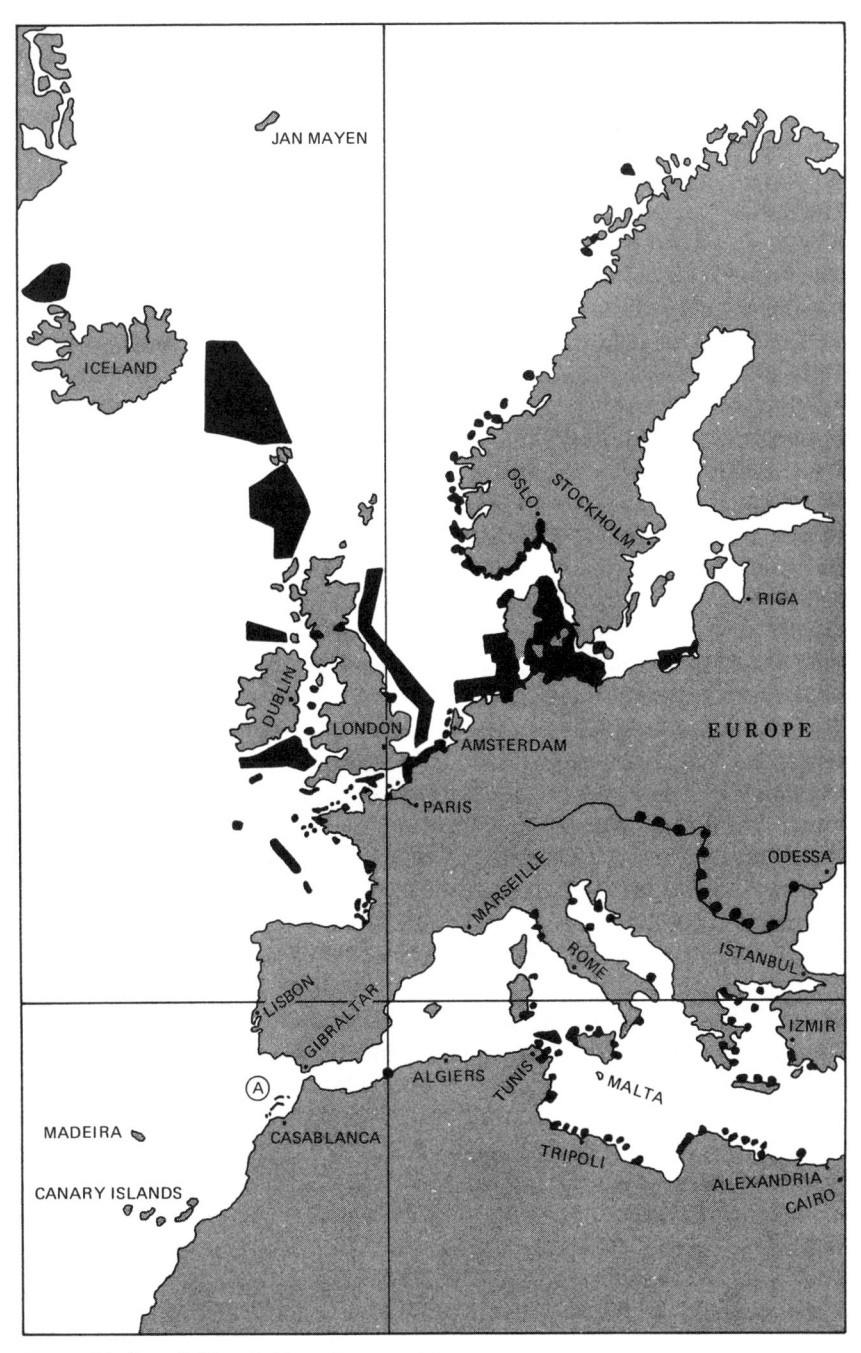

Figure 20. British Minefields in Europe 1939–45.
(American Minefields shown by (A))

247

the mines laid at Haiphong, involved both helicopters and surface mine craft working together.

The Falklands and Nicaragua

While the dramatic sinking of the HMS *Sheffield* from a single hit by an Exocet (AM 39) antiship missile brought into stark focus the present dangers of futuristic missile wars at sea, the short but intense war between Argentina and Great Britain over control of the Falkland Islands in the spring of 1982 also saw the much more mundane threat of naval mines still taken very seriously by the Royal Navy. Three days after the Argentines had captured the South Atlantic British colony on 2 April 1982, the HMS *Hermes* and HMS *Invincible*, principal combatants in the largest British task force in recent history—which eventually numbered some 110 warships, naval auxiliaries, and merchant Ships Taken Up From Trade, or STUFT—left the United Kingdom with the sole objective of retaking the Falklands. Included in the British armada were five Extra Deep Armed Team Sweep Trawlers taken up from trade, fitted with minesweeping gear and manned by Royal Navy personnel. [57, Pp. 5–6, 26]

As the Royal Navy mounted its campaign to retake the Falklands in May 1982, a solitary Argentine surface ship managed to lay a single minefield off Port Stanley, complicating the landings of Royal Marines and the continued supply of materiel and foodstuffs to the British forces and civilians on the Islands. [68, Pp. 50–51] (The Argentine ship with mines on board was actually sunk by Harrier aircraft, but this was not known by the Royal Navy until after hostilities ended.) Although the mines were World War I moored contact devices, "obsolete in comparison with modern naval mines," the U.S. Navy noted that "they still had to be swept or neutralized before British ships could operate in those waters." More ominously, the Navy also recognized that "had the Argentines placed minefields in Falkland Sound before the arrival of the British Force, the landings might have encountered significant delays." As it was, the initial amphibious landings at San Carlos Water were subjected to furious attacks by the Argentine Air Force, causing great damage to the Royal Navy; it could have been much worse had even a limited number of mines been in place.

The five commercial STUFT trawlers were outfitted with deep-water sweeping gear but were employed principally in utility and cargo roles until the hostilities ended. In all, these five little ships managed to sweep "some ten enemy buoyant mines," according to the Royal Navy's review of the Falklands campaign, although no data on the actual type and design of the mines have been provided publicly. The U.S. Navy's 1983 assessment of the "lessons learned" for the United States from this conflict is telling in its candor:

The limited mining capabilities of the Argentine Navy, if used in a timely and effective manner, could have created major problems for the British force. The U.S. Navy had learned the implications of a minor or nonexistent naval power using obsolete naval mines during the Korean War (1950–1953). The U.S. Navy MCM [Mine Countermeasures] capability built up in the 1950s has been allowed to deteriorate. Today the Navy has 21 minesweeping helicopters (RH-53D) and three active minesweepers (built in the early 1950s), plus another 18 outdated minesweepers in the Naval Reserve Force.

Major programs are underway to revitalize the U.S. Navy's MCM capability.... Unfortunately, MCM forces tend to be particularly vulnerable to budget reductions. The Navy will attempt to ensure that the MCM program is completed as now structured. [68, P. 51]

Less than two years from the time that relations between Argentina and Great Britain began to sour over who owned the Falklands, mines—or mine-like devices—were again used in a smoldering dispute between an upstart nation and, apparently, a great power intent on preserving its interests in the third world. In March 1984 the Soviet-flag oil tanker *Duhler* was seriously damaged and several of its crew wounded by an underwater explosion while at Nicaragua's Pacific-coast port of Corinto. A mine of "unknown origin" was identified as causing the damage. Another two vessels were reportedly damaged by mines (limpets?) earlier in the month, with the Sandinista government of Daniel Ortega eventually blaming *Contra* rebels in league with the United States for all incidents.

Almost from the earliest days of the first Reagan administration, the pronouncement of the Reagan Doctrine had focused on Nicaragua as a key country from which Soviet influence was to be "rolled back." Speculation had indeed focused on clandestine activities of the U.S. CIA as behind the mining in support of the administration's policies for Latin America. [76, Pp. 1239–48]

While not explicitly admitting a U.S. role in the Corinto mining, the State Department in March 1984 presented to the Congress a legal opinion that "naval mines can be a legitimate means of self-defense and have long been accepted as such by the international community." [84] Various legal experts took exception to the official U.S. position, arguing that, among other issues, a resort to mine warfare is a very serious step and usually is done only after a declaration of war, and then within the context of a blockade where official notice is given to all nation-states of the potential danger to their ships. The State Department brief also rationalized the use of mines against Nicaragua because they were weapons of harassment rather than powerful charges capable of sinking ships or killing people.

For all the conjecture, details of this incident remain highly ambiguous: What types of mines were used, and how many? Who actually laid them, when, and how? Who supplied them? What is known beyond the legal brief presented to Congress is that the United States was willing to risk international reprobation by refusing to accept the compulsory jurisdiction of the World Court in the resulting dispute between Nicaragua and the United States, rather than possibly letting all the facts come out in a public hearing. That this mine-tempest was soon overshadowed by other, more pressing events is perhaps testimony to the fact that the world was becoming ever more inured to low-level international violence, which increasingly has come to include the use of naval mines in "peacetime" acts of terrorism against the West.

The Gulf of Suez/Red Sea Crisis

In what became known as the "Mines of August" crisis, beginning with a mysterious explosion on 9 July 1984 involving the Soviet-flag merchant ship *Knud Jesperson,* within three months the masters of some nineteen ships under the flags of fifteen different countries would claim to be victims of clandestine "minefields" in the Red Sea and Gulf of Suez. [79] The suspected victims—three of which are highly questionable—are listed in table 10.

The intensive mine-hunting/sweeping efforts comprised modern and not-so-modern forces from Egypt, the United States, Great Britain, France, Italy, and the Netherlands, while the Soviet Union had three minesweepers operating for a short time in the Bab el Mandeb and off Aden. The United States sent two contingents of RH-53D Sea Stallion AMCM (Airborne Mine Countermeasures) helicopters from the Norfolk-based HM-14 squadron to the area, providing the first real-world "combat test" of the AQS-14 side-scan sonar. Flying from the amphibious transport dock *Shreveport* (LPD-12), the operations of the four U.S. helicopters in the Gulf of Suez emphasized mine *hunting,* as the goal was to locate and retrieve a mine to discover its origin. In Saudi waters off the port of Jidda, however, the three RH-53Ds operating from the USS *La Salle* (AGF-3) focused on mine*sweeping* in order to make the sea approaches to Jidda safe as quickly as possible.

During the countermeasures operations the various navies came up with literally hundreds of mine-like contacts, many of which were buried under several feet of mud and sediments. The four British mine hunters, for example, classified fifteen contacts for each four square miles of search area. Most turned out to be innocent, including parts of aircraft, oil drums, toilets, anchors, and similar debris. The Dutch mine hunters complained of the astonishing amount of junk littering the sea floor, including refrigerators! Nevertheless, most contacts had to be examined by divers for positive identification.

TABLE 10.
RED SEA/GULF OF SUEZ "MINES OF AUGUST" INCIDENTS (1984)

Date	Ship	Area
9 July	*Knud Jesperson* (Soviet)	N Gulf of Suez
27 July	*Este* (W German)	SW Gulf of Suez
	Medi Sea (Liberian)	NE Gulf of Suez
	Meiyo Maru (Japanese)	NE Gulf of Suez
28 July	*Bigorange XII* (Panamanian)	SE Gulf of Suez
	Linera (Cypriot)	SW Gulf of Suez
31 July	*Hui Yang* (Chinese)	S Red Sea/Bab el Mandeb
	Peruvian Reefer (Bahamian)	S Red Sea/Bab el Mandeb
	Valencia (Spanish)	Gulf of Suez
2 Aug	*Kriti Coral* (Panamanian)	S Red Sea/Bab el Mandeb
	Morgul (Turkish)	S Red Sea/Bab el Mandeb
	Dai Hong Dan (N Korean)	S Red Sea/Bab el Mandeb
	George Shumann (E German)	S Red Sea/Bab el Mandeb
3 Aug	*Tang He* (Chinese)	S Red Sea/Bab el Mandeb
5 Aug	*Oceanic Energy* (Liberian)*	S Central Red Sea
6 Aug	*Bastion* (Soviet)	S Red Sea/Bab el Mandeb
11 Aug	*Jozef Wybicki* (Polish)	Bab el Mandeb
15 Aug	*Theopoulis* (Greek)	S Red Sea/Bab el Mandeb
30 Sept	*Belkis I* (Saudi Arabian)	N Gulf of Suez

* Assessed by commercial surveyors and the U.S. Navy as not a mining victim, since there were no signs of an external explosion; possible "insurance scam," according to diplomatic and military sources. Two other incidents are viewed with great suspicion: *Theopoulis* and *Belkis I*.
Source: Reference 79, P. 97.

Occasionally suspicious buried contacts located in the Gulf of Suez were blown up without positive knowledge of what they were, on the assumption that they could have been mines; at times this produced a much larger explosion than anticipated, indicating that the contact had indeed been ordnance. All buried contacts were assessed as too old to be related to the ships damaged in the late summer of 1984, as a newly laid mine would not have been expected to be covered by thick sediments. The British and French detonated numerous mines, some dating to World War II, a 2,000-pound bomb, and a practice torpedo. The French alone destroyed ten mines lying on the bottom in depths of some 250 feet. The U.S. force found no mines in the Gulf of Suez, although numerous contacts had to be investigated. No mines were located or swept by the French, Dutch, or U.S. forces operating in Saudi waters.

The Royal Navy located only one modern mine, which was brought on shore, carefully, for examination. Dubbed the "99501" mine because of the markings on its case, it was too clean to have been in the water for very long. The Royal Navy assessed it to be an "export" version of a Soviet ground mine with a combined magnetic/acoustic/seismic-influence target detection device, fully packed with 750 kg of high explosive.

Piecing together all available evidence gleaned from many sources, it seems beyond doubt that the "culprit" in this case was a Libyan-flag civilian roll on/roll off ship, the *Ghat,* which had embarked military personnel, including the head of the Libyan mine-laying division, prior to entering the Suez Canal on 6 July 1984, and thereafter sailed to Assab, Ethiopia, and back, returning to the Canal on 21 July. Normally the voyage, if for peaceful, commercial purposes only, would have taken about eight days, including unloading at Assab the "general cargo"—crated goods—carried by the ship as she steamed through the Canal and the Red Sea. In this instance, however, the *Ghat* needed an additional seven days to make the round trip.

With the *Ghat* almost certainly sailing under the orders of Libyan leader Colonel Muammar Qaddafi, the motives behind her mine-laying mission were variously judged to be Qaddafi's hatred for Egyptian President Mubarak, the need to avenge recent political embarrassments, the desire to support Iran in its continuing struggle against Iraq (Mubarak having offered strong support to Iraq in the form of "volunteers" and funds), and perhaps even the desire to embarrass Saudi Arabia, keeper of the two most sacred holy places of Islam, as the annual *Hajj* was just beginning and thousands of pilgrims were expected to make the journey by sea. Given the physical configuration of the mine recovered, it was seen as a small matter for Qaddafi's henchmen to have merely rolled the "uncrated" weapons down the *Ghat*'s stern ramp, into the sea. When the *Ghat* was subsequently seized in the French port of Marseilles, an examination revealed extensive damage and warping of her stern ramp, which were assessed as possibly caused by the ramp's having been lowered in a seaway. Admittedly this was circumstantial, yet other, more incriminating "evidence" was reportedly found during the inspection. But as culpability could not be positively proved, the ship was later released from custody.

One remarkable aspect of the Red Sea/Gulf of Suez incidents was that only one new mine was located, that only some nineteen ships—if indeed all nineteen *were* mine victims—detonated explosions, and that there have been no other reports of mine incidents in the region since September 1984. The inference thus is that only about twenty mines were laid and that each one, save the one recovered by the Royal Navy, was set off by a target ship, a remarkable performance! There were some contradictory statements about the water depths in which the mine explosions occurred, with the Suez Canal Authority stating that the depths were from 200 to 300 *meters,* whereas other authorities stated that all ships damaged were in waters with depths of around 300 *feet.* The "99501" mine was located in waters only 150 feet deep. Adding to the confusion, early reports from the U.S. and Royal navies, moreover, said that intelligence sources had indicated that the mines had been loaded with only half-charge weights (how would they know?), but the "99501" mine, according to the Royal Navy's Captain Mine Countermeasures, speaking at a U.S. Naval Institute symposium in May 1985, was "fully loaded" with high explosive. [67]

Whether the charge weights in these mines were less than a full load or not, the distances of the explosions away from the target ships, from 20 to 200 meters—a factor loosely correlated with the depth of the water, with the closer explosions occurring in deeper water—and, except for one or two instances, the relatively small amount of damage actually suffered by the ships, all indicated that the mines' target detection devices were set at a high sensitivity so that they would explode beyond the serious damage radius. No ships were sunk or in imminent danger of sinking. Why all this was done, and at the "crazy" Qaddafi's bidding, has remained something of a mystery. Perhaps Qaddafi had taken a cue from the U.S. State Department's brief on the Corinto mines: use weapons of harassment rather than powerful charges capable of sinking ships or killing people.

One observer, however, has argued that Libya may have asked for and received mines from the Soviet Union to protect its coasts and ports from an anticipated amphibious attack by the United States and other anti-Arab states. [54] An operational test of the mines and Libya's capabilities to use them was necessary to ensure confidence that the feared attack from the West could be repelled. No better operational evaluation was available to Libya than to place a limited number of the mines in a well-traveled sea lane where the exact location and water depth of each mine, its charge weight, influence combinations, and sensitivity setting could be correlated with its target size, speed, and damage reports. Given the politics of the region, it was also highly likely that some terrorist group would claim responsibility for the acts (several did), and that the world's press would play up any and all such claims, thus blunting suspicion that Libya was the culprit. Moreover, the use of a civilian vessel would further cover up the trail. If high sensitivities were set for each mine's target detection device or if the explosive charges were reduced, ships would probably not be sunk or people killed. Insurance companies would pay the bills, while reports from the press, port authorities, or shipyards would constitute the test results. Indeed, the Lloyd's of London weekly shipping casualty reports provided a great deal of comprehensive information on the location of each ship "struck" by mines and the resulting damage. And during the crisis diplomatic sources reported that the Soviet leadership was "furious" at Qaddafi for his indiscriminate use of the Soviet "export" mines.

The "Mines of August" crisis, in a much broader context, served warning that mining had become an accepted terrorist tactic. Indeed, the terrorist group Islamic Jihad was quick to claim credit for the mines, and several other terrorists also clamored for recognition during the confusing days in late July and early August, when numerous ships were radioing for assistance and warning that they were "in the middle of minefields!" Given numerous narrow straits and approaches to key harbors and ports, the potential for terrorist mining, whether carried out by subnational entities or nation-states, is alarming. This crisis underscored what had been known all along: there are no easy means to prevent surreptitious mining of strategic sea areas, and once these areas have been mined, or are *thought* to have been mined, significant efforts are required to ensure the safe passage of ships.

"Obsolete" Mines in the Iran-Iraq War

The use of mines by Iranian Revolutionary Guards to confound naval and merchant traffic in the Persian Gulf came late in the Iran-Iraq conflict, which had already seen Iran's use of children "martyrs of the faith," each issued a golden key to heaven, to clear suspected Iraqi minefields on land. (Iraq, however, in 1980 apparently mined the Shatt al-Arab waterway, sinking one vessel and damaging another two.) That being so, the first public warning that naval mines might be introduced into the fighting came in a 1980 announcement by Iran, a month after the start of hostilities. On 14 October Tehran radio renewed the revolutionary government's warning that it would take "drastic action" if other Gulf states aided Iraq. The commander of the Iranian Navy was quoted to the effect that he would mine the Strait of Hormuz if other countries gave military aid to Baghdad.

Nevertheless, according to a 27 October 1980 *Time* magazine report, American officials remained confident that an Iranian threat to mine the Strait of Hormuz "was a bluff. U.S. intelligence found no evidence that Iran was manufacturing mines or acquiring them from abroad." Moreover, an Iranian mining of the Strait of Hormuz was seen as something of a suicidal gesture that would cut Tehran's own economic throat, as all Iranian oil exports, the major source of hard currency to prosecute the war with Iraq, had to transit that waterway. Perhaps sensing growing international concern, on 22 October Iran sent a message to United Nations Secretary General Kurt Waldheim affirming that it was committed to keeping the Strait of Hormuz open to international shipping. Denying "certain rumors" that it intended to blockade vital oil lanes, the message stated that Iran would "not spare any effort" to ensure freedom of navigation.

Several years later, as the "tanker war" in the Gulf reached frightening proportions and Iran responded with speedboats armed with rocket-propelled grenades to the Iraqi escalation of aircraft-launched missile attacks against "large naval targets," Kuwait requested that both the Soviet Union and the United States provide naval protection for its tankers in their passage through the region. [73, 74] Only after the Soviets were seen to be willing to assist Kuwait did the United States act in the late fall and early winter of 1986, offering to reflag *all* the Kuwaiti tankers in question, rather than share duty—and political influence and good will in the Gulf—with the Soviet Union. Eleven tankers were eventually reflagged as U.S. merchant vessels, although Kuwait also chartered three Soviet tankers that remained under the flag, and ultimate protection, of the Soviet Union.

According to some reports, as early as 1980 three ships had been mined and in 1984 two more ships struck naval mines in the headwaters of the Persian Gulf. But details of these minings were sketchy at the time, and no more incidents were reported until mid-1987; most observers acknowledged little concern about mines. The next confirmed mine attack occurred near Kuwait on 16 May 1987, the day before an Iraqi Etendard aircraft fired two Exocet missiles at the USS *Stark* (FFG-31), and was thus relegated to the shadows of U.S. and international attention after the *Stark* was nearly sunk. (There were also unconfirmed reports that another three ships had struck mines around this time.) And as this first mining victim was one of the Soviet-flag tankers chartered to Kuwait, the *Marshal Cheykov,* and the Soviet Union had little to say on the incident, the threat of mines continued to be disregarded.

Indeed, after the United States earlier in the year undertook to reflag eleven Kuwaiti tankers and organize naval escorts for them as they plied the Persian Gulf, published reports indicated that the Navy's threat assessment for the convoy mission all but ignored mines. Several early listings of possible threats to U.S.-flag ships operating in the Gulf showed that Navy planners had identified everything from Chinese-supplied Silkworm land-launched antiship missiles to shoulder-fired rocket-propelled grenades launched from speedboats. Early on there were *no* mentions of naval mines, even though a U.S. Navy ordnance disposal team had been in Kuwait since late spring helping clear Kuwaiti waters of mines. In private communications with U.S. Navy personnel familiar with the planning for the convoys, these officers admitted that the mine threat had been relegated to such a low priority that if it ever appeared on official lists, it did not remain in focus for very long.

On 18 June 1987, for example, a Pentagon spokesman called the Iranian mines an undeniable risk, but one that was relatively remote. In his assessment, Iran "had nothing that we would think of as a mine-laying capability in military terms, and no modern mines." [71] Later that summer, after U.S. forces had found themselves mired in minefields, Admiral William J. Crowe, Chairman of the Joint Chiefs of Staff, admitted to the Senate Armed Services Committee that the Navy had "overrated our intelligence" and had underestimated the dangers posed by Iranian mines as it prepared to escort U.S.-flag ships through the Persian Gulf. [72]

With supreme irony, then, during the first U.S. Navy escort operation of two reflagged tankers, on 24 July 1987 the 401,400-ton *Bridgeton* struck a mine in the Farsi Channel, about twenty miles west of Farsi Island, as it traveled north to the Kuwaiti anchorage of Al Ahmadi, an area that had previously been considered safe from any Iranian threat. (After the *Bridgeton* was hit, then–Secretary of Defense Caspar Weinberger candidly noted that "we weren't looking for mines there because we had never seen a mine in that area.") Although damage was extensive—a fifteen-by-thirty-foot hole was ripped in her one-and-one-eighth-inch plating, with the explosion driving shrapnel through several decks—the supertanker was in no danger of sinking. She proceeded under her own power, now operating as the convoy's minesweeper, with the other tanker and the three U.S. combatants falling in astern. [77]

The mine was later determined to be a 1908-vintage Russian-design M-08/39, a bottom-moored contact mine with chemical horns and a 115-kg high-explosive charge. The *Bridgeton*'s master, Captain Frank Seitz, later remarked that the damage his ship suffered "would have done a number" on a frigate. (By the end of 1987 a total of ten ships reported damage from mines sown *seemingly* haphazardly throughout the Persian Gulf. In actuality, the Iranian minelayers knew exactly what they were doing, at times even focusing on specific convoy targets by planting mines during the night in the channels expected to be used the next day.) Speculation, never publicly confirmed, focused on the mines being supplied by North Korea—the mines were said to be identical to those used against U.N. forces at Wonsan thirty-seven years earlier—or the People's Republic of China, with some analysts suggesting that even the Soviet Union might have supplied them in an attempt to better its relations with Iran's reigning mullahs. Others noted that perhaps the Iranians had obtained a few mines from North Korea and copied their design, reproducing them by the hundreds in secret arsenals; in published accounts, intelligence sources discounted these reports, however.

Now at General Quarters regarding the mine threat in the Gulf, the Joint Chiefs of Staff decided as a first step to send in elements of a squadron of AMCM helicopters to help protect the U.S. convoys. The initial movement of these helicopters and their equipment was held up, seemingly interminably, as discussions with some Gulf countries friendly to the United States—including, surprisingly, even Kuwait—regarding landing privileges for the C-5 transports foundered on concerns about raising the political visibility of the U.S. presence in the region.

As it was, the helicopters were finally brought in, but only after their transports landed at the Diego Garcia Navy Base and the helicopters transferred to the amphibious warfare ship USS *Guadalcanal* (LPH-7) for the long sea transit to the Gulf. Flying from the *Guadalcanal* with the minimum of shore-based logistics support—to keep the United States' onshore visibility at a minimum and thus not exacerbate local political sensitivities—AMCM operations using Mk 103 mechanical sweep gear and AQS-14 side-scan sonars began in mid-August. The mine-hunting sonar operations were frustrated because the AQS-14 "looks" downward to locate bottom mines, while the mines in the Gulf were buoyant devices either floating or moored on the bottom and suspended only a short distance below the surface. (Some Navy and industry observers noted, however, that a "field-engineered" reconfiguration of the AQS-14 sonar "fish" allowed it to make a good contribution to the mine-hunting efforts.)

The helicopter operations went well, but they did illuminate some of the tactical shortcomings of AMCM (short-duration sorties; high costs per hour of AMCM operations; operational constraints exacerbated by the hot, dusty, and humid weather conditions of the Gulf; and political constraints posed by host-country concerns about landing rights), as well as the operational advantages compared with surface mine countermeasures vessels (AMCM vehicles can be on the scene quickly once the decision to go is made). Thus, right from the start of planning—however belatedly in actuality—for mine clearance in the Persian Gulf, the Navy looked to both the "cutting edge" of its AMCM helicopters and the staying power of its "aging" surface mine force.

Six minesweepers—three from the Atlantic Fleet and three from the Pacific Fleet, only one of which, the *Illusive* (MSO-448), was from the active force, the other five being Naval Reserve Force Ocean Minesweepers (MSOs)—were also ordered to clear the shipping lanes and protect the protectors. Arriving in the Gulf in October 1987, by early spring of 1988 the "obsolete" 1950s-built U.S. minesweepers had identified and destroyed thirty-six mines.

Nine mines were seized on the landing craft/"minelayer" *Iran Ajr,* in a joint operation involving high-flying U.S. reconnaissance aircraft, airborne intelligence and electronic eavesdropping, and Army special operations forces operating "strike helicopters" (AH-58D Kiowa Warriors) from Navy surface combatants. [82, 83] The capture of the *Iran Ajr* on 21 September 1987, and her subsequent scuttling by the Navy after the mines on board were neutralized, points out that astute planning for mine countermeasures operations includes making the best use of intelligence from all sources, and that the best mine countermeasures effectiveness comes from preventing the mines from being put into the water in the first place. That being the case, however, the Navy's mine force still had to address the threat of the mines that somehow slipped through and were planted in the Gulf. As the commanding officer of one of the Naval Reserve Force MSOs sent to the Persian Gulf stated in the early spring of 1988, "*Inflict* [MSO-456] came to the Gulf on short notice, found mines the first day and destroyed them. I'd say we countered the threat effectively." [59, P. 2]

Not quite as threatening-looking as the World War II contact mine bristling with "horns," but much more deadly: the instrumentation section of the Soviet mine raised by the Royal Navy in the Gulf of Suez in September 1984. Barely legible are serial numbers ending in "81," indicating that the mine was of recent manufacture.

A Russian-design World War I–era M-08/39, by all accounts an "obsolete" weapon, nonetheless caused extensive damage to the reflagged tanker SS *Bridgeton* on 24 July 1987. It ripped a fifteen-by-thirty-foot hole in the tanker's one-and-one-eighth-inch steel plating, shook the bridge and superstructure, and drove shrapnel through several decks.

Close-up of the Soviet-design World War I–era contact mines found on the Iranian landing craft *Iran Ajr* on 21 September 1987, during the 1987–88 Persian Gulf crisis. Intelligence sources noted the "crude manufacturing of the mines," which belied their ultimate effectiveness.

259

U.S. Navy Explosive Ordnance Demolition (EOD) teams relax after disarming the M-08/39–type mines found on the captured *Iran Ajr*. It proved to be a relatively simple matter for Iranian Revolutionary Guards to learn of a U.S. convoy of reflagged Kuwaiti tankers, load the mines onto ships like *Iran Ajr,* and roll them off the ship—using the bow ramp, in the case of *Iran Ajr*—literally in the path of an unsuspecting convoy. In this instance, however, good intelligence and coordination of many different U.S. assets proved that the best mine countermeasures operation is one that prevents an adversary from placing the mines in the water in the first place.

Only the most heroic efforts saved the U.S. frigate *Samuel B. Roberts* (FFG-58) after she hit an "obsolete" World War I–era mine laid by Iranian Revolutionary Guards. Her back broken in the 14 April 1988 incident, the *Roberts*' repairs ultimately cost $96 million.

On 14 April 1988 the frigate USS *Samuel B. Roberts* (FFG-58) was hit by an underwater explosion while east of Bahrain. The ill-fated ship's crew knew they had stumbled into a minefield, having maneuvered past three mines before a fourth, unnoticed, struck the ship below the waterline on her starboard quarter. According to Vice Admiral William Rowden, then Commander Naval Sea Systems Command, ten crewmen were injured in the blast, which threw the ship's two gas-turbine engines off their mounts, ignited a fire in one of the engines, and sent a fireball 150 feet above the ship. A twenty-two-foot hole was ripped in the ship by the mine, said to be one of the M-08/39s that had been planted by the Iranians, which caused flooding in two adjacent compartments. The *Roberts*' keel was broken, putting her in extreme danger of sinking. Admiral Rowden tersely remarked at a Pentagon briefing called after the incident, "It required the effort of all hands to keep her afloat." The ship was later carried on a heavy-lift ship to Bath Iron Works for repairs estimated to cost $96 million. All this from an "obsolete" mine.

At the time of her near-sinking Navy sources said that there had been no reports of recently sown mines in the Gulf off Bahrain and that the mines the *Roberts* encountered may have floated into the area from an old minefield. One floating mine *had* been destroyed in the area on 9 April, less than a week before the *Roberts* was hit. Most of the mines that had been swept during early 1988 had obviously been in the water for some time, as many were covered with heavy marine vegetation.

After the Persian Gulf minesweeping operations began and up to the *Roberts* incident, the U.S. Navy found and destroyed forty-four mines, including sixteen in the first seven months of 1988. Contacted in mid-1989 for more comprehensive data, Navy sources said that no complete listing of the number of mines swept had been declassified, and, sadly, the total effectiveness of the Navy's "overage" minesweepers could not be validated. On faith, almost, these sources stated that "we did a great job, using equipment that long ago had been rejected by various pundits as obsolete and useless." There are, moreover, no reliable data on the total minesweeping efforts of the other navies—including the British, Dutch, French, and Soviet navies, in addition to the limited capabilities of the littoral states—which had mine countermeasures assets in the area. The Soviet Navy had three ocean minesweepers in the Gulf, even before the crisis escalated in the spring of 1987, raising suspicions that the Soviets knew more about Iranian mines than they were willing to admit.

Mines will continue to be used, with good results, by those individuals and countries intent on frustrating the movements of ships throughout the world. In late 1990 as this updated chapter was being prepared, mine warfare sailors and experts in the U.S. Navy were waiting for the next "Mines of August" crisis to erupt, threatening U.S. or allied interests where it was least expected. In the case of Operation Desert Shield, the U.S. response to Iraq's invasion of Kuwait in August 1990, the crisis once again erupted in the Persian Gulf. As one naval officer, well familiar with most of the postwar mine warfare vicissitudes of the U.S. Navy, expressed the situation in early 1988:

> Clearly the historical record confirms that mine warfare is an integral element of naval power. It illustrates that maritime nations are vulnerable to mining in both home and distant waters. It shows that less developed nations and rogue political groups can wage effective mine warfare. And it demonstrates a record of interwar neglect in mine warfare in the U.S. Navy that culminated in the Persian Gulf crisis of 1987 when the world's foremost sea power failed to counter with celerity the antique mines of a minor power, despite knowing beforehand that mines would certainly be laid. [78, P. 47]

The denouement of these most recent experiences may be summarized best in a new maxim that describes the Navy's pendulum-like attention to mine warfare readiness, which heightens during a crisis only to plummet after the threat goes away: "Out of sight, out of mine." But it ain't funny.

12 Into the Twenty-First Century

Prediction is, at best, a risky business, and in most cases an impossible one. It is rather easy, however, to identify changes that have come about since the last great war, or even over the course of the "limited" wars of the last forty years, which are certain to have an effect on naval operations. Nevertheless, it remains difficult to foretell what their effect will be. Experience seems to indicate that most armies and navies spend their time preparing to fight the previous war. The pugnacity and tenacity of old weapon systems in the face of new systems and threats are indeed spectacular. The Navy was successful, despite the vocal protests of numerous critics, in the proposal to resurrect and modernize the four Iowa-class (BB-61) battleships for operations in the last decades of the twentieth century and extending into the next millennium. Criticism of the Iowa-class modernization continued through 1989, focusing on the questionable use of "obsolete technology" in a modern navy. This is indeed odd, as an "obsolete technology," World War I–era mines, had created much consternation in the U.S. Navy's political and uniformed leadership just one year before.

In future wars, therefore, whether global conflict or the much more limited but much more likely low-intensity variety, the presumption must be that the threat or actual use of naval mines will play an increasing role in diplomacy or in negotiations to reach an end to the hostilities. One must not forget Clausewitz's dictum!

Mining is, of course, a supplement or even a complement to other modes of naval warfare, but no single mode is self-sufficient. Some tacti-

This chapter was prepared for the updated edition by Dr. Scott C. Truver and Dr. Gregory K. Hartmann.

cal and operational circumstances may favor one mode over another. It is therefore necessary to maintain a balance of unique and complementary capabilities to meet contingencies as they arise. If the effectiveness of mines can be maintained by healthy research and development, even in an environment of improved countermeasures, then the cost-effectiveness and military value of mining are likely to remain the same if not improve in the future. The greater damage width given to mines by providing them with mobility (e.g., Captor) will mean that the same minefield threat can be maintained with fewer weapons. However, these sophisticated weapons cost much more than their less intelligent predecessors, simply because they are more complex and are able to do much more against "tough" targets.

Image and reality: Mine warfare in a superpower navy

In order for mine warfare to be considered and properly weighted in the force levels for the nation's defense, it will be necessary for the planners and programmers to have a good understanding of mine warfare, the potential for the use of mines, and the need for effective countermeasures against the threat. The 1970 Nimrod Report [33] stated that mine warfare does not receive the attention or emphasis it should and that many persons in decision-making positions even within the Navy do not understand the characteristics or principles of mine warfare. The report speculated on the reasons for this sad state of affairs in a superpower's navy. First, military people are vehicle-oriented; mines, except for the Mk 67 Submarine-Launched Mobile Mine (SLMM), usually have no motors of their own and are not ordinarily associated with a particular vehicle. Second, mine actions are hidden or delayed—no tactical excitement or satisfying "bang" in their use!—and when something happens it is difficult to assign credit for it. Third, mine warfare does not contribute to personal career development in the U.S. Navy. Unlike the situation in the Soviet Navy, where mine warfare has tended to be a popular specialty, mine warfare in the United States is something to be avoided while, or endured for the shortest period of time before continuing, on the career path to admiral. Finally, the Nimrod Report even argued that mines are not popular in the U.S. Navy because of the lack of readiness of mine stockpiles and the deteriorating condition of the surface mine warfare force in light of advances in the threat.

All of these reasons are somewhat fabricated, and do not necessarily or sufficiently describe the status of mine warfare as the U.S. Navy enters the twenty-first century. Much has been done to heighten the Navy's awareness of the importance of mine warfare to a superpower. Yet the fact seems to be that mine warfare, in spite of its early beginnings in the United States, is in reality a relatively new thing in U.S. naval history, and that it is somewhat obscure both technically and in its operational planning, encumbered with the statistics of maritime traffic, target and damage widths, threat calculations, and so on. In short, it is a difficult discipline that has been omitted from the educational and practical experience of most of the active-duty Navy, save for the two Airborne Mine Countermeasures (AMCM) helicopter squadrons, the three or four active force surface ships, and a handful of mine readiness and training personnel (e.g., the active-duty Mine Readiness Certification and Inspections detachments) that train the land-based P-3 squadrons and attack squadrons on aircraft carriers, crews of surface mine countermeasures ships (MSO/MCM/MHC), and the resulting cadre of trained key personnel on each of these platforms. A large measure of the operational mine warfare experience and capabilities, however, remains with the Naval Reserve Force, including its active-duty personnel. And while the specific indictments cited in the Nimrod Report might be a bit contrived, the *perception* that they do indeed adequately summarize the current situation in the U.S. Navy has persisted through the 1980s and into the 1990s.

For example, when briefed on the Persian Gulf mine situation in early July 1987 and the proposal to send helicopters and mine-hunting/sweeping gear to help clear the sea lanes, one senior uniformed Surface Warfare officer asked, apparently with sincere incredulity, "We have *helicopters* to do this job?"

Then there is the criticism that there are no "real" mine warfare admirals, or any willingness on the part of the major naval warfare "unions"—Air, Surface, and particularly Submarine Warfare—to provide career paths to allow such developments. During the 1980s, three of the Commanders of the Mine Warfare Command, at the Charleston Navy Base, South Carolina, all rear admirals, were a surface warrior, a submariner, and an aviator who for all practical purposes were new to mine warfare when they assumed command. The planned rotation of the command among the surface, air, and submarine communities was actually proposed by Rear Admiral Charles F. Horne III, the third Commander of the Mine Warfare Command (CoMineWarCom), which itself had been reorganized a few years earlier, in 1975. Implemented originally by Chief of Naval Operations Admiral James Watkins and carried forward by his successor, Admiral Carlisle A. H. Trost, the objective of this rotation of the Command among the three platform communities was to involve all "unions," develop within each a greater knowledge and appreciation of mine warfare, and thus broaden, it was hoped, mine warfare's integration throughout the Navy. It is perhaps far too early to determine whether this strategy will be successful; in late 1989 the Navy was in only its fourth such planned rotation of CoMineWarCom, with the first surface warrior since Admiral Horne at the helm. There are concerns, however, that because of the Command's "nonplatform" complexion, it still may not be regarded by "hard-charging" junior and senior naval officers as desirable duty.

One vice admiral, who had served as the Director of Naval Warfare in the 1980s, challenged the belief that mine warfare sailors do not make flag rank, citing his service in an MSO during the 1950s. Questioned further on this, he admitted to being assigned to the mine force against his personal desires and getting out as soon as his first tour was over, returning to the CruDes (Cruiser-Destroyer) Navy. Why? "The mine force was seen as 'not conducive' to career development," he acknowledged, albeit reluctantly.

Isolated cases, perhaps, but they do tend to reinforce the perception of widespread institutional and personal biases against the Navy's mine forces.

Perhaps a first step to redress such a prejudicial imbalance against mine warfare in the U.S. Navy would be to emphasize mine warfare training or at least to offer all naval officers the opportunity to learn something about the subject, certainly at the Naval Academy or during ROTC or OCS, or at the very least at department head school. The structure for this was put into place as early as 1940, when the Mine Warfare School was commissioned at Yorktown, Virginia, nearly a year before Pearl Harbor and just after the German magnetic mine was discovered at Shoeburyness. Throughout the war this school trained officers and enlisted men to fill new billets in mine warfare, including countermeasures. In 1959 it was moved to Charleston and in 1972 was combined with the Fleet Training Center to form the Fleet and Mine Warfare Training Center. Courses include navigation, communications, and engineering topics, as well as mine warfare. This type of training, or at least the awareness of the importance of mines and mine countermeasures to the U.S. Navy, should be included in all naval officers' professional development curriculum.

The next objective concerns what future plans can be made for mine warfare and countermeasures, and how these plans can be brought to fruition. There are two areas of concern. First, what kind of Navy-wide organization, if that indeed is the solution, should there be for mine warfare? And second, what kind of R&D and production policies and programs should there be in mine warfare, and how should these areas interact to produce the best results?

In the postwar era, numerous changes in the Navy organization affected the way the Navy thinks about mine warfare, especially the conversion of the Navy into a single-line organization in 1963 away from the old so-called bilinear organization. Formerly, the actual operations of the Navy were nominally in the hands of the Chief of Naval Operations (CNO), and the materiel side was the responsibility of the Bureau Chiefs who reported for this to the Secretary of the Navy while attempting to meet the operational requirements established by the CNO. By 1966, with the establishment of the Naval Material Command, the Bureaus were replaced with a somewhat larger number of "Systems Commands" reporting to ComNavMat, who reported to the CNO, with OpNav now becoming much more involved in materiel work and budgets. The old Bureau of Ordnance, which in World War II was responsible only for guns, torpedoes, and mines, had expanded into missiles and aircraft weapons. In an attempt to solve weapon-aircraft interface problems, the Bureau of Naval Weapons had been created from an amalgam of the old Bureaus of Ordnance and Aeronautics. When the Systems Commands were formed in 1966, the Bureau of Naval Weapons was disintegrated back into the Ordnance Systems Command and the Naval Air Systems Command. At this time, however, the mitosis conveyed all air-delivered ordnance to NavAir, with the remaining weapons for surface ships and submarines relegated to the new organization called the Ordnance Systems Command.

In just a half-dozen years a new sea change in Washington created the Naval Sea Systems Command by combining the Ship Systems Command and the Ordnance Systems Command. Subsequently, there was a command within which ships and submarines were, however ill-suited, housed together with their weapons. The Navy could, in reality, be said to have two ordnance "commands" within NavAir and NavSea, with the solid respective backing of each set of vehicle sponsors so long as the ordnance or weapon systems could be positively identified with a specific platform. Because mines are multiplatform-oriented, they are often regarded as "orphans" or "bastards," depending on personal predilections, by the various platform sponsors. The dissolution of NavMat at the hands of the energetic Secretary of the Navy John Lehman in the early 1980s strengthened the direct linkages among the Secretariat, OpNav, and the SysComs, but it did nothing to emphasize the importance of mine warfare in the U.S. Navy.

Key questions for mines are whether, for example, air-delivered mines will be developed by various contractors as some sort of air-specific ordnance, as some other air ordnance has been, and whether there will be a reasonable commonality between parts of different mines and a common safety practice. Will the sea environment, and the influences and vulnerabilities of ships and submarines, be adequately dealt with by an aircraft-oriented weapons branch?

Raised in the mid-1970s, these concerns have been allayed somewhat through a Navy organization that has vested mine development and sponsorship within the Surface Warfare community, both in OpNav and in NavSea, but there are still doubts about overall coordination and oversight among the various platform advocates and program sponsors. For example, the oversight of mine warfare ordnance within the Office of the Assistant Chief of Naval Operations for Air Warfare (OP-05) in the early 1990s is located within the Anti-submarine Warfare Branch (OP-503), and even then it is buried with other systems and equipment in a billet ("ASW/Guns/Mines/Equipment Coordinator," OP-503G) that, some observers note, had experienced personnel "turbulence" and a lack of continuity during the 1980s. Yet OP-05 has *two* billets that focus on separate materiel—acquisition and logistics—aspects of the AMCM helicopters, and these billets have rarely gone unfilled for any significant length of time.

The Assistant Chief of Naval Operations for Surface Warfare (OP-03) does have a separate branch specifically devoted to mine warfare issues (OP-374). But that branch is bureaucratically located within the Combat Logistics, Auxiliaries, Amphibious, and Mine Warfare Division (OP-37) —the "cats and dogs" of the Surface Warfare community. The "aura of respectability" enjoyed within OP-03 by, say, the Aegis cruiser and destroyer program coordinators is not usually accorded to the mine warfare folks. For those critics of the Navy establishment who seek to "raise the consciousness" of the Navy regarding its mine warfare forces, the situation in 1990 is still seen as relegating mine warfare concerns to the backwater of Navy plans and programs, a perception that is fostered by the apparent fragmentation of mine warfare programs within the naval establishment.

Within NavAir, for example, there is a separate Program Manager for Airborne Mine Defense (PMA-210), focusing on the mine-hunting/ sweeping equipment side of the equation, with the Navy's AMCM helicopters (RH/MH-53) subordinate to another project manager, a Marine's billet, for the overall heavy-lift (H-53) program. Mine warfare programs within NavSea are divided among the Deputy Commander for Ship Design and Engineering (several branches subordinate to NavSea-05); the Amphibious, Auxiliary, Mine, Sealift Ships Directorate (NavSea-93; the mine platforms are grouped within the omnibus Tender/Repair/Mine/AD branch); the MCM/MHC Ship Acquisition Program Office (PMS-303); and the Mine Warfare Systems Project Office (PMS-407), which has responsibilities encompassing mine weapons and minesweeping/hunting equipment. Certainly there are good, practical reasons for the airborne systems to be under NavAir's aegis and, similarly, ship-related mine warfare programs to be vested in NavSea. Individuals associated with these organizations attest to the fact that "we all talk together, and in one way or another the OpNav requirements get translated into systems for the fleet." But they also admit to focused loyalties within their individual commands and branches, which tend to foster an "us and them" perspective.

Perhaps more insidious and dangerous for future mine warfare programs is the fragmentation of R&D efforts in the mine warfare laboratories, White Oak and Panama City, which come under the aegis of the Space and Naval Warfare Systems Command (SpaWar) and thus in no meaningful way are integrated with CoMineWarCom or OpNav. Most objective observers recognize that there have been, and will probably continue to be, difficulties in getting everyone convinced that they all work for the same Navy.

In short, the "platform orientation" of the Navy and its laboratories exacerbates the problem of maintaining the responsibility and competence in the materiel side of mine warfare. Current and future Navy and Defense Department reorganization, if anything, makes even more tenuous the responsibilities for mine warfare. The proposed Defense Department management, R&D, and acquisition changes (NSR-11/DMR-1) announced by Secretary of Defense Richard B. Cheney in July 1989 were seen as vesting increased responsibility and power within the office of the Secretary of Defense and his civilian advisors. That mine warfare has, more often than not, been assigned an inferior priority within the Navy may be a reason for increased concern in the future, especially as the important funding decisions may largely be made beyond the control of the uniformed experts. On the other hand, that denouement may *enhance* the future for mine warfare, as it has usually been civilians outside of the Navy, especially in key congressional committees, who have tended to force the Navy to think more seriously about its mine warfare forces and capabilities, often at the expense of "higher-priority" programs.

High-Level Command Support and Visibility

The recommendation that mine warfare policy and programs need a more "visible" home in the Navy seems to be satisfying and correct. The backing of Admiral Zumwalt for Captor and for increased cooperation with the Air Force is what a single, strong voice can do for mine warfare. Since 1979, with the wholehearted support of then-CNO Admiral Thomas B. Hayward, an aviator, mine warfare has benefited from high-level concern and interest. Following Hayward's lead, Admiral Watkins, a submariner, provided continued support, and his successor, Admiral Trost, another submariner, continued to foster mine warfare initiatives; witness the 1987–88 reorganization of the mine warfare "desk" within OpNav (discussed in more detail, below).

In 1989, the last CNO Executive Board (CEB) briefings on mine warfare had come nearly a decade previously, in 1979 and 1980, when the programs just reaching the fleet in the late 1980s were being born. In August 1989 a follow-up mine warfare CEB specifically focused on mine countermeasures was scheduled to take into account the key experiences of the intervening years, such as the "Mines of August" and Persian

Gulf mining. It is understood that the 1989 CEB recommended an increase in the numbers of new mine countermeasures surface ships currently on order, the Avenger (MCM-1) and Osprey (MHC-51) classes, in light of a reassessment of the critical requirement for U.S. port clearance early in any future conflict. Another eight "dual-mission" ocean mine hunters (for a total of twenty-two MCMs) and eight more coastal mine hunters (a new total of twenty-five MHCs) are in the Navy's future plans. More, the CEB recommended that twelve of the fourteen Avengers already authorized be kept in the active fleet and not transferred, as originally proposed, to the Naval Reserve Force after just a year in active status. Thus, even the most callous and jaded observer must admit that mine warfare has received, at least during the 1980s, command support at the highest levels of the Navy.

We cannot, however, expect every CNO to be a mine warfare champion, and yet a continuity of champions is required because it takes longer to do anything in R&D, procurement, planning, or force levels than the tour of a single officer. The first (1979) edition of this book called for the establishment of a "mine desk" at the highest level of the Navy for coordination, planning, and direction of all aspects of mine warfare—offensive, defensive, and countermeasures. This was seen as necessary because mines may use all naval vehicles—aircraft, surface ships, and submarines—and may enter into all aspects of naval operations. (As an example of this, the operations officer on the USS *Coral Sea* (CV-43), who planned the mission and led the initial Mk 52/DST mining of Haiphong in May 1972, had no previous mine warfare *planning* experience and, like so many other naval aviators, saw mine duty as unglamorous and taking scarce training time away from more important missions. Much later he admitted to having been fascinated by all aspects of the Haiphong operation and railed against the "political constraints" that forestalled using mines against the North Vietnamese until so late in the war: "All maritime traffic stopped, *immediately*, after we dropped the mines into the harbor, which greatly complicated the North's logistics support." [56]) Without the necessary emphasis *throughout the Navy,* mine warfare cannot achieve its full potential for the most fundamental missions of deterrence, sea control, and power projection.

The reorganization of the Fleet Commands, MinLant and MinPac, into a single Commander Mine Warfare Force (1971) and then Commander Mine Warfare Command (1975) was an initial step in the right direction. Although the Command does not actually "own" any mine warfare platform assets—a significant constraint, according to some analysts—CoMineWarCom today occupies a "single-node" advisory position, reporting directly to the CNO on all aspects of mine warfare. [55] That central focus is explicit in the command's formal statement of "Missions and Functions":

1. To act for the Chief of Naval Operations in all matters affecting Mine Warfare Readiness, training, tactics, and doctrine for active naval forces and corresponding Naval Reserve Programs.

2. To coordinate with Fleet Commanders in Chief, and with OPNAV, on matters affecting fleet mine warfare readiness above the level of groups, squadrons, or divisions, and other forces possessing a mining and/or mining countermeasures capability.

Two elements exist within the command: MOMAG and MineWarInsGru. MOMAG (Mobile Mine Assembly Group) is an OpNav field activity responsible for the maintenance and readiness of all the Navy's mines. With twenty-seven Reserve Force units in the United States and thirteen active detachments throughout the world, the some 600 personnel assigned to MOMAGs specialize in maintaining, assembling, and readying mines for laying. The principal mission, in addition to an ancillary training role, of the Mine Warfare Inspection Group (MineWarInsGru) is to conduct biannual readiness inspections of the 350 Navy commands and units that themselves have a mine warfare mission.

A relatively new development in 1988, the Commander Mine Warfare Command was "double-hatted" within the Office of the Deputy Chief of Naval Operations for Naval Warfare (OP-07) as the Director Mine Warfare Division (OP-72), a Division Director in OpNav with a full-time captain deputy and two full-time commanders who are immersed within the Washington naval establishment but who have organizational loyalty to CoMineWarCom/OP-72. There was significant discussion within OpNav whether CoMineWarCom should be double-hatted as a division within OP-07, the Naval Warfare Directorate, or within OP-03, the platform sponsor for mines and mine countermeasures. As the director of *all* naval warfare programs, OP-07 (previously, OP-095) has in recent years grown increasingly powerful vis-à-vis the submarine, surface, and aviation platform sponsors (OP-02, OP-03, OP-05). In the 1987–88 OpNav reorganization, for example, OP-07 remained a Deputy CNO billet as opposed to the platform sponsors, which were downgraded a bit to Assistant CNO billets. From mine warfare's perspective, then, OP-07 appeared to offer greater prospects than OP-03 for achieving specific mine warfare objectives, especially as OP-07 establishes the top-level warfare requirements to support the Navy's Maritime Strategy. [63]

This decision to make the mine warfare "desk" a Division Director billet has raised the visibility, and presumably the effectiveness, of the mine warfare community within OpNav, especially compared with the previous organization when it was buried within the Strike and Amphibious Warfare Division as a commander's billet (OP-744). But this brightening situation may be illusory, as OP-72 per se does not explicitly control fiscal resources, still the source of real leverage within OpNav.

Mine warfare's principal resource/funding sponsorship and responsibility still reside within OP-374, but again they are buried within the OpNav Surface Warfare community. The fight for resources for mine warfare programs—first getting the needed funds and then preserving them against uniformed and civilian budget cuts and raids for higher-priority programs—is extremely difficult, particularly when the principal attention of the organization is perhaps focused on saving Aegis destroyers and other "high-value" targets from various fiscal axes wielded by the civilian budgeteers in the Office of the Secretary of Defense and the Office of Management and Budget, as well as the Congress. In short, despite the "enhanced visibility" of mine warfare in recent years, much of it the result of diverse criticisms directed at current forces and existing programs, the mine warfare community can exercise little effective political leverage in the Pentagon.

Current and Future Needs

That lack of political leverage creates difficulties for what is really required for the future of mine warfare in the U.S. Navy: vigorous and competent R&D and acquisition efforts that will link the "operators" in the fleet with the technologists and scientists in Navy, industry, and academic laboratories, and the officers and civilians who establish warfighting requirements.

At the most fundamental level, this means that there must be continuing strong in-house Navy R&D effort both at the Naval Surface Warfare Center, White Oak, Maryland (for the mines), and at the Naval Coastal Systems Center, Panama City, Florida (mine countermeasures), with appropriate and direct liaison with CoMineWarCom and OP-72. Such a Navy lab base derives its competence from the record of experience with technologies, with environment, with engineering, with production, and with evaluation, the heritage of many mines and many conflicts.

The R&D problems and acquisition interests of future mine developments are exceedingly broad. They intersect just about every aspect of the physical sciences and engineering. Because no single mine is adequate for all contingencies, it is advisable, in having many, for components of different mines in a family to be interchangeable. Many mines, though different, have common problems and common parts, such as power supplies, ship counters, sterilizers, arming devices, case material and construction, camouflage, arming delays and dead-time devices, firing mechanisms, shock and countermine resistance, trajectory and water-entry/sinking considerations, compatibility with launching vehicles, safety, information processors, sweep resistance, and so on. An adequate mine warfare arsenal requires both large and small stationary and propelled mines for different water depths and different targets. The stockpile should be flexible and adaptable for different tactical or strategic purposes. The stockpile should be continuously updated to replace expenditures incurred in training or operational exercises, and obsolete weapons and components should be replaced. These ideals, however, are hardly ever achieved in the real world.

The continuing goals of mine development will be to make mines hard to sweep and hard to find; easy to build and with easy minimum assembly consistent with flexibility of use; adjustable in threat and in response to type of traffic and targets; compatible with the aircraft and submarines that will carry them; resistant to the environment and designed so as not to limit the conditions of use. New developments in the targets (e.g., submarine quieting and anechoic coatings) will have to be accommodated.

Although a highly controversial subject, one area of Navy mine warfare development that some (not all!) observers argue has not been fully exploited since the end of World War II is minefield theory, that is, the analytical methods and processes that provide information on the preferred geometry and density of a minefield given expected environmental conditions, threat characteristics, mine capabilities, and mine countermeasures characteristics. This is concerned with determining mine influence zones, target transits, and widths so that the probability of attacking penetrating targets is sufficiently high, intermine fratricide is acceptably low, and criteria for the efficiency and sufficiency of the number of mines planted are satisfied. The core issue about minefield theory is that as more complex and expensive, and hence more capable, mines have been acquired since the 1940s, some Navy planners have assumed that fewer numbers of mines need to be acquired to achieve a given level of performance.

However, the development of new theoretical approaches since the late 1960s in both the United States and especially NATO, when linked to recent assessments of target characteristics, has indicated a requirement for more, not fewer, mines. As one insider noted, somewhat ruefully, "The debate will continue into the 1990s, and what progress that has been made has so far failed to penetrate into official doctrine." Nevertheless, this assessment is not shared by all, with others noting that minefield planning theory and execution have indeed kept pace with the threat. But these individuals also acknowledge, reluctantly, that translating the theory into reality—that is, acquiring the advanced mines needed to defeat very quiet submarines, and buying sufficient numbers of these mines to satisfy operational requirements—continues to present difficult funding resource dilemmas to the Navy's leadership.

The mine *countermeasures* problem also requires continuing effort and coherent organization within the U.S. Navy. Indeed, since the mid-1970s this has been the one area in which the Navy has seen a compelling requirement to modernize. That being the case, however, the programs to meet these needs have often been delayed or drawn out. For example, the requirements for a new ocean mine countermeasures ship program, which eventually became the MCM-1 Avenger class, by most accounts the premier surface mine hunter in the world, were first discussed in 1976, with the acquisition program being proposed in 1979. But it took another three years for the first units of the MCM-1 program to become funded, after which it was mired in controversy, and only four ships had been commissioned by mid-1990. Furthermore, the original requirement of twenty-one MCM-1s had been reduced because of fiscal parsimony to only fourteen ships. In spite of the progress in state-of-the-art mine countermeasures ships through new sonars and mine hunting with sophisticated devices, it would be foolish in the extreme to relax a state of vigilance and development in the belief that all problems are now at a standoff and that the situation is static. The maintenance of an adequate state of readiness in mine countermeasures, generally, demands an integrated activity of training, procurement, development, and research going on into the future.

A 1975 discussion of the Navy's mine countermeasures capabilities concluded that minesweeping art had not advanced much beyond the World War II developments of magnetic tails and hammer boxes. [31] The author of this gloomy assessment, one Lieutenant James McCoy, USN, had served as mine warfare officer for Commander Mobile Mine Countermeasures Command and formulated all the detailed minesweeping instructions for both surface and helicopter units in Operation End Sweep. McCoy argued that a vastly increased effort in mine *hunting* was required and future attempts at ship simulation as a means of *sweeping* should be deemphasized.

McCoy's principal point was that the then-current focus on ship simulation and minesweeping actually played into the hands of mine designers, as it merely reinforced the value of sophisticated influence detonators, signal processing, ship counts, and delayed or random arming. This was seen as increasing dramatically the number of sweeps that would have to be made to ensure that the suspected field had indeed been cleared. As McCoy noted about End Sweep, however,

> If the approximately 11,000 U.S. mines planted in North Vietnam had not had sterilization and self-destruct features, and if they had contained batteries of indefinite active life, then the completion of that operation would have been measured in years rather than months. . . . Statistical techniques for evaluating sweeping success are only as good as the assumptions on which they were based.

After all, the Navy was "sweeping" its own mines and thus did not have to guess as to their properties and, in some cases, actual location. In reality, then, McCoy argued, End Sweep was not a valid test of the Navy's contemporary mine countermeasures capabilities, because all the mines had either self-destructed or self-sterilized by the time the operation was over. McCoy noted caustically that the United States would be incapable of clearing an enemy minefield in a timely manner as long as the Navy remained oriented toward fooling the mine into detonating—*sweeping*—rather than *hunting* for the mines suspected of being in place.

In the mid-1970s the United States and NATO continued to perpetuate the myth that our mine countermeasures forces were capable of defeating the threat. Clearly, they were not. A perceptive and well-informed British officer, Lieutenant Commander G. R. Spooner, RN, in 1976 reviewed and deplored the decreases in Mine Countermeasures (MCM) force levels of the previous decade. [45] In every NATO state except Turkey, Spooner noted, the number of active MCM ships had been reduced, sometimes significantly, as World War II–era ships were scrapped and NATO governments failed to replace them on anything close to a one-for-one basis. The largest force reductions came in the Royal Navy (from 101 MCM ships in 1966 to 40 in 1976) and the U.S. Navy (from 200 ships to just 3 MSOs in the active force).

Spooner compared this near-dereliction in the West with the continued mine warfare vigilance in the Warsaw Pact countries. During the same period of wholesale MCM scrapping in NATO, the Soviet Navy maintained its surface MCM force at some 300 ships, replacing with new construction the numerous vessels it had transferred to other countries. Although Poland, Romania, Bulgaria, and the German Democratic Republic each had no MCM ships in 1966, in ten years they had come to possess forty-four, twenty-two, six, and seven, respectively. Spooner's conclusion on the then-current state of affairs reads like an epitaph:

> On the mine countermeasures side it will not matter how powerful, skilled or efficient the rest of the surface or submarine forces are if they can be destroyed, or bottled up in harbor, by the laying of a few mines. Similarly, there is little point in planning a prolonged land campaign if essential supplies cannot be delivered to sustain the effort. National MCM forces must be rebuilt to a level where there are sufficient craft to allow the other maritime forces to maintain their credibility, to keep the harbors operating and to bolster the NATO forces. . . . Time will not be on the side of NATO in any future conflict, and it will be too late to start building MCM vessels, equipment, and mines and teaching mine warfare skills once mines have been laid or war declared. [45]

Despite these and other warnings, by the mid-1980s the situation had not improved significantly. Former Deputy Chief of Naval Operations for

Surface Warfare, Vice Admiral Joseph Metcalf III, during an April 1985 congressional hearing bluntly stated that "no element of our Navy is as deficient in capability against the threat as is the mine countermeasures force." The threat Admiral Metcalf described was principally the Soviet Union's ability in crisis and war to strangle naval and merchant shipping movements. [81] With a mine stockpile variously estimated at 250,000 to more than 400,000 weapons, including both the most sophisticated and the obsolete yet still effective types (as the Persian Gulf experience with the M-08/39 mine attests), the Soviets can lay these devices in depths of more than 3,000 feet from a variety of aircraft, surface ships, and submarines. [75, 85, 86, 87, 88] And as was demonstrated during the 1984 "Mines of August" crisis, the threat of mines used by terrorist states or groups intent on politico-military intimidation is very real and must be taken into account.

The Navy's Maritime Strategy documentation of the 1980s clearly articulated the critical need for highly capable mine warfare assets to conduct both mine countermeasures operations (to ensure the safety of our ships' transits) and offensive and defensive mining of important sea areas (to frustrate the movements and operations of our adversaries' ships).

Figure 21. A "Backwater" Mission? Several crises during the 1980s illustrated the Navy's penchant for high-tech weapons and the real need for highly capable mine warfare forces in "peace" and war. (Stuart Carlson, Milwaukee *Sentinel*)

And while it sought to underscore the salient importance of our allies' mine warfare contributions in a war with the Soviet Union, the Maritime Strategy admitted that we did not possess adequate forces to protect even our own coastal waters and harbors well, let alone project mine countermeasures platforms forward, into regions where battles will be fought. Indeed, the Maritime Strategy clearly underscored the Navy's strong reliance on the mine warfare assets of our allies to clear ports, harbors, and critical sea lanes in the event of war. [80]

Nevertheless, knowledgeable Navy observers—Admiral Wesley McDonald, a former Supreme Allied Commander, Atlantic, for one—predicted that NATO's mine warfare capabilities will likely dwindle to a point where their contributions to collective offense and defense will be marginal at best. Despite widespread acknowledgment within NATO of the dimensions of the Soviet threat, Admiral McDonald lamented,

> The generally accepted main threat which would be posed to this shipping by the Warsaw pact would be torpedo- and missile-armed submarines. It is less well known that Soviet mining capability and the environmental conditions of ports on both sides of the Atlantic make NATO's coastal waters and the use of deep-water approaches to them very vulnerable. Unlike the submarine threat, the mining problem has been, and continues to be, "under study" within NATO. [69]

Even if contemporary U.S. mine warfare programs like the Avenger-class (MCM-1) ocean-going mine countermeasures ship, the Osprey-class (MHC-51) coastal mine hunter, and the MH-53E Sea Dragon helicopter, as well as other mine warfare initiatives like COOP (Craft of Opportunity Program) and NOMAD (Naval Operations and Maintenance, Aviation Deck) finally bear the fruit of a rejuvenated mine countermeasures force in the mid-1990s, the Navy recognizes that "sequential operations" will have to be undertaken to deal with the massive threat posed by the Soviet Union, no matter how much *perestroika* takes place in the future. How much more difficult will be these tasks if today's mine warfare programs fail to achieve their stated objectives!

Such concerns, regrettably, are real. The much-needed Avenger program early on had been plagued with rising costs, technical problems, and schedule delays, with the first unit of the class delivered some two years behind schedule and almost twice as expensive as originally estimated in 1979. Despite these difficulties, sadly not uncommon for any new ship-acquisition program, the USS *Avenger* has met or exceeded all operational expectations since joining the fleet in 1988. And in early 1990 it appeared that the Congress, theretofore highly skeptical of the Navy's management of the surface MCM programs, was thinking about expanding the buy of the Avenger program beyond the planned fourteen ships.

Other U.S. mine warfare programs have also been the focus of controversy and recrimination. The Osprey (MHC-51) coastal mine-hunter/sweeper program for seventeen ships is a "pinch-hitter" of sorts for the ill-fated Cardinal-class (MSH) surface-effect mine hunter/sweeper. The Navy canceled the Cardinal program in 1986 after being hounded by a Congress that was convinced that no amount of funds could fix the technical problems the program had encountered. There have been reports, moreover, that technical problems also confronted the Osprey project and that its ultimate cost and schedule were "uncertain" in mid-1989. And the Department of Defense Operational Test and Evaluation Office in early 1989 reported that the new MH-53E Sea Dragon AMCM helicopter could not meet its operational requirements. There is, moreover, political wrangling with the Congress over the appropriate dimensions and platforms for COOP, and even the new AMCM minesweeping gear (ALQ-166) and mine-hunting vehicles (SLQ-48 Mine Neutralization System) have been criticized on technical and cost grounds. (See appendix B for descriptions of U.S. mine warfare platforms and systems.)

Perhaps unfairly compared with other Navy "unions," which themselves have their vocal critics, the slowly improving overall capability in U.S. Navy mine countermeasures is, however, implicated by a current lack of offensive and defensive mining assets, or at least the public *perception* of such deficiencies. While other mine initiatives of the 1960s and 1970s were never fulfilled, the Mk 60 Captor mine *was* finally procured, but only after a desultory on-again/off-again R&D program. Yet some analysts now question that weapon's full effectiveness against the new modern Soviet submarines. A similar programmatic history and assessment are given for proposals to upgrade the Mk 67 SLMM. Furthermore, some observers have criticized the size and composition of the existing mine stockpile, commenting that it is unclear whether the Navy has enough mines to satisfy initial wartime mining and carry out any meaningful minefield reconstitution.

And what is the prognosis for other, more capable mines and the platforms to deliver them? Work on a new Advanced Sea Mine (ASM) for intermediate water depths had been reinvigorated in the mid-1980s—after the untimely demise of the Mk 68 rocket-propelled rising mine PRAM (Propelled Rapid-Ascent Mine) in 1978—with a memorandum of agreement for cooperation with the United Kingdom in place. But that program had not previously received steadfast support within the United States, outside of the mine warfare community, and some observers questioned its future "viability" in a "constrained fiscal environment"—a fate not unusual for mine warfare programs.

Indeed, Britain's Ministry of Defence announced in late 1988 that it was pulling out of the co-development program for the ASM, causing the U.S. Navy to consider whether to drop the program altogether or to try to

Just entering the fleet in the late 1980s, an MH-53E Sea Dragon helicopter tests the new-design AN/ALQ-166 mine countermeasures sled. In 1989 the Defense Operational Test and Evaluation Office labeled the Sea Dragon "not operationally capable" of performing its mission.

Close-up of the U.S. Navy/EDO Corporation AN/ALQ-166 hydrofoil mine countermeasures sled, touted to be "orders of magnitude" better at countering modern magnetic influence mines compared with the "workhorse" Mk 105 AMCM sled.

USS *Avenger*, MCM-1, lead ship of a fourteen-unit program intended to provide a baseline ocean-going surface mine countermeasures capability for the U.S. Navy.

A U.S. Air Force B-52G Stratofortress releases practice mines during Team Spirit '87, the annual joint U.S./Republic of Korea exercise. Critics point out that in a "real war" the Air Force would likely find other missions for its aging Stratofortresses, and in any event the 1950s-built B-52s would soon be retired, with no other assets available to replace them.

corral the resources to carry out the effort on its own. However, the "fiscal constraints" of the revised Bush/Cheney Defense Department budget for fiscal 1990 made it practically impossible to embark on a unilateral U.S. ASM effort *and* carry out the proposed program for Captor upgrades at the same time. The Navy's dilemma was, unfortunately, which mine warfare program to forgo; there was no support from the other warfare "unions" for the reallocation of their funds to support both the ASM and the Captor upgrades. Congress may yet come to the rescue, as deliberations late in the fiscal 1990 and 1991 appropriations process illuminated a desire to provide sufficient funds to keep both programs alive.

Even before the Bush administration reordered many of the priorities in President Reagan's final budget submission in January 1989, the Navy's own *Highlights of the FY 1990/92 Department of the Navy Biennial Budget* was mute on the need for mines, out of a Navy weapons procurement request of $5.7 billion in fiscal 1990 and $6.3 billion in fiscal 1991, perhaps indicating that there are no "highlights" regarding mine warfare. A bit more digging showed that no funds were to be requested for procurement of the Mk 67 SLMM in either fiscal year; $2.9 million had been appropriated for SLMM procurement in fiscal 1988 and $11.4 million in 1989. No Captor (Mk 60) modifications were requested for fiscal 1990 and 1991, although $38 million had been appropriated in fiscal 1988. Looking to all R&D accounts, however, the situation looked a bit more sanguine: out of the total Research, Development, Test, and Evaluation (RDT&E) requests of $10.2 billion in fiscal 1990 and $9.6 billion in fiscal 1991, mine developments totaled $32.6 million and $34.5 million, respectively. Still, mine warfare programs constituted only a minuscule fraction of total Navy R&D; by mid-1989 the prospect for reaching even these thresholds looked grim.

However, now that mine warfare has an on-scene protagonist in OP-72, perhaps in the near future this might be translated into effective protection of those mine warfare things most needed from a broad naval warfare perspective, especially when the platform sponsors become myopic or parochial as budgets get squeezed even more. Perhaps, but numerous Cassandras raise their eyebrows at such optimism.

The Navy may also be constrained by the number of platforms— aircraft and submarines—that will be made available to lay mines in future conflicts. Submariners, as a caste, seem to be loath to "waste" critically small weapons load-outs on mines, preferring instead to have only "real weapons" for their primary mission on board. The Navy's sea-based tactical aircraft and land-based Maritime Patrol Aircraft (MPA) are capable of mine laying, but this is considered a secondary mission. And where MPA tactical roles and missions clearly embrace mining, there may be some severe "disconnects" among the locations of the stockpiles of mines, where the aircraft are based, and the projected minefields. In wartime some of these "disconnects" would be difficult to overcome in a timely fashion to make a meaningful contribution to naval operational plans.

Reliance on the Air Force's B-52s to help out in future mining campaigns—a critical assumption of the Maritime Strategy—may run into resistance during wartime operations, perhaps reminiscent of the Army Air Force's heated arguments in the winter of 1944–45, when Admiral Chester Nimitz tried, eventually successfully, to get General "Hap" Arnold and the 21st Bomber Command interested in Operation Starvation, the Japanese Inner Zone Mining Campaign of March–August 1945. In a future war the Air Force may also find it has more important things to do with those aircraft—assuming they are still flying—regardless of what a preexisting Navy/Air Force memorandum of understanding says. The new Air Force bombers, the B-1B and "Stealth" B-2, will almost assuredly *not* be offered for collateral-duty mine laying because of their small numbers—the B-1 fleet originally at 100 aircraft, in 1989 at 97 and presumably decreasing due to operational losses; no more than 75 of the $850 million-plus B-2s, if a highly skeptical Congress acquiesces to the administration's pleas for the $70 *billion* program—and higher-priority roles. Skeptics will likely point out, moreover, that the only reason the Air Force even offered to consider naval roles—mine laying and Harpoon antiship attack—for its B-52Gs was to maintain the force levels of its aging bombers, not out of any real interservice altruism.

Because of mine warfare's historically poor image in the view of much of the Navy and the public and the fragmentation of oversight and sponsorship, attempts during the mid-1970s to reinvigorate all aspects of the Navy's mine warfare community fell victim to other more important priorities, the politics of "Sea Plan 2000" force-level options, and a decreasing slice of the federal pie for defense spending. Cutting the fat by the Carter White House inadvertently sliced into the Navy's muscle. The reality is that, except for increasing the inventory of the Captor antisubmarine mines and marginal improvements in sensors and weapons, the really substantive mine warfare developments of the recent past were for varying reasons continually relegated to the out-years of the Defense Department's Five-Year Defense Program.

Mine warfare in the U.S. Navy *did* enjoy a renaissance in the 1980s, benefiting from the bow-wave of increased defense spending generated by the Reagan administration's quest to "make America strong again." In spite of growing pains and other difficulties encountered, the programs now under way will go far in making the Navy's mine warfare forces among the best in the world. But it is highly uncertain how much longer this rejuvenation can be sustained if needed programs are not initiated and mine warfare's public and internal Navy image remains unchanged and sponsorship fragmented.

By way of comparison, it is illuminating to survey other countries' mine warfare programs, not all of which have been completely successful. What follows is intended not to be a comprehensive analysis, but merely to show how other navies continue to emphasize mine warfare capabilities and readiness.

Great Britain: Current British plans call for fifty to sixty modern ships to be built by the end of the next decade, and conversion of another forty offshore oil field support vessels to a mine countermeasures configuration. The excellent new class of Vosper-Thornycroft single-role mine hunters has been funded through at least the fifth unit, but no funds have been forthcoming to carry out the conversion program. And, as noted previously, the Royal Navy dropped out of the joint ASM program because of funding constraints. Several British mine programs have made it into production, including the Sea Urchin mine (charge weight 350–1,200 kg), the advanced Stonefish mines (a family of modular mines, with multiple influences and advanced signal processing, warhead weights up to 600 kg, laid in depths of 15 to 600 feet), and the Dragonfish mine (an anti-invasion shallow-water mine, with acoustic and magnetic influence and an advanced microprocessor said to be "virtually immune" to countermeasures).

France: Since the late 1960s the French Navy has pursued numerous mine warfare initiatives, mines, new mine-hunting and classification sonars, remotely operated vehicles (e.g., the PAP-104), the new Tripartite countermeasures vessels (constructed in concert with Belgium and the Netherlands), and modification of older ocean minesweepers. The objective was to be able to control coastal waters and the continental shelf to a depth of approximately 250 feet. Future French efforts will seek to extend operations to depths of some 1,000 feet. The five-year plan submitted to the Parliament in May 1987 included the procurement of ten BAMOs (*Batiments Antimines Oceaniques,* or ocean mine hunters) with expanded mine-hunting and classification systems and extremely precise navigation systems. Several mines are under production, including the Thomson-Sintra sea mine (TSM3510, a multi-influence device with 850 kg of high explosive) and the TSM3550 ground mine (1,500 kg).

The Netherlands and Norway: These two countries are considering the construction of new classes of coastal mine hunters and sweepers based on a civilian surface-effect glass-reinforced plastic hull developed in Norway. (This was also used in the abortive U.S. Cardinal class [MSH-1].) Norwegian craft construction was to start in 1989 and all ships were programmed to be in service by 1996. The highly successful Tripartite mine hunter, produced in consort with Belgium and France, continued to be acquired for the Royal Netherlands Navy.

Japan: The last two Hatsushima-class MCMs were completed by early 1989, leaving the Japanese Self-Defense Force with thirty-nine MCM vessels of various kinds and seven obsolescent helicopter sweepers. The latter are to be replaced by twelve MH-53E heavy AMCM helicopters, the first four of which were authorized in the 1986 budget. This force will be augmented by a planned class of six 1,000-ton deep sea MCM ships, the first ordered in the 1989 budget, followed by eight 600-ton mine hunters. Two 490-ton hunters, glass-reinforced plastic versions of the Hatsushima design, were included in the 1987 budget.

Canada: Domestic public opinion in Canada has generated new governmental concern over the control of Arctic waters within Canadian jurisdiction, with implications for both defensive mining and mine countermeasures requirements. A mid-1980s White Paper suggested the establishment of a mine countermeasures force using the naval reserves. The British River-class MCM ships appeared most suitable for Canadian needs, but the Belgian/Dutch/French Tripartite vessels were also seen as excellent platforms. In 1988 the Canadian government programmed about $540 million for twelve MCM vessels, which, given the high cost of some off-the-shelf platforms and systems, may compel Canada to design its own multirole Maritime Coastal Defense Vessel (MCDV) that will combine minesweeping and patrol capabilities.

Australia and New Zealand: Mine countermeasures programs were receiving much attention with the publication of the Dibb Report in 1986, which indicated the increased mining threat. Two prototype Australian-designed Baydan catamaran inshore mine hunters were undergoing trials in 1988–89, and an Australian Craft of Opportunity Program (COOP, modeled after its U.S. counterpart) was showing great prospects as an inexpensive but capable form of an easily mobilized MCM force. In New Zealand, a 1987 Defense White Paper called for a greatly enhanced mine countermeasures capability. However, financial constraints were expected to limit the Royal New Zealand Navy to a COOP effort only.

Returning to the situation confronting the U.S. Navy as it enters the next century, although Admiral Metcalf's and many other people's illumination of the full dimensions of the Soviet mine threat served to heighten awareness of the need for adequate mine warfare readiness, some American analysts have discounted much of the Soviets' capabilities and their reliance on mines in a war with the United States, however unlikely that possibility may seem in the new world emerging from the rubble of the Berlin Wall. They argue that most of the massive stockpile comprises defensive mines and that the United States would "probably" not be subjected to a barrage of mines along its coastlines. But they should consider that these defensive mines will "probably" be deployed to protect the bastions for Soviet missile submarines, into which the Maritime Strategy stated the Navy intends to project its attack submarines. Who is going to ensure that the courses these U.S. SSNs will take will be mine-free or of a sufficiently low mine threat that the operational commanders will still commit to battle the relatively few submarines we have? And would they deploy several carrier battle groups to North Atlantic and Norwegian Sea operations if they were not assured that the mining threat had been neutralized, especially given the optimistic projection that the Navy will possess—at Secretary of Defense Cheney's direction of April 1989—at most only fourteen carrier battle groups for the foreseeable future?

Moreover, these analysts should not be so sanguine in their assessment that the continental United States could not be blocked for a critical period of time by even a small number of mines laid in important harbors and choke points by merchant vessels or submarines during the run-up to hostilities. The Soviets, always good students of history, will certainly have noted America's World War II experience when German submarines laid no more than 340 mines off the U.S. Atlantic coast, sinking twelve ships and sealing ports from Nova Scotia to Panama for periods of one to sixteen days.

In the end, the German mining proved to be only a minor inconvenience in the context of the entire conflict, although the very real threat of mines laid along the eastern seaboard tied up some 100 minesweepers for the duration of the war. [70, Pp. 91ff.] However, sixteen days at the beginning of the next global conflict and the war could be lost. Although the dissolution of the Warsaw Pact following the fall of the Berlin Wall in late 1989 makes a European Central Front conflict highly unlikely, the future political and military course of the Soviet Union remains an enigma. Support for continued modernization of the Soviet Navy has remained strong in the late 1980s and early 1990s, and the Soviets, by all accounts, remained convinced of the requirements for and value of mine warfare. It would be foolish in the extreme to ignore the potential threat of Soviet mines in any future conflict.

Finally, the use of mines in limited wars, low-intensity conflicts, and terrorism by states and subnational groups must not be discounted. Mines were used ineffectively by the Argentines in the Falklands conflict, but they nonetheless posed a threat and had to be taken into account by the Royal Navy. The U.S. MCM assets in both the Gulf of Suez and the Persian Gulf were effective, once on the scene, but political constraints and distance from U.S. home ports work to frustrate overall U.S. responsiveness. And reliance on our allies to help protect even joint interests in freedom of navigation is fraught with uncertainties. During the early summer of 1987, for instance, U.S. administration personnel literally went hat in hand to several NATO countries' capitals in an attempt to organize, however loosely, a combined effort to clear the Persian Gulf of mines. All initial efforts were rebuffed, and only when the British finally agreed to send an independent force did other NATO and Allied governments also agree to help.

Since then, numerous reports in the "popular" defense press have recounted the efforts of many nation-states—third-world countries and developed states alike—to acquire modern naval mines. (The Italian shallow-water Manta mine, with 170 kg of HBX-3, is a particularly hot item. Because of its fiberglass case and "stealth" shape, this mine poses great problems for all but the most sophisticated countermeasures.) Similarly, many states have seen the need to upgrade their countermeasures forces. After each mining event since the Vietnam War, there has been a noticeable increase in interest in mines and mine countermeasures, even by governments far removed from the mine-crisis area. For example, in the wake of the 1984 Red Sea–Gulf of Suez crisis, the Egyptian Navy actively sought out the Dutch, Belgian, and French builders of the Tripartite mine hunter. Kuwait, the target of much of the Iranian Revolutionary Guards' mining in 1987–88, and Saudi Arabia likewise were investigating upgrading their mine countermeasures capabilities with off-the-shelf systems and platforms. Reports circulated that even the navies of Singapore and Indonesia were shopping for advanced mine countermeasures systems in the wake of the 1987–89 Persian Gulf crisis. And, just a few short months after the last of the U.S. minesweepers deployed to the Persian Gulf in 1988–89 returned to their home ports, the United States found itself responding to Saddam Hussein's attack on Kuwait. Navy MCM vessels were quickly redeployed to the Persian Gulf to meet whatever mine threat Iraq posed to international shipping.

Final observations

Rear Admiral Horne, the third Commander of the nascent Mine Warfare Command, in 1982 published an insightful—perhaps *inciteful* is a more appropriate modifier—article that noted, "Mines may be the weapons that wait, but mine warfare will wait no more. Mine warfare in the U.S. Navy is back on course and receiving command support from the highest levels." [62] In the years since Admiral Horne wrote of the fact that mine warfare "boils down to 85 percent preparation (in peacetime) and 15 percent execution (in wartime)," the situation has come to this:

1. The Navy through mid-1990 commissioned only four of the new, and much more capable, Avenger-class (MCM-1) ships, a program some two years behind the original schedule.
2. The Cardinal MSH program was all but rejected unilaterally by an indignant Congress, thus sending the Navy back to the drawing board—and to the Italian firm of Intermarine Sarzana, builders of the Lerici-class mine hunter—for the Osprey (MHC-51) coastal mine hunter/sweeper.
3. The Department of Defense Director of Operational Test and Eval-

uation criticized the new Sea Dragon (MH-53E) AMCM helicopter as "not operationally suitable" for its mission.

4. The ASM has been all but canceled because of a lack of funding commitment.

5. The Captor and SLMM upgrade programs have been delayed, perhaps interminably.

By any measure, this is not an enviable history of accomplishment.

In late 1990 Admiral Horne offered to illuminate the intent behind his article. He noted that his phrase "back on course" did not mean or even seek to imply that the objective or liberty port had been reached, or that many of the obstacles en route had been or would be successfully navigated. [63] He did, however, want to emphasize that a well-conceived program plan for mine warfare, particularly mine countermeasures, finally existed, that a course had been laid out, and that clear objectives had been established. Beginning with the leadership of Admiral Hayward in 1979, Admiral Horne believed that mine warfare was no longer a "rudderless ship in its own fog bank since World War II, but now back on a reasonable, planned and productive course," with the personal support of every Chief of Naval Operations since then. For example, in his remarks at the CoMineWarCom change of command in July 1989, Admiral Trost stated,

> The bottom line is simple. Mine warfare may not be considered glamorous, some even call it ugly, but it works and it works well. . . . Mine warfare encompasses the entire spectrum of warfare, and it can threaten our future national security, our economic health, and that of our friends and allies. It can't be taken lightly or ignored. Too often mine warfare, particularly mine countermeasures, are [sic] considered after the fact. One hundred, even forty years ago, we could afford to do that. Today we can't.

Much effort was indeed expended in the 1980s proposing and defending mine warfare programs before Congress and justifying expenditures within the Defense Department. Yet almost in the face of the comments by Admiral Trost, it is only when some crisis somewhere in the world raises our collective consciousness, if only for a short time, to the true value of mines and mine countermeasures that the Navy's mine warfare programs and forces emerge from the gloom. As Admiral Tobin, former CoMineWarCom, noted, "Some people had forgotten about [mine warfare] . . . until the activity in the Persian Gulf jogged their memories a little bit." [61, P. 6] Professor Wayne P. Hughes of the Naval Postgraduate School has written, perhaps prophetically, that "today, as a consequence of U.S. maritime superiority, mining and mine countermeasures, which are peripheral to the study of fleet tactics on their own, are likely to dominate wartime actions on both sides in surprising ways." [65, P. 272]

The key to ensuring that the Navy's memories will remain jogged and that future surprises are avoided is to reach a broad consensus among all of the Navy's "unions" about the importance of mine warfare to a superpower's navy. Perhaps it is propitious, therefore, to recommend yet another step toward a better and more effective integration and employment of the force multipliers and warfare leverage that mine warfare offers for the twenty-first century. All the elements currently exist for a truly integrated approach to mine warfare; the problem is that they continue to have limited visibility, divided loyalties, and divergent foci.

Instead of the existing fragmented situation, a truly integrated mine warfare command for the Navy should be created, which—in both horizontal and vertical dimensions—would be totally responsible for both mines and mine countermeasures; for R&D and acquisition; for planning, technology assessment, and budgets; for instruction and training; and for administrative and operational control of surface and airborne platforms, with appropriate liaison and subordination when deployed to other operational commands, of course. This can only be accomplished with the strong support and advocacy of the Chief of Naval Operations and concurrence at the highest levels of the Navy.

Mine warfare and countermine warfare programs should be connected at the top of the Navy hierarchy, rather than remaining bureaucratically separated among diverse actors with diverse levels of motivation toward mine warfare. (Indeed, this is precisely the same prescription accepted in 1989 for the Navy's "new approach" to ensuring that its antisubmarine warfare programs remained focused on the real requirements for defeating the increased Soviet and other-country undersea threats.) This may be necessary to bring together the critical talents, skills, and funds to address the problems and opportunities of today and those that will come tomorrow. In short, an Assistant Chief of Naval Operations for Mine Warfare is needed to complement the Mine Warfare Division (OP-72) within the Office of the Deputy Chief of Naval Operations for Naval Warfare (OP-07).

If, however, the U.S. Navy reverts to business as usual and does not adequately and comprehensively address mine warfare, with all the creativity and enthusiasm its great potential in the twenty-first century deserves, we will continue to run a severe risk. The risk is that the significant momentum in mine warfare programs that was built up in the 1980s, after some thirty years of near-neglect, will not continue to keep pace with the multinational, multidimensional, and increasingly sophisticated and serious threat. Should we allow this to happen, the United States will doubtlessly be surprised yet again by its adversaries' weapons that wait.

A Mine Chronology

This chronology shows some of the technological events which were necessary for the applications in mine warfare that followed.

From Beginning to Civil War

600 B.C.	Thales—Static electricity from amber
600–1200 A.D.	Byzantine Period—"Greek fire" used—mixture of charcoal, sulphur, resins, fats, natural petroleum, and possibly saltpeter
1300	Bacon & Schwarz—Gunpowder—In 1242 Bacon "published" in a secret code the composition of a mixture of sulphur, carbon, and saltpeter that would explode if lighted by a flame. This was later called gunpowder. At Bacon's time no guns were known. Schwarz added the gun to the powder about 1300
1600	Gilbert—Investigations of electric action
1672	vonGuericke—First electric machine—rotating ball of sulphur
1672	Flint lock musket came into use
1731	Grey—Conductors and nonconductors
1733	Dufay—Two classes of bodies electrified by friction

1745	Cunaeus of Leyden—Leyden jar
1747, 1749	Franklin—Leyden jar, charge on conductor surface explained
1751	Franklin—Gunpowder fired with spark from Leyden jar
1773	Cavendish—Inverse square law of electric attraction proved
1777	Bushnell—Flintlock gunpowder keg mine
1791	Galvani—Electricity from dissimilar metals
1795	Cavallo—Gunpowder with steel filings fired at 300 feet by Leyden spark
1798–1802	Volta—Pile, battery
1800	Howard—Discovery of fulminate of mercury
1800	Davy—Carbon arc with battery
1805	Battle of Trafalgar
1807	Forsyth—Patent for fulminating powder, S,C & $KClO_3$
1812	Fulton—Proposed moored mines with gunpowder and flintlock. —Drifting mine with clock to explode —Bolo idea: two drifters loosely tied with a rope
1812	Schilling—Fired gunpowder across River Neva with insulated cable, carbon arc fuse, electro-chemical battery
1814	Thomas Shaw of Philadelphia—Percussion cap using fulminate of mercury
1820	Oersted—Magnetic field of electric current
1833	Robert Hare—Proposed that United States use controlled mines with coastal forts; underwater explosions of gunpowder
1838	Pelouze—Gun cotton made from cotton and nitric acid

1839	Pasley—Removes wreck of *Royal George* using Daniell Cell, platinum hotwire, insulated cable
1844	Colt—Battery, hot wire, moored charges, optical fire-control spotting
1846	Sobero—First nitrated glycerine
1848	Siemens—controlled minefield at Kiel Harbor in Schleswig-Holstein War; fuze (chemical)
1854	Jacobi—Russian moored contact minefield in Crimean War at Kronstadt, using chemical fuzes
1859	Unknown—Harbor of Venice controlled mines and camera oscura for fire control

Civil War through World War I

1861	Maury founds Torpedo Bureau in Richmond; "Singer" moored mines with percussion caps, spring-driven plunger released by ship impact
1862	First mine countermeasure device, a raft with grappling hooks used by the Union forces
1864	Farragut and Mobile Bay
1864	The *Hunley* delivers spar torpedo and sinks the USS *Housatonic* in Charleston, S.C.
1864	Whitehead automotive torpedo started development in Fiume
1866	Nobel invented dynamite by mixing nitroglycerin with diatomaceous earth (or kieselguhr) which absorbs three to four times its weight in nitroglycerin. The resulting mixture is safe to transport and can be detonated
1868	Hertz horn—breaking a glass tube releases electrolyte and completes firing circuit
1871	Sprengel detonates picric acid, with mercury fulminate. This was the first benzene ring explosive
1870	Mines used in Franco-Prussian War
1865–70	Mines used in Paraguay, Argentine, Brazil War

1877–78	Mines used in war between Russia and Turks
1885–90	HMS Vernon (British "Torpedo" School at Portsmouth, England) develops automatic anchor, and puts dry-cell batteries in moored mines
1888	Nobel makes Ballistite smokeless powder by mixing gun cotton and nitroglycerin
1898	U.S. BuOrd reports use of gun cotton to replace gunpowder in mines
1899	Commandante Elia design prototype of Vickers and Sautter-Harlé moored contact mine. (described by USN BuOrd mine pamphlet in 1909). Mechanical firing device never satisfactory, abandoned in World War I
1898	Spanish-American War—Spanish used mines without effect. The USS *Maine* suffered self explosion. See Rickover [42]
1902	TNT first used by Germans
1904	Russo-Japanese War (sixteen ships sunk)
1907	Hague Convention
1912	Inertial firing device developed by Russians for moored mines
1914–18	World War I: Heligoland Bight; Turks mine Dardanelles; Dover straits
1917	M-sinker (two-needle magnetic mechanism and acoustic chattering relay (experimental)
1917–18	Galvanic antenna firing mechanism and the Great North Sea Barrage. K device (Browne and Fullinwider) reduces number of mines needed from 300,000 to 100,000. New mine called Mk 6. Cost $400 each (1918 dollars)

Post–World War I

1919	Sweeping up the North Sea Mine Barrage
1919	Mine Building in Navy Yard, Washington; Lt. L. W. McKeehan, first officer in charge

1929	Mine unit joined with Experimental Ammunition Unit and became NOL
1931	First British coiled rod mu metal induction mine, M Mk 1
1939	Air-launched German magnetic needle ground mine (first operational use of influence mine)
1940	Degaussing
1940	Magnetic sweeping (L sweep)
1941	Operations research in mining
1942	U.S. sub-laid mines Mk 10, Mk 12: one enemy ship lost for every eight to twelve mines laid
1943	Magnetic Induction firing mechanism with electronic amplification
1943	Acoustic Mines—high frequency and low frequency
1944	Pressure mines first used (independently invented in Germany, U. K., and U.S. at earlier dates)
27 March 1945	First mines planted in Shimonoseki Strait (Japanese shipping cut to zero in five months)
1951	Wonsan, Korea, mined by Soviets with N. Korean help
1950s	Transistor and solid-state electronics
1966	Thin film magnetometer—a high-frequency oscillator whose frequency shifts when the ambient magnetic field shifts
May 1972	Aerial mining of Haiphong Harbor stopped all ship movement
Jan 1973	U.S. agrees to clean up mines as condition of peace
May 1982	Argentina uses obsolete WWI-era mines to block access to Port Stanley
July–Sept 1984	Libya plants Soviet "export" mines in Gulf of Suez and Red Sea
1987–88	Iranian Revolutionary Guards sow minefields in Persian Gulf. USS *Roberts* (FFG-58) nearly sunk after striking WWI-era mine in April 1988

The future—New and old mines will continue to be used in all forms of warfare, from low-intensity conflict to global war.

B U.S. Navy Mine Warfare Platforms and Systems

Airborne and surface platforms

MH-53E Sea Dragon Helicopter

The E series of the H-53 is the heaviest-lift helicopter in service in the West, with the MH-53E Sea Dragon being a minesweeping variant with larger fuel capacity.

The H-53E series is similar to the earlier CH-53A/D Sea Stallion helicopters in U.S. Navy and Marine Corps service. The airframe follows the pattern of the earlier CH-53 versions, having an unobstructed cabin with lowering rear ramp behind a blunt nose and flight deck and under the power train. The cabin is spacious enough to accept seven pallets measuring 3 ft 4 in × 4 ft; seating for fifty-five troops is also provided. A single-point cargo hook below the fuselage can carry an 18-ton sling load, although 16 tons is more typical.

The MH-53E Sea Dragon is a U.S. Navy multipurpose aircraft employed for vertical replenishment and Airborne Mine Countermeasures (AMCM) operations. In the latter role it can tow a variety of sweep gear, including the AN/ALQ-166 Lightweight Magnetic Sweep, Mk 103 mine cable cutters, Mk 104 acoustic countermeasures, Mk 105 hydrofoil sled, Mk 106 acoustic sweep suite for the sled, and the AN/SPU-1W Magnetic Orange Pipe (MOP) for countering shallow-water mines. The greatly en-

This appendix was prepared for the updated edition by Dr. Scott C. Truver.

Sources: USNI Military Database, 1990; Norman Friedman, *World Naval Systems* (Annapolis: Naval Institute Press, 1989); Norman Polmar, *Ships and Aircraft of the U.S. Fleet* (Annapolis: Naval Institute Press, 1987); *Jane's Weapon Systems 1987–88* (London: Jane's, 1988).

larged fuel sponsons are made from composite materials with a total capacity of 3,200 gal. The tow boom is capable of a 30,000-lb tow tension load. The maximum useful load for an influence minesweeping mission is 26,000 lb. It is fitted with the Hamilton-Standard FCC-105 Automatic Flight Control System (AFCS), Doppler, OMEGA navigation systems, and a RAYDIST minesweeping navigation system. The MH-53E has a four-hour mission capability (plus reserves) and is fitted for night operations. The first flight of the MH-53E variant was on 1 September 1983; the first delivery to a U.S. Navy squadron (HM-14) was on 1 April 1987. The planned total of thirty-two MH-53Es was fully funded with the FY 1989 appropriation. The unit cost is about $36 million in FY 1989 dollars.

The biggest change from the earlier CH-53D versions is in the power train. A third T64 turboshaft engine is fitted behind the main rotor mast; it is fed by an intake located to the left of the rotor mast. The main transmission gearbox, located directly below the main rotor hub, is linked to the three engines by drive shafts that extend forward and outward to the outboard engines and to the rear for the third engine. On either side of the fuselage are large fuel sponsons; the main gear units fold into the rear of the sponsons. The nose gear folds forward. Additional auxiliary fuel tank pylons can be fitted to the sponsons; the 2,650-gal drop tanks increase total fuel capacity by 127 percent. The aircraft is also fitted with a retractable in-flight refueling probe.

To absorb the increased power, a seventh blade was added to the fully articulated, titanium and steel main rotor hub. Each blade has a wider chord and greater length, a titanium spar, Nomex core, and fiberglass-epoxy skinning. The seven blades are power-folded. The tail section was altered several times during development, with the final arrangement having the tail rotor pylon canted 20° off vertical to port; the large four-blade tail rotor is fitted at the top. A long, strut-braced gull-wing horizontal stabilizer extends to starboard. The tail pylon power-folds to starboard, reducing overall height by 9 ft 9 in.

The initial operational capability (CH-53E) was in February 1981; first flight on 1 March 1974 (YCH-53E); December 1980 (CH-53E); and September 1983 (MH-53E). Approximately ninety-five CH-53Es and fifteen MH-53Es were in service with the U.S. Navy and Marine Corps in late 1989. The first delivery of the S-80-M-1 export version of the MH-53E was made to the Japanese Self-Defense Force in January 1989.

After the first few years of operations, the main transmission gearbox of the CH-53E was inspected and the bull gears replaced in 1987 after tear-down inspections revealed excessive wear. The entire CH-53E fleet was grounded until the replacements could be made. The aircraft was also restricted from certain maneuvers with a sling load because of the heavy vibrations induced. A Sikorsky Technical Evaluation Program (TEP) conducted in 1987–88 apparently turned up no major design flaws. In April 1989, a report from the Director of Operational Test and Evaluation for the U.S. Department of Defense stated that the MH-53 was not operationally suitable for its mine-hunting mission. The aircraft was criticized for its high cockpit noise level, problems with the cables that tow the sled, imprecision in its navigation system, and other defects. The report noted that more than 80 percent of the Sea Dragon's flights had been aborted or delayed, although the time span covered by the report was not clear.

CHARACTERISTICS

Manufacturer	Sikorsky Aircraft Co., Stratford, Conn.
Crew	3 (2 pilots, 1 crew chief) + MCM personnel
Engines	3 General Electric T64-GE-416 turboshaft
maximum power	4,380 shaft hp each
internal fuel capacity	1,017 gal
Weight	
empty	33,226 lb
maximum payload	
internal	30,000 lb
sling load	32,000 lb
Dimensions	
rotor diameter	79 ft
length	
fuselage	73 ft 4 in
overall	99 ft 1.2 in
height	27 ft 9 in
cabin	
length	30 ft
width	7 ft 6 in
height	6 ft 6 in
disc area	4,902 sq ft
Performance (at 56,000 lbs)	
maximum speed	170 kts
cruise	150 kts
climb rate	2,500 ft/min
service ceiling	18,500 ft
hovering ceiling	11,500 ft in ground effect
	9,500 ft out of ground effect
range	approx 450 nm with 20,000-lb payload, external fuel
self-ferry	1,000 nm

RH-53D Sea Stallion Helicopter

The CH-53 is a heavy-lift helicopter in wide military use in the United States and abroad. Developed specifically for the U.S. Navy and Marine Corps, it is employed for assault, vertical replenishment, and, in the RH-53A and D variants, Mine Countermeasures (MCM) operations. Forty-four CH/HH-53B/C are being modified for special operations and are known as the Pave Low III series.

The semimonocoque fuselage has a blunt, almost stepless nose with tall "cheek" windows as well as a split windshield and side windows. The large, unobstructed cabin is roughly square in cross-section—7 ft 6 in wide and 6 ft 6 in high—for most of its length. The original payload requirement was for thirty-eight fully equipped troops, twenty-four litters, and four attendants, or 8,000 lb of cargo. In later variants, up to fifty-five troops can be carried and the external hoist was ultimately rated at 20,000 lb. The after part of the cabin tapers up to the tail boom and holds the stern ramp, which is lowered for ground loading or air drops. The ramp opening is wind-shielded by small panels fitted to the fuselage sides below the tail boom.

Large sponsons are mounted on the lower aft fuselage sides; they have an airfoil cross-section. The sponsons hold the two-wheel main landing gear struts, which retract forward behind the fuel tanks; additional fuel tanks are carried on semipermanent pylons mounted outboard of the sponsons. The two-wheel nose gear retracts forward.

The six-blade, fully articulated rotor has aluminum blades with power folding; later variants have been retrofitted with composite blades having titanium leading edges. The main rotor mast rises out of a long centerline hump that contains the Solar T62T-12 Auxiliary Power Unit (APU). The two T64 turboshaft engines are shoulder-mounted on either side of the centerline fairing. The inlets are protected by prominent D-section debris screens. The four-blade antitorque rotor is mounted on the left side at the top of the slender tail boom; the long-span, unbraced horizontal stabilizer extends to starboard opposite the antitorque rotor hub. The tail boom has a strut-braced skid and can be folded to starboard for shipboard stowage.

In 1990 the Navy had fifteen RH-53As (ex–CH-53As) in service as airborne minesweepers. Thirty units of the RH-53D were delivered to the Navy for the AMCM role. Similar to the CH-53D but powered by two 4,330-shp T64-GE-415 turboshafts, the RH-53Ds are fitted with minesweeping gear including a Mk 105 towed hydrofoil minesweeping sled, the AN/SPU-1W Magnetic Orange Pipe (MOP), and the Mk 104 acoustic minesweeping gear. (Six additional units were sold to Iran before the Revolution of 1979; they have ceased to be operational, although reports circulated that one of the Iranian RH-53Ds was seen towing some minesweeping gear during the 1987–88 Persian Gulf mine crisis.) Eight

RH-53D helicopters were launched from the U.S. aircraft carrier USS *Nimitz* (CVN-68) on 25 April 1980 as part of the ill-fated effort to rescue hostages held in the American Embassy in Tehran; seven of these aircraft were lost in the raid. Eight RH-53Ds were flown to Diego Garcia in the Indian Ocean in July 1987 by C-5 Galaxy transport. The helicopters were then placed on board the amphibious carrier *Guam* (LPH-9) for mine-sweeping operations in the Persian Gulf.

Initial operational capability (CH-53A) was in 1966; the first flight was on 15 October 1964. It is no longer in production, as the H-53E variant is now being procured. The Navy's RH-53Ds are being transferred to the Reserve Force as the MH-53Es are delivered to the two active AMCM squadrons, HM-15 and HM-16.

CHARACTERISTICS (CH-53D)

Manufacturer	Sikorsky Aircraft Co., Stratford, Conn.
Crew	3 (2 pilots, 1 crewman) + MCM personnel
Engines	2 General Electric T64-GE-413 turboshaft
maximum power	3,925 shaft hp each
internal fuel capacity	622 gal
Weight	
empty	23,628 lb
maximum takeoff	50,000 lb
Dimensions	
rotor diameter	72 ft 3 in
length	67 ft 2 in
height	24 ft 11 in
disc area	4,100 sq ft
Performance (at 50,000 lb)	
maximum speed	171 kts
cruise	151 kts
climb rate	2,180 ft/min
service ceiling	21,000 ft
hovering ceiling	13,400 ft in ground effect
	6,500 ft out of ground effect
range	886 nm

Avenger (MCM-1) Mine Countermeasures Ships

The fourteen ships of this class are relatively large mine counter-measures ships intended to locate and destroy mines that cannot be countered by conventional minesweeping techniques or minesweeping helicopters. They are to be deployed to coastal waters, choke points, and critical overseas areas. The basic MCM design is similar to previous MSO (Ocean Minesweeper) classes. The hull is constructed of fiberglass-sheathed wood (laminated oak framing, Douglas fir planking, and deck sheathing with reinforced fiberglass covering). These materials, combined with a degaussing system, enable this class to maintain a very low magnetic signature. One or two AN/SLQ-48 Mine Neutralization System (MNS) vehicles are carried in addition to conventional sweep gear.

The ships are fitted with the AN/SSN-2 precise navigation system. The ships may be fitted with the NAUTIS-M mine countermeasures command and control system beginning in 1991 if initial trials on the USS *Pluck* (MSO-464) and later trials on this class are successful. The MCMs are initially fitted with the AN/SQQ-30 variable-depth sonar. This equipment, an upgraded AN/SQQ-14, is limited in certain environmental conditions and will be succeeded in later ships by the AN/SQQ-32.

All ships have four very-low-magnetic diesel engines for propulsion; electrical power for minesweeping gear is provided by gas turbines. A bow thruster of 350-hp is fitted for precise maneuvering. Maximum mine-hunting speed is 5 kts. The Congress has directed that MCMs 10–14 will have American-made diesel engines.

The lead unit of the class, *Avenger*, was ordered on 29 June 1982 and laid down on 3 June 1983, the first large minesweeper under construction for the U.S. Navy since the USS *Assurance* (MSO-521) was completed twenty-five years earlier. The *Avenger* was almost two years behind her original contract schedule and the *Defender* more than fifteen months behind contract schedule: MCMs 2–4 were to have been completed in July through December 1988, but this has been postponed to 1989 for MCMs 2 and 3 (commissioned in September and August 1989, respectively), and to 1990 for MCM-4.

In February 1989, the Navy agreed to pay Marinette Marine $65 million for the completion of the three ships being built at that shipyard. Marinette claimed that the Navy had made numerous design changes, sometimes resulting in the removal of work already completed according to the original design. In February 1989, Peterson Builders was awarded a $185 million contract to build MCMs 9, 10, and 11. MCMs 12–14 were ordered from Peterson in December 1989. The Avenger class has suffered several design and construction problems. The first two ships were fitted with American engines from existing stocks; they were installed improperly, and subsequent tests revealed a potential fire hazard from lubricating oil leakage through the turbocharger into the exhaust stack. (This engine design had been blamed for a series of fires in previous minesweepers.) The Italian engines planned for the later MCMs then failed in their endurance tests.

Under the original plans, after approximately one year in active service the first eight units were to join the Naval Reserve Force (NRF) to replace the aging Agile and Aggressive class. *Avenger* was scheduled to enter the NRF on 30 September 1989. At the August 1989 CNO Executive Board (CEB) on Mine Countermeasures, however, the decision was made to keep at least twelve of the fourteen ships of the class in the active fleet, and to seek the funds to acquire eight additional units. These eight "follow-on" MCMs are expected to be a modified design using fiberglass/resin hulls, not wood. The unit cost is $113,833,333 (FY 1990 dollars). The final three ships of the fourteen-unit class were requested under the FY 1990 program.

CHARACTERISTICS

Displacement	1,312 tons full load
Dimensions	
length	212 ft 10 in waterline
	224 ft 4 in overall
beam	39 ft
draft	11 ft 6 in
Propulsion	MCMs 1, 2: 4 diesels (Waukesha L-1616); 2,600 bhp; 2 shafts
	MCMs 3–9: 4 diesels (Isotta-Fraschini ID36 SS-6V-AM); 2,280 bhp; 2 shafts
Performance	
speed	13.5 kts
Manning	72 (5 off + 67 enl)
Combat Systems	
radars	SPS-55 surface search
sonars	SQQ-30 mine detecting in MCMs 1–9
	SQQ-32 mine detecting in later ships

Number	Name	FY	Launched	Commissioned	Status (Jan 90)
MCM-1	Avenger	82	1985	1987	Atlantic
MCM-2	Defender	83	1987	1989	Atlantic
MCM-3	Sentry	84	1986	1989	Atlantic
MCM-4	Champion	84	1989	(1990)	Building
MCM-5	Guardian	84	1987	1989	Atlantic
MCM-6	Devastator	85	1988	(1990)	Building
MCM-7	Patriot	85	(1990)	(1991)	Building

(*continued*)

(continued)

Number	Name	FY	Launched	Commissioned	Status (Jan 90)
MCM-8	Scout	85	(1989)	(1991)	Building
MCM-9	Pioneer	85		(1992)	Building
MCM-10	Warrior	86		(1992)	Building
MCM-11	Gladiator	86		(1992)	Building
MCM-12–14		90			Building

Osprey (MHC-51) Coastal Mine Hunters

The seventeen units of this coastal mine-hunter class are based on the Italian-designed Lerici-class coastal mine hunter. The ship will have a Glass-Reinforced Plastic (GRP) monohull single-skin structure and will be equipped with the AN/SLQ-48 mine neutralizing system. Mechanical and influence minesweeping systems are being developed independently. The ships will be capable of coastal mine clearance operations of up to five days without replenishment. The Osprey class is a scaled-up version of the original Lerici design, with a full load displacement of 790 tons versus the 520 tons of the Italian craft and heavier minesweeping equipment.

In FY 1984 the U.S. Navy had contracted with Bell Aerospace Textron (now Textron Marine Systems) to design and construct the first of fourteen Cardinal-class Minesweeper Hunters (MSHs). They were to be of GRP monohull structure, but with Surface Effect Ship (SES) and other complex technology. Tests showed that this design would not withstand explosive charges, and redesign efforts failed; that contract was ended in November 1986, largely at the direction of the U.S. Congress.

In August 1986 a contract was issued to Intermarine USA (established by Intermarine Sarzana of Italy, builders of the original Lerici design) to study possible adaptations of the Lerici craft to carry U.S. combat systems and electronics. Intermarine was chosen in May 1987 to construct the lead ship of this class. The first ship was laid down in May 1988 and was scheduled for delivery in April 1991; shortly after construction began Intermarine informed the Navy that this was an optimistic schedule and that the completion date would be later than planned; a new target date of July 1991 has been set. On 17 February 1989 Intermarine was awarded a $55.3 million contract to build the second unit, the MHC-52, scheduled for completion by 30 April 1992.

In mid-1989 the Navy delivered a Request for Proposals (RFP) for a second-source builder for the third unit, which was awarded to Avondale Industries in October 1989. Intermarine USA and Avondale will compete for the remaining fourteen units. The Italian GRP technology is being transferred to Intermarine USA, as it will be to Avondale. Another result of the August 1989 CEB was the decision to plan for an additional eight units of the Osprey class, several of which may be built to a modified design.

Each ship will operate for one year with an active crew before transferring to the Naval Reserve Force (NRF). The average unit cost for the lead ships is $76,766,666 (FY 1990 dollars); $120.1 million of the FY 1986 funds allocated to the canceled MSH program were designated for construction of the lead MHC ship.

CHARACTERISTICS

Displacement	851 tons full load
Dimensions	
length	190 ft overall
beam	36 ft 1 in
draft	9 ft 6 in
Propulsion	2 diesels (Isotta-Fraschini ID 36 SS 6V-AM); 1,160 bhp; 1 shaft
	2 180-bhp hydraulic motors for mine hunting
	1 180-hp bow thruster
Performance	
speed	12 kts
range	2,500 nm at 12 kts
Manning	51 (4 off + 47 enl)
Combat Systems	
guns	1 30-mm or 40-mm
radars	1 SPS-64(V)9 surface search
sonars	SQQ-32 mine detecting

Number	FY	Commissioned	Status (Jan 90)
MHC-51	86	(1991)	Building
MHC-52	89	(1992)	Building
MHC-53	89	(1992)	Authorized
MHC-54	90	...	Planned
MHC-55–57	91	...	Planned
MHC-58–61	92	...	Planned
MHC-62–65	93	...	Planned
MHC-66–67	94	...	Planned

Craft of Opportunity Program (COOP) Minesweepers

As originally conceived, seventeen former Naval Academy seamanship training craft (YPs) and five former fishing craft were to be adapted for the Craft of Opportunity Program (COOP) to provide mine countermeasures capabilities in U.S. ports. The first COOP units were in place in 1985, with the last scheduled to become active in 1989–90. Three of the civilian craft have been identified, while the former YP class originally comprised YPs 654–675. They will use trawls, side-scan towed sonar, and Remotely Operated Vehicles (ROV) to survey harbor bottoms to detect mines. The YP radar is a Raytheon 1220. A modified AN/SQQ-14 sonar is being considered for later installation in these craft.

Tests with the first YPs, however, showed them to be unstable in their COOP configuration. In 1989 the Navy decided to kill the YP conversion program. This prompted some non-Navy initiatives, supported by a few vocal members of Congress, for a new-construction program to build "Super-COOPs" based on the YP-683–class design for the Naval Reserve Force. This was rejected by the Navy. No decision regarding this aspect of the COOP initiative had been made by the spring of 1990.

Each COOP craft will have a nine-man enlisted operating crew, with four Reserve Force crews assigned. These crews (designated Blue, Gold, Red, and Green) will alternate weekends operating the craft. In addition, each craft will have two active enlisted personnel assigned plus an administrative staff of two officers and two enlisted men.

Acme-Class Ocean Minesweepers

The two ships of this class are improved versions of the Agile and Aggressive MSO classes, being slightly larger and slower. They are fitted as mine division flagships. These ships were the newest minesweepers in the U.S. fleet until delivery of the USS *Avenger* (MCM-1) in 1987. Four ships were built to this design for the U.S. Navy (MSOs 508–511) and seven for allied navies (MSOs 512–518; since discarded). The U.S. ships are operated by the Naval Reserve Force.

CHARACTERISTICS

Displacement	682 tons light
	818 tons full load
Dimensions	
length	173 ft overall
beam	36 ft
draft	14 ft
Propulsion	4 diesels (Packard 1D-1700); 2,800 bhp; 2 shafts

(continued)

CHARACTERISTICS (continued)

Performance
  speed   14 kts
  range   3,300 nm at 10 kts

Manning   82 (5 off + 52 enl active + approx 25 reserve)

Combat Systems
  guns   2 .50-cal machine guns (twin)
  radars   SPS-53L surface search in MSO-509
                       SPS-53E surface search in MSO-511
  sonars   SQQ-14 mine detecting

Number	Name	FY	Launched	Commissioned	Status (Jan 90)
MSO-509	Adroit	54	1955	1957	Reserve
MSO-511	Affray	54	1956	1958	Reserve

Agile- and Aggressive-Class Ocean Minesweepers

These eighteen ocean minesweepers and the previous Acme class are the survivors of the massive U.S. minesweeper construction programs started during the Korean War after the shock of extensive North Korean use of Soviet-supplied mines. They are fitted for sweeping contact, magnetic, and acoustic mines, but they lack a good capability for countering modern mines. Five of these ships, including the *Enhance* and the *Esteem*, were deployed to the Persian Gulf as part of the buildup of U.S. naval mine warfare capability to counter the Iranian mine threat to shipping.

These ships are of lightweight wooden construction with laminated timbers; the fittings and machinery are of stainless steel (nonmagnetic) and bronze. Magnetic items are reduced to a minimum. These ships originally had the AN/UQS-1 mine detecting sonar; most have been refitted with the more capable AN/SQQ-14 sonar. The latter is a variable-depth sonar, lowered on a rigid rod from the hull forward of the deck structure.

The *Pluck* was test ship in the mid-1970s for the AN/SSN-2 precise navigation system fitted in the subsequent Mine Countermeasures (MCM) and Coastal Mine Hunter (MHC) classes. In September 1988 she was fitted with the NAUTIS-M mine warfare command and control system for possible later installation on the MCM-1 and MHC-51 classes. The *Fidelity* has been test ship for the AN/SQQ-30 sonar.

As built, these ships had one 40-mm gun and two .50-cal machine guns. Most were rearmed with provisions for mounting a 20-mm mount forward, originally a twin-barrel Mk 24 and subsequently a single-barrel Mk 68 gun. The smaller 20-mm mount was required in modernized ships to permit installation of the larger, retractable SQQ-14 sonar in the forward hull. Most now have a .50-cal machine gun.

Originally, fifty-eight ships of this class were built for the U.S. Navy (MSOs 421–449, 455–474, 488–496); twenty-seven more were built for allied navies (MSOs 450–454, 475–487, 498–507), with the MSO-497 being canceled. The remaining ships will be stricken as the Avenger and Osprey classes are completed.

In FY 1968 a program was begun to modernize the existing MSOs. New engines, communications, and sonar were installed and improved sweep gear provided. However, increasing costs and shipyard delays caused the program to be halted after only thirteen ships had been fully modernized: MSOs 433, 437, 438, 441–443, 445, 446, 448, 449, 456, 488, and 490. Subsequently, some of these features, especially the improved sonar, were fitted to several additional ships.

All MSOs originally were classified as minesweepers (AM with the same hull numbers); they were changed to MSO in 1955. Of the surviving ships, three are in active Navy service and the remainder are Naval Reserve Force ships. The active ships serve as mine countermeasures test platforms as well as performing operational minesweeper training and exercises. The *Fidelity* was based at the Navy's Coastal Research Center in Panama City, Florida; the two other active ships are based at Charleston, South Carolina.

CHARACTERISTICS

Displacement	716 tons light
	853 tons full load
Dimensions	
length	172 ft overall
beam	35 ft
draft	14 ft
Propulsion	4 diesels (Packard, except Waukesha L-1616 in MSOs 433, 437–438, 441–443, 448–449, 456, 488, 490)
Performance	
speed	15.5 kts
range	3,300 nm at 10 kts
Manning	active ships: 81 (6 off + 75 enl)
	NRF ships: 82 (5 off + 52 enl active + approx 25 reserve)

(continued)

307

CHARACTERISTICS (continued)

Combat Systems
  guns   2 .50-cal machine guns (1 twin)
  radars   SPS-53E/L surface search
  sonars   UQS-1 or SQQ-14 mine detecting

Number	Name	FY	Launched	Commissioned	Status (Jan 90)
MSO-427	Constant	51	1953	1954	NRF Pacific
MSO-433	Engage	51	1953	1954	NRF Atlantic
MSO-437	Enhance	51	1952	1955	NRF Pacific
MSO-438	Esteem	51	1952	1955	NRF Pacific
MSO-439	Excel	51	1953	1955	NRF Pacific
MSO-440	Exploit	51	1953	1954	NRF Atlantic
MSO-441	Exultant	51	1953	1954	NRF Atlantic
MSO-442	Fearless	51	1953	1954	NRF Atlantic
MSO-443	Fidelity	51	1953	1955	Decommissioned
MSO-446	Fortify	51	1953	1954	NRF Atlantic
MSO-448	Illusive	51	1952	1953	Atlantic
MSO-449	Impervious	51	1952	1954	NRF Atlantic
MSO-455	Implicit	52	1953	1954	NRF Pacific
MSO-456	Inflict	52	1953	1954	NRF Atlantic
MSO-464	Pluck	52	1954	1954	NRF Pacific
MSO-488	Conquest	53	1954	1955	NRF Pacific
MSO-489	Gallant	53	1954	1955	NRF Pacific
MSO-490	Leader	53	1954	1955	Atlantic
MSO-492	Pledge	53	1955	1956	NRF Pacific

Minesweeping Boats

MSB 29 Type. A single minesweeping boat of this class was built; she is based at Charleston, South Carolina.

CHARACTERISTICS

Displacement	80 tons full load
Dimensions	
length	82 ft overall
beam	19 ft
draft	5 ft 6 in
Propulsion	2 diesels (Packard); 600 bhp; 2 shafts

(continued)

CHARACTERISTICS (continued)

Performance	
speed	12 kts
Manning	11 (2 off + 9 enl)
Combat Systems	
guns	several machine guns can be fitted
radars	1 navigation
sonars	none

MSB 5 Type. These are wooden-hulled minesweeping boats that were designed to be carried to assault areas aboard amphibious ships. They were used extensively in the Vietnam War, being armored and armed for sweeping rivers and channels in that conflict. The seven surviving MSBs form Mine Division 125 at Charleston, South Carolina. A few others are employed by the Navy as utility and training craft.

This class originally included MSBs 5–54, less the MSB-24, which was not built, and the MSB-29, which was built to an enlarged design in an effort to improve seakeeping qualities. All were completed 1952–56. MSBs 1–4 were former Army minesweepers built in 1946.

The MSBs that served in the Vietnam War were armed with several machine guns, with a raised gun tub fitted aft. In August 1987, four MSBs were put aboard the *Raleigh* (LPD-1) amphibious transport dock, headed for minesweeping duties in the Persian Gulf.

CHARACTERISTICS

Displacement	30 tons light
	44 tons full load
Dimensions	
length	57 ft 3 in overall
beam	15 ft 10 in
draft	4 ft 4 in
Propulsion	2 diesels (Packard); 600 bhp; 2 shafts
Performance	
speed	12 kts
Manning	6 or 7 (enl)
Combat Systems	
guns	several machine guns can be fitted
radars	1 navigation
sonars	none

Mine countermeasures systems/mine-hunting sonars

Oropesa (Wire) Sweeps

The Oropesa (wire) sweeps are of various types, all of which are designed to deal with moored buoyant mines close to the surface. The United States, like other countries, is developing a deep sweep to cut loose the cables of rising mines, which may be more than 1,000 ft below the surface. The O Size 1 is the largest wire sweep for ocean minesweepers (MSOs and MCMs); when streamed with 300-fathom wires at a maximum speed of about 8 kts (sweep depth of 5–40 fathoms), the swept path for a double sweep is 500 yds wide (250 yds for a single sweep). The O Size 4/modified is for coastal sweepers (MSC/MHC); with 300 fathoms of wire streamed on each side, at 8 kts, and a wire depth of 2–20 fathoms, the swept path is 420 yds wide. The O Size 5 is a small wire sweep for launches and boats (MSL/MSB); this uses explosive cutters to cut the mooring cables because the tow speed of the craft is too low for the saw/abrasive action of the sweep wires to cut the mine mooring cables.

Mk 103

This helicopter-towed moored-mine sweeping system consists of a tow wire, sweep wires (with explosive cutters), floats, a depressor, otters, and float pendants. A recent development is the AN/37U, a controlled-depth, rapidly deployable helicopter wire sweep for use in greater depths, presumably against deep-moored rising mines. In the late 1980s this deep sweep was being adapted for shipboard use, as a modular single-ship sweep.

A Mk 2

The A Mk 2 acoustic sweep for both helicopters and surface vessels consists of parallel pipes or bars, towed broadside-on, in different variants for distinct frequencies and towing speeds from 4 to 20 kts. The A Mk 2(g) was used by helicopters during the End Sweep operation in North Vietnam.

A Mk 4

The A Mk 4 is a towed hammer box, originally intended for small craft in coastal waters but later extended to all types of surface minesweepers. The A Mk 4 consists of a hydraulically driven hammer in a streamlined/hydrodynamic box, driven by a gasoline engine on board the minesweeper.

A Mk 6

The A Mk 6 is a very low-frequency acoustic sweep. The eccentric mechanism is usually towed some 1,200 yds astern of a coastal minesweeper. The A Mk 6(b) uses a diaphragm driven by pistons connected to an eccentric driven by an electric motor. The A Mk 6(g) was used initially in Operation End Sweep but was later replaced by the A Mk 2(g).

Mk 104

This airborne acoustic mine countermeasures system consists of a cavitating disk within a venturi tube, driven by a water turbine. Originally developed by American General, the Mk 104 is towed behind a helicopter; minimum sweep water depth is 30 ft.

AN/ALQ-141, AN/SLQ-35

This acoustic sweep is also called Double Alfa: the ALQ-141 for helicopters is Alfa One; the SLQ-35, a ship-towed version, is Alfa Two. Presumably covering a different portion of the acoustic spectrum from the ALQ-160, the ALQ-141 was described in 1979 as an electronic (acoustic) counter to deep-water rising mines, manufactured by Westinghouse Oceanic Division.

AN/ALQ-160

This is a helicopter-towed acoustic sweep using a stable towed body. No details about this system have been made public.

M Mk 1

The M Mk 1 is a magnetic sweep consisting of no more than a magnetized iron rail towed from a minesweeper. Although it was not standard after World War II, a device of this type was used in Vietnam to clear the Cua Viet River of magnetic/acoustic mines. The MOP (Magnetic Orange Pipe) is the "modern" variant of the M Mk 1.

M Mk 4

The M Mk 4 electromagnetic sweeps are of various types: the M Mk 4(m) is a small two-ship closed loop; the M Mk 4(i) is a two-ship single-catenary small-craft sweep (the short legs of which are connected by an insulated cable forming the catenary, while the long legs stream aft); and the M Mk 4(v) is a post–World War II variant of the M Mk 4.

M Mk 5

The M Mk 5 electromagnetic sweeps are of various types: the M Mk 5(c) is the standard large-craft sweep, consisting of two buoyant cables connected through a 650-ft short leg, with long legs streaming aft another 900 ft (in some arrangements two or more ships could sweep together, pulsing their tails synchronously to increase ship counts); and the M Mk 5(a) is a straight-tail electrode sweep (also called an I-sweep).

M Mk 6

The M Mk 6 electromagnetic sweeps are of various types: the M Mk 6(a) is called a J-sweep, with the long legs curving around to meet a diverted line towed by the ship and connected to an Oropesa (wire sweep) and a kite; the M Mk 6(b) is a single-ship closed-loop sweep, using Oropesa floats to keep the legs apart; and the M Mk 6(h) is a closed-loop sweep with two legs meeting at a diverter line streamed from an Oropesa.

AN/SLQ-37

This modular magnetic/acoustic combined sweep is being procured for the Avenger-class (MCM-1) mine hunters. No information on this system is available in the public domain.

SPU-1W (MOP)

The SPU-1W Magnetic Orange Pipe (MOP) is a magnetized pipe (30 ft × 10 in) filled with styrofoam for buoyancy. A modern version of the World War II–era iron-rail sweep M Mk 1(m), the MOP was conceived as a precursor sweep mechanism for the helicopters used during Operation End Sweep. The mines laid in Vietnamese waters were so sensitive that they would have destroyed the Mk 105 sleds towed by the helicopters. The MOP was also used to sweep waters too shallow for the hydrofoil sleds. A helicopter could tow as many as three MOPs in tandem to increase the ship counts.

Mk 105

The Mk 105 is a helicopter-towed minesweeping hydrofoil sled manufactured by EDO. It carries a gas turbine generator for magnetic sweeping. First approved for U.S. Navy use in 1970, the Mk 105 was first used during Operation End Sweep in 1973, and also in the Suez Canal sweeps in 1974–75. The sled is typically towed at 20–25 kts, about 450 ft behind the helicopter, with the gas turbine providing power to the twin magnetic tails (conventional open-electrode magnetic sweeps some 600 ft long). The sled becomes foil-borne at about 13 kts, and minimum water depth is 12 ft. The Mk 105 is being succeeded by the ALQ-166.

Mk 106

The Mk 106 is a helicopter-towed acoustic/magnetic sweep, consisting of the Mk 105 sled and a Mk 104 attached to one of the magnetic tails. The Mk 106 was used extensively during the 1984 "Mines of August" sweeping operations in the Red Sea and Gulf of Suez.

AN/ALQ-166

The ALQ-166, manufactured by EDO, is a magnetic minesweeping hydrofoil sled that is replacing the Mk 105 in U.S. Navy service. The

ALQ-166 is a monohull and thus can carry a heavier load in the same dimensions as the catamaran Mk 105; it is intended for use with the MH-53E Sea Dragon helicopters, taking advantage of their greater pulling capacity compared with the RH-53D Sea Stallion helicopters. The new 1,200-hp Avco-Lycoming gas turbine engine on board the ALQ-166 is about three times more powerful than the engine on the Mk 105 and can drive a more powerful electromagnetic sweep generator. The ALQ-166 reportedly quadruples the net sweep rate of the earlier system and can operate across a greater speed spectrum (6–30 kts). The sled-mounted AQS-182 measures the local magnetic environment so that the sweep track spacing can be optimized.

AN/UQS-1

This mine-hunting sonar of the 1950s remains in service on board most ex-U.S. minesweepers but has been replaced in U.S. service by the SQQ-14. The UQS-1 is a short-pulsed electronically scanned sonar, with a frequency of 100 kHz (10-kW pulses). Range scales are 200, 500, and 1,000 yds; the precision is 5 yds at 200 yards, 25 at 500, and 50 at 1,000. The beam is 20° × 10° (vertical), but it rotates and can scan up and down to cover angles between $+5°$ and $-50°$, with bearing precision of 30'. The UQS-1 can scan over a 90° sector or over all 360°. It is not stabilized and cannot provide the mine image required for classification. About 150 UQS-1s were built during 1951–54; the SQS-15 was a variable-depth version.

AN/SQQ-14

The AN/SQQ-14 is a dual-frequency (80 kHz search/350 kHz classification) mine detection and classification sonar for minesweepers. (It is also built under license in Italy; as the P2072 it equips the Italian Lerici-class mine hunters.) The SQQ-14 is basically a UQS-1 fitted with CXRP, a copy of the British Type-193 high-resolution sonar, for classification. It is intended primarily for use in shallow water against bottom mines. It is lowered from under the hull by a flexible cable that consists of 18-in sections connected by universal joints, a result of the inherent instability of variable-depth bodies at speeds greater than 5 kts, which ruined the accuracy of the beams. (In later systems, e.g., the SQQ-30, a great deal of effort has gone into stabilizing the variable-depth body using fins.) Maximum depth is thought to be 45 m. The search sonar has a 1.5° beam (100° field of view); azimuth resolution is 1.5° and range resolution is 1 m (1.0-msec pulses, 15 kW, 126-dB source level). The classification sonar has a 0.3° beam (18° field of view); azimuth and range resolutions are 0.3° and 3 cm, respectively. Both the search and classification sonars can be tilted between $+10°$ and $-30°$ (elevation), and there is also a side-scan mode. Initial operational capability was in 1960. The system was manufactured by General Electric. A towed version, SQQ-35, was under development in the late 1980s.

AN/SQQ-30

The AN/SQQ-30 is a solid-state mine detection and classification sonar developed from the AN/SQQ-14. Manufactured by General Electric, the SQQ-30 entered Navy service in 1983. Unlike the SQQ-14, it is cable-lowered from under the minesweeper, and thus it can be operated at greater depths and handled more easily. The SQQ-30 will be succeeded in service by the AN/SQQ-32.

AN/SQQ-32

The AN/SQQ-32 is a mine detection and classification sonar manufactured by Raytheon/Thomson CSF. As it becomes available, this sonar will be installed in the newer Avenger-class (MCM-1) minesweepers, instead of the less capable AN/SQQ-30; it will be better at discriminating between genuine mines and other objects, and it will be able to identify objects with near-picture quality. The SQQ-32 displays search and classification information simultaneously, using separate search and classification transducers in a stable, variable-depth body. Multibeam operation increases the sonar's search rate. The SQQ-32 can also be used from the hull in shallow water. Initial operational capability was planned for 1989.

AN/SQS-37

The AN/SQS-37 is a hand-held mine clearance sonar for divers. It uses continuous FM transmissions or passive detection of a sonar beacon, employing a 12-in conical reflector that forms a 9° beam at 70 kHz. In search mode, frequency is swept over 30 kHz (between 50 and 90 kHz); there are three range scales (20, 60, and 120 yds). Source level is 72 dB. The reflected beam is combined with a sample of the transmitted beam to form a different frequency in the audio range; the lower the tone, the closer the object. Using the set to detect 40 kHz marker beacons, the sonar indicates the direction but not the range of the beacon.

Mk 24/Klein 590

This lightweight side-scanning sonar for mine hunting and channel conditioning can be used on board virtually any unmodified commercial helicopter or small craft. It has been adopted for the U.S. Navy's Craft of Opportunity Program (COOP) as well as for the seven surviving U.S. minesweeping boats, four of which were sent to help clear the Persian Gulf in 1987 and 1988. The Mk 24 operates at 100 kHz (beam width 0.75°), but Klein also offers an optional very high-resolution sonar at 500 kHz. Range is selectable in thirteen increments between 25 and 600 m, and the variable-depth fish can be operated down to 2,300 m. (An optional fish can be operated down to 12,000 m.)

AN/SLQ-48 MNS

The AN/SLQ-48 Mine Neutralization System (MNS) is an unmanned mine-hunting submersible built by Honeywell. The vehicle takes its power from the launching ship, proceeding at a speed of 6 kts to the target detected/classified by the ship's sonar. The vehicle (12.5 ft long × 3 ft wide) carries a small high-definition sonar and an acoustic transponder that enables the vehicle to be tracked by the shipboard sonar. There is also a low-light-level television for examining a target. The vehicle carries MP-1 (Mk 26-0) cable cutters and an explosive mine-destruction charge (MP-2, Mk 57-0). Propulsion is provided by two 15-hp motors, and there are two side thrusters for exact positioning above a target. The umbilical cable length is 3,500 ft. Two consoles on board the ship monitor and control the vehicle's operation.

Buried Mine-Hunting System

In the late 1980s this system was a new-development effort for the U.S. Navy, in conjunction with the Federal Republic of Germany and Great Britain. It is to use both magnetic and acoustic sensors to detect mines buried in sediments too deep for conventional mine-hunting sonars. The ASQ-81 MAD (Magnetic Anomaly Detector), on a catamaran, was used to locate buried mines during Operation End Sweep.

AN/AQS-14

The AN/AQS-14 is an active-controlled helicopter-towed mine-hunting sonar, developed for retrofit into the RH-53D Sea Stallion helicopter; it will also be used in the new MH-53E Sea Dragon helicopters. Manufactured by Westinghouse Oceanic Division, it is a multibeam side-looking sonar with electronic beam forming, all-range focusing, and an adaptive processor. The system consists of three parts: a stabilized underwater vehicle, electromechanical tow cable, and airborne electronic console. The 3-m-long underwater vehicle can be maintained at a fixed point above the sea floor or below the surface, and the thin, coaxial cable is armored and nonmagnetic. The display is in the form of two continuous moving television-like window pictures. The system's initial operational capability was in 1984 (RH-53D), and its first operational use came in August of that year during the Red Sea/Gulf of Suez mine crisis. Westinghouse describes the AQS-14 as suitable for installation on board hovercraft and other surface effect ships.

AN/AQS-20

The AN/AQS-20 is an advanced airborne mine-hunting system, using a sonar detection system. No details about the AN/AQS-20, formerly called the Advanced Minehunting Reconnaissance System, are available in the public domain.

ALARMS

Northrop's Airborne Laser Radar Mine Sensor (ALARMS) employs a downward-looking blue-green laser pulsed at 10 kHz. Northrop developed this in response to reports of moored mines in the Persian Gulf during 1987; first tests of the system were run in August 1988. The copper-vapor laser (a Larsen 500 Ladar built by Optech Systems) scans in a 682-shot circular pattern. The mine targets are detected in real time, the display showing the mines directly beneath the aircraft carrying the system. Northrop claims that search rates at least ten times the expected visual rates are possible. The mine targets are displayed as squares, the color of the square indicating the depth of the target. Maximum operating depth is said to be 100 ft. In late 1989 the system was still under development by the Navy. (A similar system built by Kaman Aerospace, called Magic Lantern, was also under development in the early 1990s.)

Naval mines

Mk 25 Ground Mine

This was the standard U.S. World War II aircraft-laid bottom mine, which was also used during the Vietnam War, according to some sources. In 1980, new flight gear was being issued to permit this mine to be carried at high speeds. Mk 25s were probably provided to U.S. allies postwar.

CHARACTERISTICS

Weight	
case	2,000 lb
warhead	1,274 lb TPX or 1,200 lb HBX
Dimensions	
length	7 ft 9 in
diameter	22 in
Warhead	conventional TPX or HBX high explosive
Sensors/Fire Control	
Mod 1	acoustic
Mod 2	pressure
Mod 3	acoustic

Mk 36 Ground Mine

The Mk 36 was an aircraft-laid bottom influence mine with a slanted nose for optimum underwater trajectory. The basic version was acoustic, but there was also a magnetic influence version. The mine was probably exported to U.S. allies after World War II.

CHARACTERISTICS

Weight	
case	1,082 lb (acoustic); 1,008 lb (magnetic)
warhead	638 lb TPX
Warhead	conventional HBX high explosive
Sensors/Fire Control	
Mod 1	acoustic or magnetic
Mod 2	low-frequency acoustic
Mod 3	pressure

Mk 39 Ground Mine

The Mk 39 was a high-altitude aircraft-laid magnetic bottom mine, delivered without any parachute. Total weight was 2,000 lb, including 800 lb of high explosive.

Mk 49 Ground Mine

This is a submarine-laid bottom mine, which can be launched in depths as great as 125 ft.

CHARACTERISTICS

Weight	
case	2,000 lb (magnetic)
	1,890 lb (acoustic)
	1,960 lb (magnetic/pressure)
warhead	1,180 lb
Warhead	conventional HBX-3 high explosive
Sensors/Fire Control	
Mod 0	magnetic
Mod 1	acoustic
Mod 2	magnetic/pressure

Mk 51 Harbor Defense Mine

This controlled mine for harbor defense is usually laid in groups of thirteen weapons. The Mk 51 is fired either from shore or by magnetic induction. Associated equipment may include a pair of hydrophones attached to anchors on the sea floor. This weapon was probably exported to U.S. allies postwar.

CHARACTERISTICS

Weight	
case	6,200 lb
warhead	3,275 lb
Dimensions	
length	5 ft 2 in
diameter	37 in
Warhead	conventional TNT high explosive
Sensors/Fire Control	command fired from shore or magnetic induction

Mk 52 ASW Ground Mine

The first of the postwar series of antisubmarine mines, the Mk 52 is an older modified aircraft bomb that has been deployed in an aircraft-laid bottom mine configuration by the U.S. Navy. Several mods vary principally in their firing mechanisms. The Mk 52 was part of the U.S. Long Range Mine Program begun in 1948, was released for production in January 1954, and became operational in 1961. In FY 1960, each of the 2,500 Mk 52 mines cost $4,900. As disclosed to the North Vietnamese in 1973 as part of Operation End Sweep, the Mk 52 Mod 2 mine used a two-look reverse-polarity magnetic induction fuze, with selectable magnetic sensitivities, interlook, dead periods, intership dead periods, ship counts, and preset self-sterilization/self-destruction time.

CHARACTERISTICS

Weight	
overall	1,130 lb (Mod 1) to 1,235 lb (Mod 6)
warhead	625 lb
Dimensions	
length	5 ft 1 in
diameter	19 in
	2 ft 9 in over fins
Performance	
maximum depth	150 ft (Mods 1, 3, 5, 6)
	600 ft (Mod 2)

(continued)

CHARACTERISTICS (continued)

Warhead	conventional HBX-1 explosive
Sensors/Fire Control	
Mod 1	acoustic
Mod 2	magnetic
Mod 3	pressure/magnetic
Mod 5	acoustic/magnetic
Mod 6	acoustic/magnetic/pressure
Mod 7	dual-channel magnetic
Mod 11	magnetic/seismic

Mk 55 ASW Ground Mine

The Mk 55 is a U.S. modified aircraft bomb that is employed as an aircraft-laid antisubmarine bottom mine in shallow waters. It has appeared in several mods, which differ primarily in their firing mechanisms. It is essentially the 2,000-lb equivalent of the Mk 52. The Mk 55 was released for production in November 1955 and became operational in 1961. In FY 1960, each of the 1,400 Mk 55s procured cost about $5,500.

CHARACTERISTICS

Weight	
overall	2,039–2,128 lb
warhead	1,269 lb
Dimensions	
length	6 ft 7 in
diameter	23.4 in
Performance	
maximum depth	150 ft (Mods 3, 5, 6)
	600 ft (Mods 2, 7)
Warhead	conventional HBX-1 explosive
Sensors/Fire Control	
Mod 1	acoustic influence
Mod 2	magnetic influence
Mod 3	pressure/magnetic
Mod 5	acoustic/magnetic
Mod 6	acoustic/magnetic/pressure
Mod 7	dual-channel magnetic
Mod 11	magnetic/seismic

Mk 56 ASW Mine

The Mk 56 is an aircraft-laid moored U.S. mine that was specifically designed for use against high-speed, deep-operating submarines. It has a nonmagnetic stainless steel case and is fitted with a magnetic firing mechanism. The Mk 56 was released for production in 1960 and became operational in 1966. The delay was due to lack of funding for production at the time.

CHARACTERISTICS

Weight	
overall	2,135 lb
warhead	357 lb
Dimensions	
length	9 ft 5 in
diameter	23.4 in
Performance	
maximum depth	1,200 ft
Warhead	conventional HBX-3 high explosive
Sensors/Fire Control	magnetic dual-channel using total field magnetometer as detector

Mk 57 ASW Mine

The Mk 57 is an improved submarine-laid moored U.S. mine for use against high-performance submarines (e.g., second-generation Soviet nuclear submarines). It has a fiberglass case (unlike the Mk 56, which has a nonmagnetic stainless steel case) and a magnetic firing mechanism. The Mk 57 was released for production in 1960 and became operational in 1964. As with the Mk 56, the delay was caused by a lack of funding for production.

CHARACTERISTICS

Weight	
overall	2,059 lb
warhead	340 lb
Dimensions	
length	10 ft 8 in
diameter	21 in

(continued)

CHARACTERISTICS *(continued)*

Performance	
maximum depth	approx 1,148 ft
Warhead	conventional HBX-3 high explosive
Sensors/Fire Control	magnetic influence using total field magnetometer as detector

Mk 36 Destructor

The Destructor series of U.S. Navy aircraft-laid bottom mines began as conversions of Mk 80–series bombs during the Vietnam War. The conversion included the installation of a modular arming kit containing an arming device, explosive booster, and magnetic-influence firing device. The Mk 36 is a modified Mk 82 500-lb bomb, with an initial operational capability in the early 1960s. More than 4,000 Mk 36 conversion kits were procured. Although no longer in production, it remains in service with the U.S. Navy. The Mk 36 went through several mods, the principal difference lying in the type of tail assembly and the firing mechanism.

CHARACTERISTICS

Weight	
low-drag tail	564 lb
warhead	192 lb
Dimensions	
length	7 ft 5 in
diameter	15 in
Performance	
maximum depth	300 ft
Warhead	conventional H-6 explosive
Sensors/Fire Control	
Mk 75 Mod 0	Destructor modification kit
Mods 4, 5, 6	magnetic/seismic firing mechanism

Mk 40 Destructor

The Mk 40 is a modified Mk 83 1,000-lb bomb, with an initial operational capability in the 1960s. Although no longer in production, it remains in service with the U.S. Navy.

CHARACTERISTICS

Weight	
low-drag tail	1,064 lb
warhead	445 lb
Dimensions	
length	9 ft 9 in
diameter	22.5 in
Performance	
maximum depth	300 ft
Warhead	conventional H-6 explosive
Sensors/Fire Control	
Mk 75 Mod 0	Destructor modification kit
Mods 4, 5	magnetic/seismic firing mechanism

Mk 41 Destructor

The Mk 41 is a modified Mk 84 2,000-lb bomb, with an initial operational capability in the 1960s. Although no longer in production, it remains in service with the U.S. Navy.

CHARACTERISTICS

Weight	
fixed conical tail	2,030 lb
warhead	990 lb
Dimensions	
length	12 ft 5 in
diameter	25 in
Performance	
maximum depth	300 ft
Warhead	conventional H-6 explosive
Sensors/Fire Control	
Mk 75 Mod 0	Destructor modification kit
Mods 4, 5, 6	magnetic/seismic firing mechanism

Mk 60 Captor

The Mk 60 Captor (Encapsulated Torpedo) is the U.S. Navy's principal antisubmarine mine. It is normally laid in deep water by aircraft or submarine. Captor acoustically detects moving submarines while ignoring surface ships and upon detection launches a Mk 46 Mod 4 acoustic-homing lightweight torpedo. This mode of operation utilizes Reliable Acoustic Path (RAP) propagation and hence will be effective in almost any depth. Lifetime in the water can be from several weeks to months.

The Captor concept was conceived about 1960 at the Naval Surface Weapons (now Warfare) Center at White Oak, and development began in 1961. The Tentative Operational Requirement was issued in November 1962, and the name Captor was selected in May 1964. At that time, it was thought that Captor would reduce ASW mine barrier costs by a factor of 100 and mine numerical requirements for ASW barriers by a factor of 400. The first production contract was awarded to Goodyear Aerospace (now a division of Loral) in 1972. Technical evaluation began in February 1974 and operational evaluation the following June. Initial operational capability was reached in September 1979, and the weapon was approved for service use in February 1980.

Reliability problems caused a suspension of Captor production in 1980, with resumption of production in FY 1982. Budget constraints cut Captor procurement in FY 1985 through FY 1987 from a planned 1,568 to 450. The last authorization of the weapon was in FY 1986 at a unit cost of U.S. $377,000. In FY 1987, 493 Captors were to have been requested, but the budget when submitted had none of the weapons included. Programs to upgrade Captor's performance against newer, quieter, and more sophisticated submarines have also suffered from a lack of funding commitment.

CHARACTERISTICS

Weight	
overall	2,321 lb
warhead	98 lb high explosive
Dimensions	
length	12 ft 1 in
diameter	21 in mine
	12.75 in Mk 46 torpedo
Performance	
maximum depth	3,000 ft
maximum range	1,093 yd detection range
Warhead	conventional high explosive
Sensors/Fire Control	
guidance	passive acoustic monitoring, switching to active once target is identified as a

(continued)

CHARACTERISTICS *(continued)*

submarine; monitoring is not continuous but turns on and off according to programmed schedule; the Mod 1 conversion improves target detection capability

Mk 62 Quickstrike

The U.S. Navy's Quickstrike series of aircraft-laid bottom mines are conversions of Mk 80–series bombs and are follow-ons to the Destructor series. The conversion includes the installation of a modular arming kit containing an arming device, explosive booster, and magnetic-influence firing device. These bombs can be dropped either on land for use against vehicles or in shallow water. All have standard thick-walled bomb casings, and their tails are adaptable to parachutes. These weapons use two alternative Target Detection Devices (TDDs): Mk 57 (magnetic/seismic) and Mk 58 (magnetic/seismic/pressure). These TDDs are to be replaced by the newer and more sophisticated Mk 70 and Mk 71 TDDs. The great advantage of the Quickstrike program is that the United States need not stockpile large numbers of mines. Instead, it can stockpile large numbers of specialized TDDs. The development of the Quickstrike series was justified on two grounds. First, the existing mines, Destructors and Mk 52 ground mines, were compromised in Vietnam. Second, they were beginning to become difficult to maintain, and funding constraints would not permit the one-for-one replacement of existing mine stockpiles with modern devices. Quickstrike's design emphasizes ease of maintenance, preparation, and use. None of the Quickstrike series has received a substantial percentage of the funding originally envisioned for this program, however. All were approved for service use in the early 1980s (Mk 62 Quickstrike in 1980).

The Mk 62 is a modified Mk 82 500-lb bomb (and may be an alternative designation for the DST-36; see Destructor series, above).

CHARACTERISTICS

Weight	
fixed conical tail	531 lb
low-drag tail	570 lb
warhead	192 lb
Dimensions	
length	7 ft 5 in
diameter	10.8 in

(continued)

CHARACTERISTICS *(continued)*

Performance	
maximum depth	300 ft
Warhead	conventional H-6 explosive
Sensors/Fire Control	TDD Mk 57 or 58

Mk 63 Quickstrike
The Mk 63 is a modified Mk 83 1,000-lb bomb.

CHARACTERISTICS

Weight	
fixed conical tail	985 lb
low-drag tail	1,105 lb
warhead	450 lb
Dimensions	
length	9 ft 5 in
diameter	14 in
Performance	
maximum depth	300 ft
Warhead	conventional H-6 explosive
Sensors/Fire Control	TDD Mk 57 or 58

Mk 64 Quickstrike
The Mk 64 is a modified Mk 84 2,000-lb bomb.

CHARACTERISTICS

Weight	2,000 lb
Dimensions	
length	12 ft 8 in
diameter	18 in
Performance	
maximum depth	300 ft
Warhead	conventional H-6 explosive
Sensors/Fire Control	TDD Mk 57 or 58

Mk 65 Quickstrike

Although considered part of the Quickstrike series of bomb-to-mine conversions, the Mk 65 aircraft-laid bottom mine is substantially different from the Mk 84 2,000-lb bomb that is the basis of both the Mk 64 and Mk 65. The Mk 65 has a thin-walled, modular mine-type case and several modifications to the arming mechanism, nose, and tail. The Mk 65 was approved for service use in 1983. Mod 0 uses the Mk 57 TDD; Mod 1, the Mk 58 TDD. Manufactured by Aerojet Tech Systems, and under license to the Italian firm of Misar, the Mk 65 is no longer in production. A new family of microprocessors developed by Misar will reportedly be used to replace the Mk 58 multisensor TDD in the Mk 65.

CHARACTERISTICS

Weight	2,000 lb
Dimensions	
length	9 ft 2 in
diameter	
body	20.9 in
across fins	29 in
Performance	
maximum depth	300 ft
Warhead	conventional HBX high explosive
Sensors/Fire Control	TDD Mk 57 or 58

Mk 67 SLMM

The Submarine-Launched Mobile Mine (SLMM) is a self-propelled torpedo-like mine that permits covert mining by submarines in waters that are inaccessible to other means of delivery. It is also a shallow-water bottom mine for use against surface ships. The SLMM consists of a modified Mk 37 Mod 2 torpedo with the wire guidance equipment removed and the warhead replaced by a mine. The Mk 67 Mod 2 was approved for service use in 1983; at that time a requirement for some 2,400 weapons was noted; only approximately 890 units were procured through FY 1985. The SLMM received a total of $14.3 million in funding in FY 1988 and 1989. An Extended-Range SLMM has been proposed, using a Honeywell Marine Systems NT-37E torpedo.

CHARACTERISTICS

Weight	
overall	1,759 lb
warhead	529 lb
Dimensions	
length	13 ft 5 in
diameter	
mine	21 in
torpedo	19 in
Propulsion	electric motor
Performance	
maximum depth	approx 328 ft
range	approx 10,000 yd
Warhead	Mk 13 high explosive
Sensors/Fire Control	Mk 70 magnetic/seismic TDD
	Mk 71 magnetic/seismic/pressure TDD

Universal Laying Mine

The Universal Laying Mine (ULM) is a new-development project begun in the mid-1980s. Engineering development for the weapon was completed during FY 1986, and some 3,600 units were procured that year at a cost of $3.7 million. An additional 2,400 weapons were to have been procured in FY 1988 and 1989, but none was subsequently requested because of fiscal constraints. (The mine has been included in the same program element as the Quickstrike series; no mark number has been announced.) Details on the mine have not been released to the public, but some report that the mine is a replacement for the Mk 56/57 moored antisubmarine mines. The ULM may also be associated with a new parachute package (Mk 16) for 500-lb mines.

Advanced Sea Mine

Also called the Continental Shelf Mine, the Advanced Sea Mine (ASM) until late 1988 was a joint U.K.-U.S. development effort for a mine to operate in medium water depths, down to about 600 ft. For the U.S. Navy, the ASM is to fulfill the requirement of the Intermediate Water Depth Mine (IWDM), itself replacing the abortive Propelled Rapid-Ascent Mine (PRAM, ex-68) of the 1970s. In the 1984 competition for ASM, Marconi offered a Captor-like mine based on the Stingray lightweight torpedo;

BAe offered a rocket-propelled weapon similar in principle to PRAM; and Ferranti offered a cluster of ten or twelve rockets firing upward from a box launcher. The resulting two ASM contenders are called Crusader (British Aerospace/Plessey/Honeywell) and Hammerhead (Marconi/Loral), both of which are rising vehicles intended to deal with submarines passing over the continental shelf. Both mines will be deployable by submarine torpedo tubes, aircraft, and surface ships. The details that have been released suggest either buoyant or propelled devices employing passive sonar for detection and classification and an active sonar for prosecution, much like the deep-water Captor mine. As of mid-1988, the ASM was scheduled to reach the fleet in the mid-1990s. However, in late December 1988 the U.K. Ministry of Defence canceled its participation in the program (thought to be approximately 20 percent of the total cost). The U.S. Navy then provisionally canceled the ASM as part of the Cheney reassessment of the final Reagan Department of Defense budget in April 1989, although congressional interest in restoring the program remained high.

C The Pressure Drop Under a Moving Ship

Assume the ship has a very large beam so that all flow is from front to back, nothing sideways. Then consider a slice of unit width. Let ship velocity with respect to bottom be v. Hold ship still and move bottom and water toward ship with velocity v. Ignore boundary layer on ship. See figure.

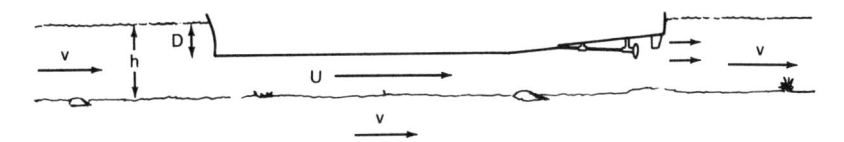

Then $hv = (h - D) U$ where $U > v$.

The velocity of the water under the ship with respect to the ship is

$$U = \frac{hv}{h - D}.$$

With respect to the bottom the water velocity is

$$U - v \text{ or } \frac{Dv}{h - D}.$$

Designating the hydrostatic pressure on the bottom away from the ship as p, and under the ship as p', Bernoulli's theorem says,

$$p + 0 = p' + \frac{1}{2}\rho\left(\frac{Dv}{h-D}\right)^2,$$

or the pressure drop equals

$$\Delta p = \frac{1}{2}\rho\left(\frac{D}{h-D}\right)^2 v^2.$$

If h is large then the change in pressure is negligible. If h is of the order of 2D, then

$$\Delta p = \frac{1}{2}\rho v^2,$$

the pressure change on the bottom in cgs units. If in this depth of water, v is 4 kt, then $\Delta p \sim 8$ inches of water. Hence this is the maximum suction for a ship of infinite beam. In practice, the suction is 1 to 2 inches at 4 kt, being reduced by the finite width of the ship, and variations in the draft of the ship and in the water depth. For example, if h = 3D, then

$$\Delta p = \frac{\rho v^2}{8},$$

or one quarter as much. (If v is 4 kt (or 204 cm/sec) and $\rho = 1$, then $\Delta p = 5200$ dynes/cm² or 2 inches of water, or in practice, less.)

A surface effects ship will have a much smaller draft when underway. However, the higher velocity will help to restore the pressure drop, though for a shorter time.

Bibliography

In the classified literature, which gradually becomes superseded, there are many parts or sections which are unclassified. Not the least of these is the historical section (chapter 2) of the Mine Advisory Commitee's Nimrod report. [33] The Mine Advisory Committee, established in 1951 but superseded by the Naval Studies Board in 1974, has produced eleven studies on subjects related to mine warfare. These reports contain information ranging from broad operational or feasibility studies to such specialities as precise navigation or the development of better power sources for mines. In 1977, the Naval Studies Board (National Academy of Sciences) published a "Historical Bibliography of Sea Mine Warfare," prepared by Patterson and Winters. This bibliography of over two thousand documents was compiled during the preparation of the Nimrod historical section already referred to, and is a complete list of the unclassified literature at that time. To it as a source material should be added all the papers that were once classified but that now need not be and that could be declassified if they were first identified.

One way to attack this problem would be to look at the over 600 papers presented in the annual Minefield Conferences at White Oak since 1958. An index of these papers up to 1973 was printed at NOL in October of that year under the title "The Naval Minefield Community, Cumulative Indexes of Communications I through XVI." Although most or many of these papers could be declassified, they are generally not available, either being bound in with other classified papers or else existing only on microfilm in their place of origin. It is to be hoped that this work will not disappear today because it was so valuable that it could not be released yesterday.

1. *American Institute of Physics Handbook (5-214)*. New York: McGraw-Hill Book Co., 1957.
2. Bergmann, Klaus. "Minenabwehrsystem 'Troika'." *Wehrtechnik*, November 1976.
3. Bitter, F. *Magnets, The Education of a Physicist*. Garden City, New York: Doubleday, 1959.
4. Bleil, D. F., Ed. *Natural Electromagnetic Phenomena Below 30 Kc/s*. New York: Plenum Press, 1964.
5. Boehe, Rolfe. "Westeuropäische Minenjäger aus GFK." *Marine Rundschau*, June 1977.
6. Boyd, J. Huntly. "Nimrod Spar: Clearing the Suez Canal." Annapolis: U.S. Naval Institute *Proceedings*, February 1976.
7. Cagle, M. W., and Manson, F. A. *The Sea War in Korea*. Annapolis: U.S. Naval Institute, 1957.
8. Cole, R. H. *Underwater Explosions*. Princeton: Princeton University Press, 1948.
9. Corse, C. D. Jr. *Introduction to Shipboard Weapons*. Annapolis: Naval Institute Press, 1975.
10. Cowie, J. S. *Mines, Minelayers and Minelaying*. New York: Oxford University Press, 1949.
11. Dorset, P. F. *Historic Ships Afloat*. New York: Macmillan Co., 1967.
12. Duncan, R. C. *America's Use of Sea Mines*. Silver Spring: U.S. Naval Ordnance Laboratory, White Oak, 1962.
13. Fleet and Mine Warfare Training Center, Charleston, S.C. "Fundamentals of Mine Warfare Planning." (CONFIDENTIAL) Course No. A-2A-0010. Charleston: 10 June 1975.
14. Fleet and Mine Warfare Training Center, Charleston, S.C. "Surface Mine Warfare Familiarization." (CONFIDENTIAL) Courses No. A-2E-0017 and A-2E-0033. Charleston: 1 October 1973.
15. Foreign Policy Association, Eds. *A Cartoon History of United States Foreign Policy 1776–1976*. New York: William Morrow & Company, Inc., 1975. (See epigraph on half-title page.)
16. Hartmann, G. K. *Mine Warfare, History and Technology*, NSWC/WOL TR 75-88. Silver Spring: Naval Surface Weapons Center, 1 July 1975.
17. Hartmann, G. K. *Wave Making by an Underwater Explosion*, NSWC/WOL MP 76-15. Silver Spring: Naval Surface Weapons Center, September 1976.
18. Hartmann, G. K., and Hill, E. G., Eds. *Compendium on Underwater Explosion Research*. 3 vols. Washington: Office of Naval Research, 1950.

19. Hezlet, Sir Arthur, Vice Admiral. *The Submarine and Sea Power.* New York: Stein & Day, 1967.
20. Hezlet, Sir Arthur, Vice Admiral. *Aircraft and Sea Power.* New York: Stein & Day, 1970.
21. Hoffmann, Roy F. "Offensive Mine Warfare: A Forgotten Strategy?" Annapolis: U.S. Naval Institute *Proceedings*, May 1977.
22. Ito, M., with Roger Pineau. *The End of the Imperial Japanese Navy.* New York: Macfadden-Bartell, 1965.
23. Johnson, E. A. and Katcher, D. A. *Mines Against Japan.* Washington: Naval Ordnance Laboratory, White Oak, 1973 (Stock Number 0856-00038).
24. Keegan, J. *The Face of Battle.* New York: Viking Press, 1976.
25. Ledebur, Gerhard Frhr von. *Die Seemine.* Munich: J. F. Lehmann's Verlag, 1977.
26. Lindsey, G. R. "Tactical Anti-submarine Warfare: The Past and the Future." London: Adelphi Papers, International Institute for Strategic Studies, 1970.
27. Lott, A. S. *Most Dangerous Sea: A History of Mine Warfare and an Account of U.S. Navy Mine Warfare Operations in World War II and Korea.* Annapolis: U.S. Naval Institute, 1959.
28. Lundeberg, P. K. *Samuel Colt's Submarine Battery, the Secret and the Enigma.* Smithsonian Studies in History and Technology, No. 29. Washington: Smithsonian Institution Press, 1974.
29. Mahan, A. T. *The Influence of Sea Power upon History, 1660–1783.* Boston: Little, Brown & Co., 1890.
30. McCauley, Brian. "Operation End Sweep." Annapolis: U.S. Naval Institute *Proceedings*, March 1974.
31. McCoy, James M. "Mine Countermeasures: Who's Fooling Whom?" Annapolis: U.S. Naval Institute *Proceedings*, July 1975.
32. Meacham, J. S. Commander. "Four Mining Campaigns: An Historical Analysis of the Decisions of the Commanders." Newport: *Naval War College Review*, Vol. 19, No. 10, June 1967.
33. Mine Advisory Committee, National Academy of Sciences, National Research Committee, Project Nimrod. "The Present and Future Role of the Mine in Naval Warfare." Chapter 2, *History of Mine Warfare*, by Andrew Patterson, Jr. Washington: 1970.
34. Morison, S. E. *The Battle of the Atlantic, September 1939–May 1943.* Vol. I. Boston: Little, Brown & Co., 1947.
35. Naval Research Advisory Committee. "Basic Research in the Navy." A Report to the Secretary of the Navy. Vol. 1. Washington: 1 June 1959.
36. Peck, T. *Round-shot to Rockets: A History of the Washington Navy Yard and U.S. Naval Gun Factory.* Annapolis: U.S. Naval Institute, 1949.
37. Preston, A. "The Circe class, Minehunters par excellence." United Kingdom: *Navy International*, January 1977.

38. Preston, A. *Submarines, the History and Evolution of Underwater Fighting Vessels.* London: Octopus Books, Ltd., 1975.
39. Price, Alfred. *Aircraft versus Submarine, the evolution of the antisubmarine aircraft 1912 to 1972.* Annapolis: Naval Institute Press, 1973.
40. Read, J. *Explosives.* New York: Penguin Books Ltd., 1943.
41. Richardson, L. F. *Statistics of Deadly Quarrels.* Edited by Quincy Wright and C. C. Lienau. Pittsburgh: Boxwood Press, 1960.
42. Rickover, H. G. *How the Battleship Maine was Destroyed.* Washington: Naval History Division, Dept. of the Navy, 1976. (Stock Number 008-046-00085-9).
43. Ruge, F. *Im Küstenvorfeld* Wehrwissenschaftliche Berichte, Vol. 15. Munich: J. F. Lehmanns Verlag, 1974.
44. Shortley, George. "Operations Research in Wartime Naval Mining." *Operations Research Journal.* Vol. 15, No. 1, January–February 1967.
45. Spooner, G. R. "Minewarfare—Policy or Palsy?" London: *Royal United Services Institute for Defence Studies Journal*, 1976.
46. Tuchman, B. *The Guns of August.* New York: Macmillan Co., 1962.
47. Urick, R. J. *Principles of Underwater Sound for Engineers.* New York: McGraw-Hill Book Co., 1967.
48. U.S. Office of Naval Records and Library. "The Northern Barrage and Other Mining Activities." Washington: 1920. Published under the direction of the Hon. Josephus Daniels, Secretary of the Navy.
49. U.S. Strategic Bombing Survey. "The Offensive Mine Laying Campaign Against Japan." 1946. Washington: Reprinted by Dept. of the Navy, Headquarters Naval Material Command, 1969.
50. Wineland, W. C. *Delay as a Measure of Minefield Effectiveness.* Naval Ordnance Laboratory Technical Report 69-206, November 1969.
51. Winters, Albim, and Patterson. *A Brief History of Mine Warfare.* Unpublished.
52. Wood, A. B. Memorial Number, *Royal Naval Scientific Service Journal.* Vol. 20, No. 4, July 1965.
53. Zumwalt, E. R. Jr. *On Watch.* New York: Quadrangle/The New York Times Book Co., 1976.
54. Blanton, Cole, Lieutenant, USN. "Mines of August Revisited." U.S. Naval Institute *Proceedings,* November 1987, pp. 16, 21–22.
55. Campbell, Guy R., Captain, USN, Chief of Staff, CoMineWarCom. Personal communication, May 1988.
56. Eikel, Harvey A., Captain, USN (Ret.). Personal communication, July 1989.
57. *The Falklands Campaign: The Lessons.* London: HMSO, Cmnd. 8758, December 1982.

58. Friedman, Norman. *Naval Weapons*. Annapolis: U.S. Naval Institute Press, 1983.
59. Giusti, James R. "Sweeping the Gulf." *Surface Warfare,* March/April 1988, pp. 2ff.
60. Greer, William, and Bartholomew, James, Commander, USN. "The Psychology of Mine Warfare." U.S. Naval Institute *Proceedings,* February 1986.
61. Heine, Kenneth A. "Sweeping Ahead." *Surface Warfare,* March/April 1988.
62. Horne, Charles F., III, Rear Admiral, USN. "New Role for Mine Warfare." U.S. Naval Institute *Proceedings,* November 1982.
63. Horne, Charles F., III, Rear Admiral, USN (Ret.). Personal communications, August 1989–September 1990.
64. Howard, Michael, and Paret, Peter, Eds. Preparatory "Note of 10 July 1827" to *On War.* 2d ed. Princeton: Princeton University Press, 1984.
65. Hughes, Wayne P., Jr., Captain, USN (Ret.). *Fleet Tactics: Theory and Practice.* Annapolis: Naval Institute Press, 1986.
66. Hunt, Lee M., Naval Studies Board, National Research Council. Personal communication, March 1988.
67. Lawson, Max, Captain, RN, Captain Mine Countermeasures. Statement at the May 1985 U.S. Naval Institute Professional Seminar, "Terrorism at Sea." U.S. Naval Academy. Transcripts (mimeo).
68. *Lessons of the Falklands, Summary Report.* Washington, D.C.: Department of the Navy, February 1983.
69. McDonald, Wesley, Admiral, USN. "Mine Warfare: A Pillar of Maritime Strategy." U.S. Naval Institute *Proceedings,* October 1985, pp. 46–53.
70. Melia, Tamara Moser. *Damn the Torpedoes: A Short History of U.S. Naval Mine Countermeasures.* Washington, D.C.: Naval Historical Center, 1989.
71. *New York Times,* 19 June 1987, p. A10.
72. *New York Times,* 30 September 1987, p. A10.
73. O'Rourke, Ronald. "Gulf Ops." U.S. Naval Institute *Proceedings/ Naval Review,* May 1989, pp. 42–50.
74. O'Rourke, Ronald. "The Tanker War." U.S. Naval Institute *Proceedings/Naval Review,* May 1988, pp. 30–34.
75. Polmar, Norman. *Guide to the Soviet Navy.* Annapolis: Naval Institute Press, 1987. Soviet mines section.
76. Rouarch, Claude. "The Naval Mine, As Effective a Weapon As Ever." *International Defense Review,* 9/1984, pp. 1239–48.
77. Seitz, Frank C., Captain, U.S. Merchant Marine. "SS *Bridgeton*: The First Convoy." U.S. Naval Institute *Proceedings/Naval Review,* May 1988, pp. 52–57.
78. Tarpey, John F., Captain, USN (Ret.). "A Minestruck Navy Forgets Its History." U.S. Naval Institute *Proceedings,* February 1988.

79. Truver, Scott C. "The Mines of August: An International Whodunit." U.S. Naval Institute *Proceedings/Naval Review,* May 1985, pp. 95–117.
80. Truver, Scott C. "Weapons That Wait . . . and Wait . . . ." U.S. Naval Institute *Proceedings,* February 1988, pp. 31–40.
81. Truver, Scott C., and Thompson, Jonathan S. "Navy Mine Countermeasures: Quo Vadis?" *Armed Forces Journal International,* April 1987, pp. 70–74.
82. "A U.S. Ambush in the Gulf." *Newsweek,* 5 October 1987, pp. 24–27.
83. "U.S. Navy Seizes Iranian Mine Ship." *Washington Post,* 23 September 1987, pp. A1/A26.
84. "U.S. Reasons for Mining Challenged." *Washington Post,* 14 April 1984, p. A14.
85. USNI Military Database. Soviet mines section.
86. Vego, Milan. "Soviet MCM Tactics." *NAVY International,* April 1987, pp. 241–44.
87. Vego, Milan. "Soviet Navy: Mines and Their Platforms." *NAVY International,* July 1986, pp. 431–36.
88. Vego, Milan. "The Soviet View of Mine Warfare." *NAVY International,* March 1986, pp. 176–80.

Index

Abel, Sir Frederick, and gunpowder improvement, 37
Absorption of sound, 94
Acoustic correlation, and passive localization, PUFFS, 206
Acoustic mines, 295
Acoustic signals, complexity of, 94
Advanced Sea Mine (ASM), 278, 327
Aegean Sea, 243
Aerial land mining, 188
Aggressive-class minesweepers, 306
Agile-class minesweepers, 306
AH-58D Kiowa Warrior, 257
Aircraft for mine laying: B-52, 182; P6M, B-29, 179; TBF & PBY, 73
Air Systems Command, 180
Akula-class submarines, 235
Al Ahmadi, 256
Alarms, 316
Albemarle, CSS, 33
Alexander I, Czar, 27
Alfenol, magnetic material, 218
Allied losses to mines, in World War I, 15
ALQ-166, 278–79, 281, 296
AM (alternating magnetic field), 97
American Revolution, 14, 242, 245
Anechoic room, 92
AN/ALQ-141, 311
AN/ALQ-160, 311
AN/ALQ-166, 280, 312–13
AN/AQS-14, 257, 315
AN/AQS-20, 316

AN/SLQ-35, 311
AN/SLQ-37, 312
AN/SLQ-48, 301, 303, 315
AN/SQQ-14, 301, 305–7, 313–14
AN/SQQ-30, 301, 306, 313
AN/SQQ-32, 301, 314
AN/SQQ-35, 314
AN/SQS-37, 314
AN/SSN-2, 301, 306
AN/UQS-1, 306, 313
Arab-Israeli War, 244
Archimedes: basic research and technology, 217; defense of Syracuse, 200; principle, 96
Arctic Ocean, 230; and influence of ice on mining, 208; and strategic submarines, 207
Argentine Air Force, 248
Argentine Navy, 249
Argonaut, USS, 60
Armstrong, John, accident paths, RAP (relative accident probability), 216
Army Air Corps: large planes, 6; and mining in Pacific, 164
Army Coast Artillery, 105
Arnold, Henry H., 283; and mine blockade of Japan, 73
Assab, 252
Atanasoff, John V., and Loch Ness pressure sweep, 91–92
Atlantic Study (1958), 185
Atomic Energy Commission, 22
Attenuation of sound, 95

337

Austrian-Prussian War, 242
Automatic anchor, 38
Avenger-class, 270, 274, 277, 281, 287, 302

Bab el Mandeb, 250
Bacon, Roger, and composition of gunpowder, 24, 291
Baltic Sea, 235
BAMO (*Batiments Antimines Oceanique*), 284
Bangkok, in World War II, 72
Bardeen, John, and pressure effect, 91
Basic research in the Navy, 112
Bastion (Soviet ship), 251
Battle of the Atlantic, 68
Beach, Dr. Eugene, developed Destructor and Firing Mechanism Mk 42, 187
Belkis I (Saudi Arabian ship), 251
Bell Laboratories, 114
Bell Report on Contracting, 219
B-52 Stratofortress, 182, 281, 283
Bigorange XII (Panamanian ship), 251
Bismanol, 114
Bittinger, Charles, and mine camouflage, 134
Black Sea, 46–47
Blockade, uses of, 3
B-1B bomber, 283
Bordentown, New Jersey, 18
Bosporus, 210
Bouvet (French battleship), lost south of Kephoz, 47
Bowdoin College, 205
Brattain, Walter H., 91
Breslau, German light cruiser, 46
Bridgeton, USS, 256, 260
Britannic, HMS, 243
British Elia mine, 37
Brooklyn, USS, at Mobile Bay, 36
Brown Bess, flintlock musket, 18
Browne, Ralph C., and electrolytic firing device, 50
B-2 Stealth bomber, 283
Bucknell, C., and sprocket wheel sweep evader, 209
Buehler and Wang, 111
Bureau of Ordnance, 267; and countermeasures cognizance, 64–65; and Gun Club, 57; and Mk 6 mine and North Sea Barrage, 48
Bureau of Weapons, 180
Buried mine-hunting system, 315
Burns, Robert, 173
Bushnell, David, and keg mines at Philadelphia, 17, 230, 292
Byron, Lord, and the law of the sea, 13

Camera oscura, 32
Camouflage, for mines, 134
Cape Fear River, 33
Cape Hatteras, 239
Cap Gris Nez, 48
Captor (Mk 60) mine, 33, 103, 232, 235, 264, 269, 278, 282–83, 288, 322–23; barrier, 195, 205; countermeasures, 147; plans, 181, 185
Carderock, Maryland Ships R&D Center, 129
Cardinal-class, 278, 284, 287
Casablanca, no German mines at, 70
CAT guided catamaran (Sperry), 158
Cavallo, Tiberius, firing gunpowder at a distance, 26, 292
Central Intelligence Agency (CIA), 229, 231
Central Powers, mine losses, World War I, 15
C-5 Galaxy, 300
Charleston, South Carolina, 33, 266
Cheney, Richard B., 269, 282, 286
Cherbourg Harbor, 71
Chesapeake Bay, 239
CH-53, helicopter for minesweeping, 152, 299
CIA. *See* Central Intelligence Agency
Circé, French mine hunter, equipment, 157
Clausewitz, Karl von, 229, 263
CNO Executive Board, 269
Coastal shelf, 197
Coles, J.S., and Undersea Warfare Committee, 134
Collinson, Peter, of the Royal Academy, 24
Colt, Samuel, 293; and defensive controlled mines, 30, 200, 230
Command and control, of mines and minefields, 8
Confederate mines: Singer mine, 33; where located, 33
Conquest, USS, 308
Constant, USS, 308
Constantinople, 47
Continental shelf, 129
Control mine Mk 2, 201
Convoys, reasons for past successes of, 164–65
Coolidge, USS, 243
COOP (Craft of Opportunity Program), 277–78, 285, 305, 314
Coral Sea, USS, 270
Corinto, 231, 245, 249
Cormorant, USS, 57
Countermeasure equipments: AN/SQS-15 sonar, 135; AN-UQS-1

sonar, 140; CAT, remote controlled, 157; DUBM-21A sonar, 157
Countermeasures for Captor, 147
Cowie, J.S., message to mine conference, 182
Crane Depot, pyrotechnics in Indianapolis, 179
Crécy, Battle of, 24
Crimean War, 28, 242
"Critical mass" of laboratories, 128
Crowe, William J., 255
Cua Viet River, 311
Cunaeus of Leyden, 25, 292
Cyclops II, 186

Dai Hong Dan (North Korean ship), 251
Damage: by explosions, 98–100; zones of ships, 101
Danube, mining of, 202, 235, 240
Dardanelles, 47, 230, 245, 294
Davids, underwater manned vessels, 34
Davy, Humphry, and carbon arc, 27, 292
Deddario, Representative, on uses of technology, 109
Deep-Sea Mine Study (1957), 232
Deep-water mine, Captor, 103
Defense Intelligence Agency, 188
Defense minefield: controlled, 31; Key West, 68
Degaussing: early British, 55; at NDL, 63, 89
Delaware River, 17
Deperming, 88
Depth charge, shock, 99
Desert Shield. *See* Operation Desert Shield
Design disclosure, 228
Destructor Mk 75: Firing Mechanism Mk 42, 187; mining at Haiphong, 188, 202
Detection devices, 86
Detector (ordnance) Mk 9, Mk 10, 138
Deterrence, 13, 191, 198
Diego Garcia, 256, 300
Dixon, George, 35
Dorothea, British brig sunk by Fulton, 29
Dover Straits, mining with surface control, 172, 230, 234, 245, 294
Drayton (captain of USS *Brooklyn*), 36
DRV (Democratic Republic of Vietnam), 153
Duhler (Soviet-flag oil tanker), 249
Duncan, Robert C., 57, 234, 236, 238

EDD (Electromagnetic Discontinuity Descriminator), 137
Egyptian Army at Suez, 155
Einstein, Albert, 15
Electrolytic firing device, K-device, 48, 50
Elia mine, 37, 294
Elizabeth I, Queen, 231
Ellingson, Herman, 206
Elmen, Gustaf, magnetic materials lab at NDL, 114
Engage, USS, 308
English Channel. *See* Hague Convention
Enhance, USS, 306, 308
EOD, Explosive Ordnance Demolition, 133, 155
Este (West German ship), 251
Esteem, USS, 306
Evaluation: independence of, 225; of mine effectiveness, 5; of weapons vs. vehicles, 223
Excel, USS, 308
Exocet missile, 248, 255
Exploit, USS, 308
Explosions, contact vs. noncontact, 69
Explosives: DATB, 180; fulminate of mercury, 34; gun cotton in Mk 2 naval defense mine, 36; HBX, 139; nitroglycerine, 24; TNT, 37
Exultant, USS, 308

Falkland Islands, 229–30, 248–49, 286
Farraday, Michael, 26, 88
Farragut, David G., 4, 35–36, 293
Fearless, USS, 308
Ferret (British destroyer), and laying of oscillating mines, 207
Fidelity, USS, 308
Fido, Mine Mk 24, first homing torpedo, 210
Finland, Gulf of, Soviet-moored minefield in, 42, 235
Firing mechanisms: Hertz horn described, 37; inertial switch and electric battery, 38; K-device, 50
Firing width, 102, 193
Fiume, 33, 293
Flanders, German submarine ports in, 47
Fleming, George, and keg mine, 18
Flight gear, high-speed, 179
Focke, Alfred B.: and damage estimates, 99; and Mine Advisory Committee, 133
Folkestone minefield, 48
Fortify, USS, 308

339

Fort Lauderdale, Florida, 135
Fort Monroe, Virginia, control mines at, 200–1
Fort Morgan, Alabama, 36
Franke, William B., 112
Franklin, Benjamin, firing gunpowder by electricity, 24–25, 292
French-Austrian War (1859), 32
Fullinwider, Simon Peter, 51
Fulminate of mercury, 34, 292
Fulton, Robert, mine proposals of, 28–29

Gallant, USS, 308
Gallipoli, 47, 171
Galvani, 25, 292
Gas globe, underwater explosion of, 99
Gates, Thomas S., Jr., 112
Geomagnetic frequency spectrum, 115
George Shumann (East German ship), 251
German-Austrian War, 37
German magnetic needle mine, 63
German minefields: at Casablanca, not used, 70; at Salerno and Anzio, 70; off U.S. Atlantic Coast, 69–70
Ghat, 252
Gibraltar, Straits of, 41
Glass Reinforced Plastic (GRP): boats for mine hunting, 156; ships: HMS *Wilton*, 156; *Viksten* (Swedish), 156
Goeben (German battle cruiser), 46
Grant's electricity machine, 31
Gravimeter, 96
Ground influence mines, 46
Groverman, W. H., 182
Guadalcanal, USS, 256
Guam, USS, 300
Gun factory, 57
Gunpowder, history and uses of, 24, 291–92

Hague Convention (1907), roles for mine warfare, 41, 238, 292
Haiphong: in Vietnam War, 8, 148, 151–54, 230, 243–45, 270; World War II mining of, 72, 295
Halifax, 239
Hamlet, 16
Hare, Robert, 30, 292
Hartford, USS (Farragut's flagship), 36
Harvard Underwater Sound Laboratory, 135
Hatsushima-class ships, 284
Havana Harbor, 38
Hayward, Thomas B., 269, 288

Helicopter sweeping, 81
Heligoland Bight, 43, 294
Hermes, HMS, 248
Hertz horn, 37, 293
Herzfeld, Karl, 81
High-speed ships, hydrofoil, surface effect, SWATH, 97
Himly, Karl, 28
Hiroshima, 16, 73
Hitler's "secret weapon," 61
Ho Chi Minh Trail, 255
Hooper, Edward, and ordnance sciences, 217–19
Hormuz, Strait of, 254
Horne, Charles F. III, 265, 287–88
Housatonic, USS, sunk in Charleston, 34–35, 293
Howard, Edward Charles, 34, 292
Howe, Lord, "Battle of the Kegs," 19
H-2 mine, at Dover, 37
Hughes, Wayne P., 288
Hui Yang (Chinese ship), 251
Humboldt, Baron Alex von, 27
Hunley, and sinking of USS *Housatonic*, 35, 293
Hussein, Saddam, 287
Hydrofoils, 97

Iceland, 245
Illusive, USS, 257, 308
Impervious, USS, 308
Implicit, USS, 308
Inchon Harbor, 236
Indigenous mines, 210
Induced magnetization of ships, 87
Inertial switch, 38
Inflexible, HMS, damaged at Kephoz, 47
Inflict, USS, 257, 308
In-house laboratories: development of, 222; management of, 220
Inland Sea, 230
Interdiction, 11
Invincible, HMS, 248
Iowa-class battleships, 263
Iran Ajr (Iranian landing craft), 259–60
Iranian Revolutionary Guards, 229, 234, 237, 254, 295
Iran-Iraq War, 230, 254–62
Irresistible, HMS, sunk in Dardanelles (1915), 47
Islamic Jihad, 253
Iso-damage contours, 99

Jacobi, Moritz von, 27–28, 293
James River, 33

340

Jane's Fighting Ships, on minesweepers, layers, 141, 145
Japan: delay in campaign against, 6; ships sunk, 16
Jellico, Admiral, 43
Jidda, 250
Johnson, Ellis, operations research, 175, 234
Josef Wybicki (Polish ship), 251
Joy, C. Turner, on the lesson of Wonsan, 78
Jutland, Battle of, 43

Kattegat, rocket-laid mines in, 210
Kegs, Battle of the, 19
Kephoz barrier, 46
Keyes, Admiral, plan for Dover Straits, 48
Key West minefield, 68
Kiel, University of, 28
King Kong, first ordnance locator, 136
Kipling, Rudyard, on minesweeping, 122
Kissinger, 84
Knott Apparatus Company (Cambridge, Massachusetts), 50
Knud Jesperson (Soviet-flag merchant ship), 250–51
Kongo, Japanese battleship sunk in Formosa Strait, 226
Korea, moored and ground mines in, 15, 78, 80
Kriti Coral (Panamanian ship), 251
Kuwait, 254–56; Iraq's invasion of, 262

Laboratories and Centers, 213
Lance, Bert, on maintenance, 109
Land mining: by air, 188; terradynamics of, 202
La Salle, USS, 250
Latvian coast, 42
Leader, USS, 308
Le Duc Tho, peace protocol with Dr. Kissinger, 84
Lee, W. A., 175
Lehman, John, 267
LeMay, Curtis, 78, 240
Leningrad, 235
Lerici-class mine hunters, 303
Leyden jars, 25, 292
Libya, 253
Limpet mines, 209
Linera (Cypriot ship), 251
Little, A. D. & Co., 112
Liverpool, magnetic sweeping of (1940), 65
Lobbies and funding, 85

Loch Ness Monster, 92
Long Range Mine Program, 184; origin and objectives of, 177; producibility of, 224; shortcomings of, 221; underfunding of, 222
Lord Byron, and law of the sea, 13
Lord Howe, 19

M-08/39 mine, 259–61
MAC (Mine Advisory Committee), 5
MacArthur, Douglas, 236
McCoy, James, 274–75
McDonald, Wesley, 277
McKeehan, L. W.: and Bureau of Ordnance, 176; and Edwards Street Laboratory, Yale, 133; first commander, Mine Building, 57, 294
McKinney, Chester, ARL, University of Texas, Austin, 133
McLean, William B., on role of submersibles, 11
McNamara line in Vietnam, remote sensors, 198
MAD (Magnetic Anomaly Detector), 138; Magnetic Airborne Detector, limit of range of, 116
Madison, James, 29
Magic Lantern, 316
Magnetic extrapolator, 89
Magnetic mine M-sinker, earliest ground mine, 43
Magnetic Orange Pipes (MOPs), 296, 299, 312; for helicopter towed sweeps, 152
Magnetization of ships: induced, 87; permanent, 87
Magnetometer, thin film, 84
Mahan, Alfred T., view on offensive mining, 7
Maine, USS, 38, 294
Makaroff, admiral lost with *Petropavlovsk,* 40
Mansfield amendment, 109
Manta mine, 287
Marseilles, 252
Marshal Cheykov, 255
Marshall Islands, 73
Martin, James M., 69
Martin P6M, 179
Maury, Matthew Fontain. and torpedo bureau in Richmond, Virginia, 32, 293
Medi Sea (Liberian ship), 251
Meiyo Maru (Japanese ship), 251
Menneken, Carl, and Mine Advisory Committee, 133
Metcalf, Joseph III, 276, 285
Mexican-American War, 31

MH-53 Sea Dragon helicopter, 277–79, 285, 287, 296–97, 300, 313
Middle East, 14
Millikan, Mrs. Robert A., on the definition of a physicist, 107
Millikan, Robert A., 107
Minas Basin, Nova Scotia, 208
Mine: Mk 6, development authorized, 48–49; Mk 11, sub-laid moored, 58, 60
Mine Advisory Committee, reports and members of, 81, 133, 232
Mine Building in Washington Navy Yard, 57
Mine campaign against Japan, 71
Mine chronology, 291–95
Mine conferences, 81
Mine countermeasure forces, NATO and Eastern Europe, 271
Mine countermeasure systems: policy, 66; Sea Nettle, 139; Shadowgraph, 140; Turtle, 139
Mine Defense Laboratory, becomes Coastal Systems Laboratory, 127–29
Mine detecting and classifying sonar, 135–40
Mine Division, 125, 309
Minefields: calculating threat of, 193–94; Dover, 47; Gulf of Finland, 42; Hanoi, 84; Heligoland, 43; North Sea Barrage, 53; Straits of Otranto, 55; U.S. Atlantic Coast, 69
Mine firing influences, 66
Mine hunter: (French) *Circé*, 157; Poisson Autopropulsée, 157
Mine laying, by submarine, 12–13
Mine locator, sonar. *See* Countermeasure equipments
Mines, odd: Mk 2 Control, 201; Mk 19 Penstock, 209; Mk 24, homing torpedo, 210; Mk 29, towed explosive hose, 210; Mk 49, free fall, 210
Mines, pressure distinguished from acoustic, 91
Mine spotting, 134
Minesweeping in Vietnam. *See* Operation End Sweep
Minesweeping policy, 66
Mine warfare: defensive, offensive, 6–7; defined, 3; effectiveness of, 15; reasons for neglect of, 6, 8; statistical assessment of value of, 4
Mine Warfare Command, 239, 265, 270
Mine Warfare Force, 270

Mine Warfare Inspection Group, 271
Mine Warfare Operations Research Group ("Morgue"), 67
Mine Warfare School, Yorktown, Virginia, 266
Mine warfare system, management of, 120
Mining operations, Kephoz, Ostende, Zeebrugge, 171
Mk 2, 37, 310
Mk 4, 310
Mk 6, 15, 48–50, 310–12
Mk 10, 71, 239
Mk 11, 58
Mk 12, 72, 239
Mk 24, 68, 307, 314
Mk 25, 66, 177, 316
Mk 27, 203–4
Mk 36, 66, 153, 177, 232, 243, 317, 321
Mk 39, 317
Mk 40, 243, 321
Mk 41, 243, 322
Mk 46, 323
Mk 49, 317
Mk 51, 317
Mk 52, 243, 318
Mk 55, 319
Mk 56, 320
Mk 57, 320, 326
Mk 58, 326
Mk 60. *See* Captor (Mk 60) mine
Mk 62, 324
Mk 63, 325
Mk 64, 325–26
Mk 65, 326
Mk 67, 233, 235, 264, 278, 282, 326
Mk 68, 278, 307
Mk 103, 257, 296, 310
Mk 104, 299, 311
Mk 105, 280, 296, 312
Mk 106, 312
M Mk 1, 311
M Mk 4, 311
M Mk 5, 311
Mobile Bay, 33, 293
Mobile mine. *See* Mk 67
MOMAG (Mobile Mine Assembly Group), 271
Morgul (Turkish ship), 251
M-sinker, 294
Musket, flint lock, 291
Muzzey, David: and Nitinol, 111; and solion pressure detector, 93

Nagasaki, 73
NATO, mine countermeasure efforts, 156, 275, 277

Nautilus: first nuclear submarine, 10; Fulton's submarine, 28
Naval Air Systems Command, 267
Naval Gun Factory, 227
Naval Material Command, 267
Naval Mine Depot, Yorktown, readiness of, 176, 178
Naval missions, 9–10
Naval Ordnance Laboratory: as center for mine development, 15; change of name of, 15, 223; conferences at, 167; and development of Lulu, 216; 50th anniversary of, 57; and magnetic extrapolator, 89; and magnetic materials laboratory, 111; and Nova Scotia, recording temperature in, 208; origins of, 57
Naval Sea Systems Command, 267
Naval Torpedo Station, Newport, 36
Nelson, Lord, 30
Neva River, 27, 192, 292
New Ironsides, USS, damaged in Charleston, 34
New Orleans, 35
Newton, Isaac, 167
Nicaragua, 249–50
Nicholas I, Czar, and Committee on Underwater Experiments, 27
Nickel-titanium alloy, 111
Nimbus Moon, 244
Nimbus Star, 244
Nimitz, Admiral, 283; and mine blockade of Japan, 73
Nimitz, USS, 300
Nimrod, 5; reasons for study, 183, 232, 264–65
Nobska study (1956), string mine, 185, 205
NOMAD (Naval Operations and Maintenance Aviation Deck), 277
Normandy beach, mines at, 70
North Sea Barrage, 48, 230, 245, 294; effects of, 53; and Norway sector, 55
North Vietnam, 84
Northwest Passage, 207
Norwegian Sea, 239
Nova Scotia, 208, 286

"Obsolete" mines, 254–62
Ocean, HMS, sunk in Dardanelles (1915), 47
Oceanic Energy (Liberian ship), 251
Ocean vs. air transport, 9
Oersted, magnetic field of electric current, 292
Offensive mining, 7
Okinawa, 73

Old Point Comfort, Virginia, 200
OMEGA navigation system, 297
Operation Desert Shield, 262
Operation End Sweep, 148, 149, 154, 244, 274–75, 312
Operations Research: beginnings of, 66–67; functions of, 162–63
Operation Starvation: mines against Japan, 73; results, 73, 78, 240–41, 283
Operation Yo-Yo, 235
Ordnance, definitions of, 22
Ordnance locators, Mk 15, Mk 9, EDD, SMSD, King Kong, Queen gear, 136–38
Ordnance Systems Command, organization of, 180, 260
Orkney Islands, 55, 194, 245
Oropesa sweeps, 310
Ortega, Daniel, 231, 249
Orwell, George, 42
Oscillating mine, 207
Osprey-class, 270, 277, 287, 303–4
Otranto, Straits of, 55
Ottoman Empire, 47
Ouvry, John, and magnetic mine recovery, 61. *See also* A. B. Wood
Oyster mine, 67, 131, 238

Palau Atoll, 73, 231, 234, 245
Panama, 239
Panama City, Florida, 269, 272
Parabonal, magnetic material, 64
Paraguayan War, 242
Pasley, Charles, underwater demolition, 25–26, 293
Passive ranging, 206
Pasteur, Louis, 217
Patterson, Andrew, Jr., 134
Pearl Harbor, 60, 266
Pebble, Project, 232
Pelouze, gunpowder, 36, 292
Pensacola, Florida, 218
Perestroika, 277
Permanent magnetization of ships, 87
Permeability, magnetic, 87
Persian Gulf, 14, 229, 233–34, 237, 254–55, 259, 262, 286, 287
Perth, Australia, 72
Peruvian Reefer (Bahamian ship), 251
Philadelphia, 17
Philippines, 38
Phipps, Tom, 122
Piney Point, Maryland, 136
Pioneer, CSS, 34
Platform systems, assignment of missions to, 213–14
Pledge, USS, 308

343

Plessy Type 193, mine-hunting sonar, 161
Plowman, Joseph, 18
Pluck, USS, 301, 308
Plymouth, Massachusetts, 33
Poisson Autopropulsée (PAP), 157
Polaris, 12–13
Porpoise-class submarines, 56
Port Arthur, 245
Portsmouth Harbor, 26
Port Stanley, 245, 248, 295
Potomac River, 31
PRAM (Propelled Rapid-Ascent Mine), 278
Pressure mine, Oyster, 67, 131
Pressure sweeping, Eggcrate, XMAP, 132
Price, R. S., 38
Prime-contractor concept, for Captor, 221
Principles of mine warfare, 3
Probability in decision making, 173
Project General, 210
PUFFS (Passive Underwater Fire-control Feasibility System), 205–6

Qaddafi, Muammar, 229, 231, 252–53
Queen magnetometer, 136–37
Quickstroke, 188

Raleigh, USS, 309
Ramsey, G. B., 69
RAYDIST minesweeping navigation system, 297
Red Sea, 229, 231, 250, 287, 315
Reich, Eli, 226
Reliability requirement, 215
Reliable Acoustic Path (RAP), 323
Research and Development (R&D): competition in, 213; mission assignments and motivation in, 214; relevance of, 212
RH-53 helicopter, 300, 313, 315
Richardson, L. F., 15
Rickover, Admiral Hyman, 38
Rocket-laid mines, in Kattegat, 210
Rolfe Committee, on mine readiness, 168, 177–78
Rowden, William, 261
Rowzee, Charles, and Destructor Adaption Kit and FM, 187
Royal George, HMS, 293
Russo-Japanese War, 40, 245, 294
Russo-Turkish War, 242

Safety of weapons: goals for, relative accident probability of, 216; opinions on, 223
Saigon, 209

St. Julian's Creek, Virginia, 52
St. Petersburg, Russia, 27, 42
St. Vincent Earl, and criticism of William Pitt, 7
Samuel B. Roberts, USS, 229, 261–62
Sanders, Admiral, CoMinLant (1955), 168, 177
Sautter-Harlé, 294
Schilling, Baron Pavel L'vovich, demonstration at the Neva, 27, 192, 292
Schleswig-Holstein War, 28
Schwarz, Berthold, and handgun prototype, 24
Sea Nettle, 139
"Sea Plan 2000," 283
Sea Systems Command. *See* Naval Sea Systems Command
Secrecy in countermeasures, 127
Seine River, 27
Seitz, Frank, 256
Seventh Fleet, 256
Shadowgraph, 140
Shark and Captor, 204–5
Shatt al-Arab, 254
Shaw, Thomas, and percussion cap, 34, 292
Sheffield, HMS, 248
Shimonoseki Strait, 73, 245, 295
Ships' influence fields, UEP and AM, 97
Ships R&D Center, Carderock, 129
Shockley, William, 91
Shock-waves: damage by, 98; relevance of, 112–13
Shoeburyness, German mine exposed at, 46, 61
Shreveport, USS, 250
Siemens, Werner Von, and design of galvanic mines at Kiel (1848), 28, 293
Signatures of ships, 88
Singer, Confederate moored mine, 33
Smith, A. E., 78
Smithsonian Institution, 36
SMSD Submarine Mine Surface Detector, 136
Sney, H. G., 100
Sobero, and nitrated glycerine, 293
Solion, pressure detector, 93
Solomons, Maryland, 127
Sömmering, Samuel Thomas, and insulated wire, 26–27
Sonar, for mine hunting, 135–40, 158
Souchon, Admiral, German alliance with Turkey (World War I), 46
Soviet devices and nomenclature, 56
Soviet Navy: mine stockpiles of, 276; mining tradition, 6; modernization

of, 286; and NATO, 275, 277; submarine force, 12; surface MCM force, 275
Spanish-American War, 242, 294
Sperrbrecher, 101
Spofford, Captain, and countermeasures at Wonson, 78
Spooner, G. R., 275
Sprengle, 293
Sprocket wheel sweep evader, 209
Stark, USS, 255
Sterilizer, 8
Stratofortress bomber (B-52), 182, 281, 283
Stratton Report, 197
String mine, 204–5
STUFT (Ships Taken Up From Trade), 248
Sturtevant, USS, 68
Submarine forces, composition of, 10
Submarine minelayers: *Argonaut*, 58, 182; British E class, Russian Krab, German UC, 181; in North Sea, 43; *Porpoise*-class, 56
Submersible cargo ships, 11
Subroc, 250
Suez, sweeping of, 155, 229, 231, 244–45, 251, 258, 286–87, 295
Suez Canal Authority, 252
Surface effect ships, 97
Surveillance: mine technology, 198; systems, 195
SWATH (small waterplane area twin hull), 97
System for mine warfare, 119–20

Tamura, Admiral, and Japanese countermeasures, 16
Tang He (Chinese ship), 251
Tannenburg, Battle of, 42
Team Spirit, 281
Technology, definitions of, 108
Tecumseh, USS, 35
Tehran, 254
Television, underwater, 135
Theopoulis (Greek ship), 251
Third Reich, end of, 171
Thomson-CSF, DUBM20A, mine-hunting sonar in *Circé*, 157
Thomson-Sintra sea mine, 284
Tinian, 73
TNT, 241
Tobin, Admiral, 232–33, 288
Torpedo, Mk 48, 228
Torpedo Bureau, 32, 34, 293
Trafalgar, Battle of, 29, 292
Transport of the future, 9
Trident submarines, 199
Trinadad, 239

Troika (German), minesweeping and hunting, 156
Trost, C. A. H., 235, 239, 265, 269, 288
Truk, 234
Truman Proclamation: on law of the sea, 197
Turner, Richmond Kelly, 175
Turtle: Bushnell's submarine, 25; mine location and destruction, 139
Twenty-first Bomber Command, Pacific Theater, 78, 240, 283

UEP, underwater electric potential, 97
Universal laying mine, 327
Urakaze, Japanese destroyer sunk with *Kongo*, 226

Valencia (Spanish ship), 251
Venice, Harbor of, 32
Venturi tube, 131
Vernon, HMS, 38, 294
Vickers Company, 37, 294
Vietnam, 15
Volta, 25, 292
Voltaire, 155

Waldheim, Kurt, 254
Wang and Buehler, 111
War of 1812, 30
Warsaw Pact, 277, 286
Washington, George, 17
Watkins, James, 265, 269
Weinberger, Caspar, 256
Weiss and Warner, minefield simulation, 168
West German Navy, Mk 52 mine procurement, 228
Wheatstone, Sir Charles, 26
Whitehead torpedo, in Fiume, 32, 293
White Oak Laboratory, 15, 213, 269, 272, 323. *See also* Naval Ordnance Laboratory
Wonsan, 230–31, 245, 256; delay at, 173; landing at, 80–81
Wood, A. B., on magnetic mine recovery, 62–63

XMAP experimental device, 132

Yazoo River, 33
Yenisei, Japanese minelayer, lost to mines, 40

Zumwalt, Elmo R., Jr., 269; on helicopter sweeping, 130, 133; and Project 60, 146–47; on "unions," 214

345

About the Authors

Gregory K. Hartmann, Ph.D., is a former technical director of the U.S. Naval Ordnance Laboratory in White Oak, Maryland. His tenure there covered an important period during which the Laboratory developed SUBROC, the Mk 48 torpedo, and the Navy's first nuclear weapons. Dr. Hartmann served in the Bureau of Ordnance during World War II, and before that taught physics at the University of New Hampshire and at Brown University, where he earned his Ph.D. in the science of ultrasonics. He has been honored with the National Civil Service Award and Distinguished Civilian Service Award from the Navy and the Department of Defense. His list of publications is impressive for its length as well as for the variety of subjects on which he is recognized as an authority.

Scott C. Truver, Ph.D., is the director of national security studies at Information Spectrum, Inc., a studies and analysis firm specializing in defense, national security, and international security programs and issues. He holds the first doctoral degree conferred in the field of marine policy studies. He has undertaken numerous analytical projects for clients in government and industry, is the author of *The Strait of Gibraltar and the Mediterranean*, part of the *International Straits of the World* series, and has written articles for such publications as the U.S. Naval Institute *Proceedings, International Defense Review,* and *Naval Forces,* among others. He has lectured on naval and defense issues in the United States and abroad. In addition to his responsibilities at Information Spectrum, he assists the Naval Institute's professional seminar program and the major Fleet and Industry Conference and Exhibition held in San Diego, California.

The **Naval Institute Press** is the book-publishing arm of the U.S. Naval Institute, a private, nonprofit professional society for members of the sea services and civilians who share an interest in naval and maritime affairs. Established in 1873 at the U.S. Naval Academy in Annapolis, Maryland, where its offices remain today, the Naval Institute has more than 100,000 members worldwide.

Members of the Naval Institute receive the influential monthly magazine *Proceedings* and discounts on fine nautical prints, ship and aircraft photos, and subscriptions to the quarterly *Naval History* magazine. They also have access to the transcripts of the Institute's Oral History Program and get discounted admission to any of the Institute-sponsored seminars regularly offered around the country.

The Naval Institute's book-publishing program, begun in 1898 with basic guides to naval practices, has broadened its scope in recent years to include books of more general interest. Now the Naval Institute Press publishes more than forty new titles each year, ranging from how-to books on boating and navigation to battle histories, biographies, ship and aircraft guides, and novels. Institute members receive discounts on the Press's more than 375 books.

Full-time students are eligible for special half-price membership rates. Life memberships are also available.

For a free catalog describing the Naval Institute Press books currently available, and for further information about U.S. Naval Institute membership, please write to:

Membership & Communications Department
U.S. Naval Institute
Annapolis, Maryland 21402

Or call, toll-free, (800) 233-USNI. In Maryland, call (301) 224-3378.

THE NAVAL INSTITUTE PRESS
WEAPONS THAT WAIT
Updated Edition
Set in Primer
by Brushwood Graphics, Inc., Baltimore, Maryland

Printed on 50-lb. Hi Brite Vellum
and bound in Joanna Arrestox A
by The Maple-Vail Book Manufacturing Group,
York, Pennsylvania

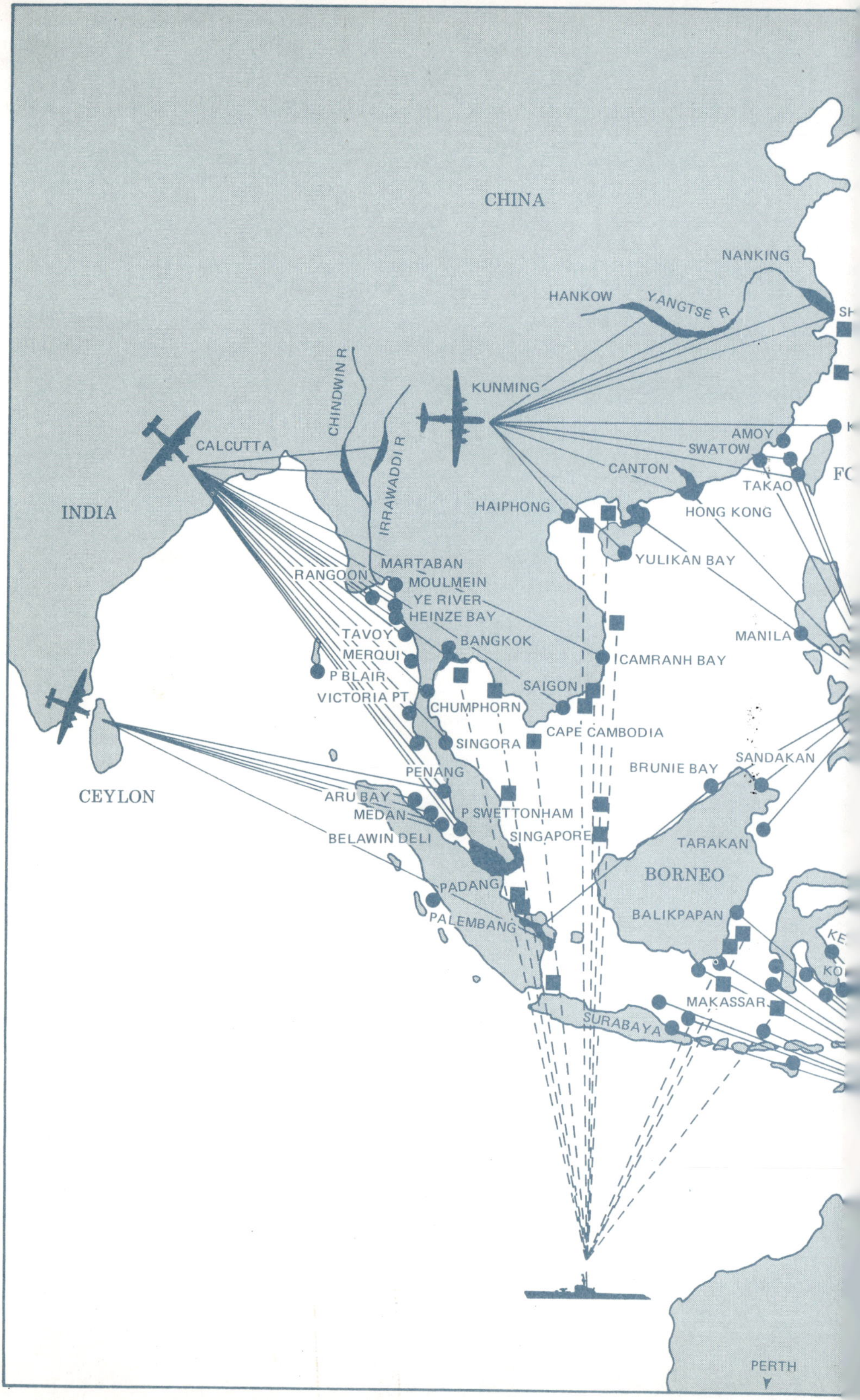

CHINA
NANKING
HANKOW
YANGTSE R
KUNMING
CHINDWIN R
IRRAWADDI R
CALCUTTA
INDIA
AMOY
SWATOW
CANTON
TAKAO
HAIPHONG
HONG KONG
YULIKAN BAY
MARTABAN
RANGOON
MOULMEIN
YE RIVER
HEINZE BAY
TAVOY
MERQUI
BANGKOK
MANILA
CAMRANH BAY
P BLAIR
SAIGON
VICTORIA PT
CHUMPHORN
CAPE CAMBODIA
SINGORA
BRUNIE BAY
SANDAKAN
CEYLON
PENANG
ARU BAY
MEDAN
P SWETTONHAM
BELAWIN DELI
SINGAPORE
TARAKAN
BORNEO
PADANG
BALIKPAPAN
PALEMBANG
MAKASSAR
SURABAYA
PERTH